COMPUTERS AND INTRACTABILITY

A Guide to the Theory of NP-Completeness

A SERIES OF BOOKS IN THE MATHEMATICAL SCIENCES
Victor Klee, Editor

COMPUTERS AND INTRACTABILITY

A Guide to the Theory of NP-Completeness

Michael R. Garey / David S. Johnson

BELL LABORATORIES
MURRAY HILL, NEW JERSEY

W. H. FREEMAN AND COMPANY
New York

Library of Congress Cataloging in Publication Data

Garey, Michael R.
Computers and Intractability.

Bibliography: p.
Includes index.
1. Electronic digital computers--Programming.
2. Algorithms. 3. Computational complexity.
I. Johnson, David S., joint author. II. Title.
III. Title: NP-completeness.
QA76.6.G35 519.4 78-12361
ISBN 0-7167-1044-7
ISBN 0-7167-1045-5 pbk.

AMS Classification: Primary 68A20
Computer Science: Computational complexity and efficiency

Printed in the United States of America

Twenty-fourth Printing 2003

Contents

Preface

Few technical terms have gained such rapid notoriety as the appelation "NP-complete." In the short time since its introduction in the early 1970's, this term has come to symbolize the abyss of inherent intractability that algorithm designers increasingly face as they seek to solve larger and more complex problems. A wide variety of commonly encountered problems from mathematics, computer science, and operations research are now known to be NP-complete, and the collection of such problems continues to grow almost daily. Indeed, the NP-complete problems are now so pervasive that it is important for anyone concerned with the computational aspects of these fields to be familiar with the meaning and implications of this concept.

This book is intended as a detailed guide to the theory of NP-completeness, emphasizing those concepts and techniques that seem to be most useful for applying the theory to practical problems. It can be viewed as consisting of three parts.

The first part, Chapters 1 through 5, covers the basic theory of NP-completeness. Chapter 1 presents a relatively low-level introduction to some of the central notions of computational complexity and discusses the significance of NP-completeness in this context. Chapters 2 through 5 provide the detailed definitions and proof techniques necessary for thoroughly understanding and applying the theory.

The second part, Chapters 6 and 7, provides an overview of two alternative directions for further study. Chapter 6 concentrates on the search for efficient "approximation" algorithms for NP-complete problems, an area whose development has seen considerable interplay with the theory of NP-completeness. Chapter 7 surveys a large number of theoretical topics in computational complexity, many of which have arisen as a consequence of previous work on NP-completeness. Both of these chapters (especially Chapter 7) are intended solely as introductions to these areas, with our expectation being that any reader wishing to pursue particular topics in more detail will do so by consulting the cited references.

The third and final part of the book is the Appendix, which contains an extensive list (more than 300 main entries, and several times this many results in total) of NP-complete and NP-hard problems. Annotations to the main entries discuss what is known about the complexity of subproblems and variants of the stated problems.

The book should be suitable for use as a supplementary text in courses on algorithm design, computational complexity, operations research, or combinatorial mathematics. It also can be used as a starting point for seminars on approximation algorithms or computational complexity at the graduate or advanced undergraduate level. The second author has used a preliminary draft as the basis for a graduate seminar on approximation algorithms, covering Chapters 1 through 5 in about five weeks and then pursuing the topics in Chapter 6, supplementing them extensively with additional material from the references. A seminar on computational complexity might proceed similarly, substituting Chapter 7 for Chapter 6 as the initial access point to the literature. It is also possible to cover both chapters in a combined seminar.

More generally, the book can serve both as a self-study text for anyone interested in learning about the subject of NP-completeness and as a reference book for researchers and practitioners who are concerned with algorithms and their complexity. The list of NP-complete problems in the Appendix can be used by anyone familiar with the central notions of NP-completeness, even without having read the material in the main text. The novice can gain such familiarity by skimming the material in Chapters 1 through 5, concentrating on the informal discussions of definitions and techniques, and returning to the more formal material only as needed for clarification. To aid those using the book as a reference, we have included a substantial number of terms in the Subject Index, and the extensive Reference and Author Index gives the sections where each reference is mentioned in the text.

We are indebted to a large number of people who have helped us greatly in preparing this book. Hal Gabow, Larry Landweber, and Bob Tarjan taught from preliminary versions of the book and provided us with valuable suggestions based on their experience. The following people read preliminary drafts of all or part of the book and made constructive comments: Al Aho, Shimon Even, Ron Graham, Harry Hunt, Victor Klee, Albert Meyer, Christos Papadimitriou, Henry Pollak, Sartaj Sahni, Ravi Sethi, Larry Stockmeyer, and Jeff Ullman. A large number of researchers, too numerous to mention here (but see the Reference and Author Index), responded to our call for NP-completeness results and contributed toward making our list of NP-complete problems as extensive as it is. Several of our colleagues at Bell Laboratories, especially Brian Kernighan, provided invaluable assistance with computer typesetting on the UNIX® system. Finally, special thanks go to Jeanette Reinbold, whose facility with translating our handwritten hieroglyphics into faultless input to the typesetting system made the task of writing this book so much easier.

Murray Hill, New Jersey
October, 1978

MICHAEL R. GAREY
DAVID S. JOHNSON

COMPUTERS AND INTRACTABILITY

A Guide to the Theory of NP-Completeness

1

Computers, Complexity, and Intractability

1.1 Introduction

The subject matter of this book is perhaps best introduced through the following, somewhat whimsical, example.

Suppose that you, like the authors, are employed in the halls of industry. One day your boss calls you into his office and confides that the company is about to enter the highly competitive "bandersnatch" market. For this reason, a good method is needed for determining whether or not any given set of specifications for a new bandersnatch component can be met and, if so, for constructing a design that meets them. Since you are the company's chief algorithm designer, your charge is to find an efficient algorithm for doing this.

After consulting with the bandersnatch department to determine exactly what the problem is, you eagerly hurry back to your office, pull down your reference books, and plunge into the task with great enthusiasm. Some weeks later, your office filled with mountains of crumpled-up scratch paper, your enthusiasm has lessened considerably. So far you have not been able to come up with any algorithm substantially better than searching through all possible designs. This would not particularly endear you to your boss, since it would involve years of computation time for just one set of specifications, and the bandersnatch department is already 13 components behind schedule. You certainly don't want to return to his office and report:

"I can't find an efficient algorithm, I guess I'm just too dumb."

To avoid serious damage to your position within the company, it would be much better if you could prove that the bandersnatch problem is *inherently* intractable, that no algorithm could possibly solve it quickly. You then could stride confidently into the boss's office and proclaim:

"I can't find an efficient algorithm, because no such algorithm is possible!"

Unfortunately, proving inherent intractability can be just as hard as finding efficient algorithms. Even the best theoreticians have been stymied in their attempts to obtain such proofs for commonly encountered hard problems. However, having read this book, you have discovered something almost as good. The theory of NP-completeness provides many straightforward techniques for proving that a given problem is "just as hard" as a large number of other problems that are widely recognized as being difficult and that have been confounding the experts for years. Armed with these techniques, you might be able to prove that the bandersnatch problem is NP-complete and, hence, that it is equivalent to all these other hard problems. Then you could march into your boss's office and announce:

"I can't find an efficient algorithm, but neither can all these famous people."

At the very least, this would inform your boss that it would do no good to fire you and hire another expert on algorithms.

Of course, our own bosses would frown upon our writing this book if its sole purpose was to protect the jobs of algorithm designers. Indeed, discovering that a problem is NP-complete is usually just the beginning of work on that problem. The needs of the bandersnatch department won't disappear overnight simply because their problem is known to be NP-complete. However, the knowledge that it is NP-complete does provide valuable information about what lines of approach have the potential of being most productive. Certainly the search for an efficient, exact algorithm should be accorded low priority. It is now more appropriate to concentrate on other, less ambitious, approaches. For example, you might look for efficient algorithms that solve various special cases of the general problem. You might look for algorithms that, though not guaranteed to run quickly, seem likely to do so most of the time. Or you might even relax the problem somewhat, looking for a fast algorithm that merely finds designs that meet *most* of the component specifications. In short, the primary application of the theory of NP-completeness is to assist algorithm designers in directing their problem-solving efforts toward those approaches that have the greatest likelihood of leading to useful algorithms.

In the first chapter of this "guide" to NP-completeness, we introduce many of the underlying concepts, discuss their applicability (as well as give some cautions), and outline the remainder of the book.

1.2 Problems, Algorithms, and Complexity

In order to elaborate on what is meant by "inherently intractable" problems and problems having "equivalent" difficulty, it is important that we first agree on the meaning of several more basic terms.

Let us begin with the notion of a problem. For our purposes, a *problem* will be a general question to be answered, usually possessing several *parameters,* or free variables, whose values are left unspecified. A problem is described by giving: (1) a general description of all its parameters, and (2) a statement of what properties the answer, or *solution,* is required to satisfy. An *instance* of a problem is obtained by specifying particular values for all the problem parameters.

As an example, consider the classical "traveling salesman problem." The parameters of this problem consist of a finite set $C = \{c_1, c_2, \ldots, c_m\}$ of "cities" and, for each pair of cities c_i, c_j in C, the "distance" $d(c_i, c_j)$ between them. A solution is an ordering $<c_{\pi(1)}, c_{\pi(2)}, \ldots, c_{\pi(m)}>$ of the given cities that minimizes

$$\left[\sum_{i=1}^{m-1} d(c_{\pi(i)}, c_{\pi(i+1)})\right] + d(c_{\pi(m)}, c_{\pi(1)})$$

This expression gives the length of the "tour" that starts at $c_{\pi(1)}$, visits each city in sequence, and then returns directly to $c_{\pi(1)}$ from the last city $c_{\pi(m)}$.

One instance of the traveling salesman problem, illustrated in Figure 1.1, is given by $C = \{c_1, c_2, c_3, c_4\}$, $d(c_1, c_2) = 10$, $d(c_1, c_3) = 5$, $d(c_1, c_4) = 9$, $d(c_2, c_3) = 6$, $d(c_2, c_4) = 9$, and $d(c_3, c_4) = 3$. The ordering $<c_1, c_2, c_4, c_3>$ is a solution for this instance, as the corresponding tour has the minimum possible tour length of 27.

Algorithms are general, step-by-step procedures for solving problems. For concreteness, we can think of them simply as being computer programs, written in some precise computer language. An algorithm is said to *solve* a problem Π if that algorithm can be applied to any instance I of Π and is guaranteed always to produce a solution for that instance I. We emphasize that the term "solution" is intended here strictly in the sense introduced above, so that, in particular, an algorithm does not "solve" the traveling

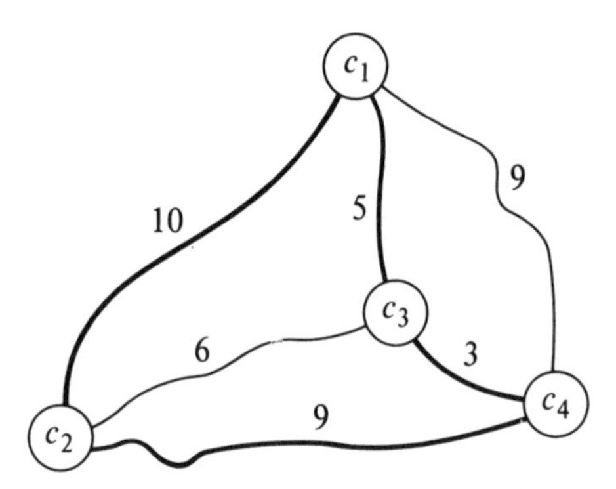

Figure 1.1 An instance of the traveling salesman problem and a tour of length 27, which is the minimum possible in this case.

salesman problem unless it always constructs an ordering that gives a minimum length tour.

In general, we are interested in finding the most "efficient" algorithm for solving a problem. In its broadest sense, the notion of efficiency involves all the various computing resources needed for executing an algorithm. However, by the "most efficient" algorithm one normally means the fastest. Since time requirements are often a dominant factor determining whether or not a particular algorithm is efficient enough to be useful in practice, we shall concentrate primarily on this single resource.

The time requirements of an algorithm are conveniently expressed in terms of a single variable, the "size" of a problem instance, which is intended to reflect the amount of input data needed to describe the instance. This is convenient because we would expect the relative difficulty of problem instances to vary roughly with their size. Often the size of a problem instance is measured in an informal way. For the traveling salesman problem, for example, the number of cities is commonly used for this purpose. However, an m-city problem instance includes, in addition to the labels of the m cities, a collection of $m(m-1)/2$ numbers defining the inter-city distances, and the sizes of these numbers also contribute to the amount of input data. If we are to deal with time requirements in a precise, mathematical manner, we must take care to define instance size in such a way that all these factors are taken into account.

To do this, observe that the description of a problem instance that we provide as input to the computer can be viewed as a single finite string of symbols chosen from a finite input alphabet. Although there are many different ways in which instances of a given problem might be described, let us assume that one particular way has been chosen in advance and that each problem has associated with it a fixed *encoding scheme*, which maps problem instances into the strings describing them. The *input length* for an instance I of a problem Π is defined to be the number of symbols in the description of I obtained from the encoding scheme for Π. It is this number, the input length, that is used as the formal measure of instance size.

For example, instances of the traveling salesman problem might be described using the alphabet $\{c, [,], /, 0, 1, 2, 3, 4, 5, 6, 7, 8, 9\}$, with our previous example of a problem instance being encoded by the string "$c[1]c[2]c[3]c[4]//10/5/9//6/9//3$." More complicated instances would be encoded in analogous fashion. If this were the encoding scheme associated with the traveling salesman problem, then the input length for our example would be 32.

The *time complexity function* for an algorithm expresses its time requirements by giving, for each possible input length, the largest amount of time needed by the algorithm to solve a problem instance of that size. Of course, this function is not well-defined until one fixes the encoding scheme to be used for determining input length and the computer or computer model to be used for determining execution time. However, as we shall see, the particular choices made for these will have little effect on the broad distinctions made in the theory of NP-completeness. Hence, in what follows, the reader is advised merely to fix in mind a particular encoding scheme for each problem and a particular computer or computer model, and to think in terms of time complexity as determined from the corresponding input lengths and execution times.

1.3 Polynomial Time Algorithms and Intractable Problems

Different algorithms possess a wide variety of different time complexity functions, and the characterization of which of these are "efficient enough" and which are "too inefficient" will always depend on the situation at hand. However, computer scientists recognize a simple distinction that offers considerable insight into these matters. This is the distinction between polynomial time algorithms and exponential time algorithms.

Let us say that a function $f(n)$ is $O(g(n))$ whenever there exists a constant c such that $|f(n)| \leqslant c \cdot |g(n)|$ for all values of $n \geqslant 0$. A *polynomial time algorithm* is defined to be one whose time complexity function is $O(p(n))$ for some polynomial function p, where n is used to denote the input length. Any algorithm whose time complexity function cannot be so bounded is called an *exponential time algorithm* (although it should be noted that this definition includes certain non-polynomial time complexity functions, like $n^{\log n}$, which are not normally regarded as exponential functions).

The distinction between these two types of algorithms has particular significance when considering the solution of large problem instances. Figure 1.2 illustrates the differences in growth rates among several typical complexity functions of each type, where the functions express execution time in terms of microseconds. Notice the much more explosive growth rates for the two exponential complexity functions.

Time complexity function	Size n					
	10	20	30	40	50	60
n	.00001 second	.00002 second	.00003 second	.00004 second	.00005 second	.00006 second
n^2	.0001 second	.0004 second	.0009 second	.0016 second	.0025 second	.0036 second
n^3	.001 second	.008 second	.027 second	.064 second	.125 second	.216 second
n^5	.1 second	3.2 seconds	24.3 seconds	1.7 minutes	5.2 minutes	13.0 minutes
2^n	.001 second	1.0 second	17.9 minutes	12.7 days	35.7 years	366 centuries
3^n	.059 second	58 minutes	6.5 years	3855 centuries	2×10^8 centuries	1.3×10^{13} centuries

Figure 1.2 Comparison of several polynomial and exponential time complexity functions.

Even more revealing is an examination of the effects of improved computer technology on algorithms having these time complexity functions. Figure 1.3 shows how the largest problem instance solvable in one hour would change if we had a computer 100 or 1000 times faster than our present machine. Observe that with the 2^n algorithm a thousand-fold increase in computing speed only adds 10 to the size of the largest problem instance we can solve in an hour, whereas with the n^5 algorithm this size almost quadruples.

These tables indicate some of the reasons why polynomial time algorithms are generally regarded as being much more desirable than exponential time algorithms. This view, and the distinction between the two types of algorithms, is central to our notion of inherent intractability and to the theory of NP-completeness.

The fundamental nature of this distinction was first discussed in [Cobham, 1964] and [Edmonds, 1965a]. Edmonds, in particular, equated poly-

Size of Largest Problem Instance
Solvable in 1 Hour

Time complexity function	With present computer	With computer 100 times faster	With computer 1000 times faster
n	N_1	$100\ N_1$	$1000\ N_1$
n^2	N_2	$10\ N_2$	$31.6\ N_2$
n^3	N_3	$4.64\ N_3$	$10\ N_3$
n^5	N_4	$2.5\ N_4$	$3.98\ N_4$
2^n	N_5	$N_5+6.64$	$N_5+9.97$
3^n	N_6	$N_6+4.19$	$N_6+6.29$

Figure 1.3 Effect of improved technology on several polynomial and exponential time algorithms.

nomial time algorithms with "good" algorithms and conjectured that certain integer programming problems might not be solvable by such "good" algorithms. This reflects the viewpoint that exponential time algorithms should not be considered "good" algorithms, and indeed this usually is the case. Most exponential time algorithms are merely variations on exhaustive search, whereas polynomial time algorithms generally are made possible only through the gain of some deeper insight into the structure of a problem. There is wide agreement that a problem has not been "well-solved" until a polynomial time algorithm is known for it. Hence, we shall refer to a problem as *intractable* if it is so hard that no polynomial time algorithm can possibly solve it.

Of course, this formal use of "intractable" should be viewed only as a rough approximation to its dictionary meaning. The distinction between "efficient" polynomial time algorithms and "inefficient" exponential time algorithms admits of many exceptions when the problem instances of interest have limited size. Even in Figure 1.2, the 2^n algorithm is faster than the n^5 algorithm for $n \leqslant 20$. More extreme examples can be constructed easily.

Furthermore, there are some exponential time algorithms that have been quite useful in practice. Time complexity as defined is a *worst-case* measure, and the fact that an algorithm has time complexity 2^n means only that at least one problem instance of size n requires that much time. Most problem instances might actually require far less time than that, a situation that appears to hold for several well-known algorithms. The simplex algorithm for linear programming has been shown to have exponential time complexity [Klee and Minty, 1972], [Zadeh, 1973], but it has an impressive record of running quickly in practice. Likewise, branch-and-bound algorithms for the knapsack problem have been so successful that many consider it to be a "well-solved" problem, even though these algorithms, too, have exponential time complexity.

Unfortunately, examples like these are quite rare. Although exponential time algorithms are known for many problems, few of them are regarded as being very useful in practice. Even the successful exponential time algorithms mentioned above have not stopped researchers from continuing to search for polynomial time algorithms for solving those problems. In fact, the very success of these algorithms has led to the suspicion that they somehow capture a crucial property of the problems whose refinement could lead to still better methods. So far, little progress has been made toward explaining this success, and no methods are known for predicting in advance that a given exponential time algorithm will run quickly in practice.

On the other hand, the much more stringent bounds on execution time satisfied by polynomial time algorithms often permit such predictions to be made. Even though an algorithm having time complexity n^{100} or $10^{99}n^2$ might not be considered likely to run quickly in practice, the polynomially solvable problems that arise naturally tend to be solvable within polynomial time bounds that have degree 2 or 3 at worst and that do not involve extremely large coefficients. Algorithms satisfying such bounds *can* be considered to be "provably efficient," and it is this much-desired property that makes polynomial time algorithms the preferred way to solve problems.

Our definition of "intractable" also provides a theoretical framework of considerable generality and power. The intractability of a problem turns out to be essentially independent of the particular encoding scheme and computer model used for determining time complexity.

Let us first consider encoding schemes. Suppose for example that we are dealing with a problem in which each instance is a graph $G = (V,E)$, where V is the set of vertices and E is the set of edges, each edge being an unordered pair of vertices. Such an instance might be described (see Figure 1.4) by simply listing all the vertices and edges, or by listing the rows of the adjacency matrix for the graph, or by listing for each vertex all the other vertices sharing a common edge with it (a "neighbor" list). Each of these encodings can give a different input length for the same graph. However, it is easy to verify (see Figure 1.5) that the input lengths they determine differ at most polynomially from one another, so that any algorithm having polynomial time complexity under one of these encoding schemes also will have polynomial time complexity under all the others. In fact, the standard encoding schemes used in practice for any particular problem always seem to differ at most polynomially from one another. It would be difficult to imagine a "reasonable" encoding scheme for a problem that differs more than polynomially from the standard ones. Although what we mean here by "reasonable" cannot be formalized, the following two conditions capture much of the notion:

(1) the encoding of an instance I should be concise and not "padded" with unnecessary information or symbols, and

(2) numbers occurring in I should be represented in binary (or decimal, or octal, or in any fixed base other than 1).

If we restrict ourselves to encoding schemes satisfying these conditions, then the particular encoding scheme used should not affect the determination of whether a given problem is intractable.

Encoding Scheme	String	Length
Vertex list, Edge list	V[1]V[2]V[3]V[4](V[1]V[2])(V[2]V[3])	36
Neighbor list	(V[2])(V[1]V[3])(V[2])( )	24
Adjacency matrix rows	0100/1010/0010/0000	19

Figure 1.4 Descriptions of the graph $G = (V,E)$ where $V = \{V_1, V_2, V_3, V_4\}$ and $E = \{\{V_1, V_2\}, \{V_2, V_3\}\}$, under three different encoding schemes.

Encoding Scheme	Lower Bound	Upper Bound
Vertex list, Edge list	$4v + 10e$	$4v + 10e + (v+2e)\cdot\lceil \log_{10} v \rceil$
Neighbor list	$2v + 8e$	$2v + 8e + 2e\cdot\lceil \log_{10} v \rceil$
Adjacency matrix	$v^2 + v - 1$	$v^2 + v - 1$

Figure 1.5 General bounds on input lengths for the three encoding schemes of Figure 1.4 for graphs $G = (V,E)$ with $|V| = v$, $|E| = e$. Since $e < v^2$, these show that the input lengths differ at most polynomially from each other. ($\lceil x \rceil$ denotes the least integer not less than x.)

Similar comments can be made concerning the choice of computer models. All the realistic models of computers studied so far, such as one-tape Turing machines, multi-tape Turing machines, and random-access machines (RAMs), are equivalent with respect to polynomial time complexity (for example, see Figure 1.6). One would expect any other "reasonable" model to share in this equivalence. The notion of "reasonable" intended here is essentially that there is a polynomial bound on the amount of work that can be done in a single unit of time. Thus, for example, a model having the capability of performing arbitrarily many operations in parallel would not be considered "reasonable," and indeed no existing (or planned) computer has this capability. At any rate, so long as we restrict ourselves to the standard models of realistic computers, the class of intractable problems will be unaffected by the particular model used, and we can make our choice on the basis of convenience without sacrificing the applicability of our results.

Simulated machine B	Simulating machine A		
	1TM	kTM	RAM
1-Tape Turing Machine (1TM)	—	$O(T(n))$	$O(T(n)\log T(n))$
k-Tape Turing Machine (kTM)	$O(T^2(n))$	—	$O(T(n)\log T(n))$
Random Access Machine (RAM)	$O(T^3(n))$	$O(T^2(n))$	—

Figure 1.6 Time required by machine A to simulate the execution of an algorithm of time complexity $T(n)$ on Machine B (for example, see [Hopcroft and Ullman, 1969] and [Aho, Hopcroft, and Ullman, 1974]).

1.4 Provably Intractable Problems

Now that we have discussed the formal meaning of "intractable problem," it is appropriate that we briefly survey the current state of knowledge about the existence of intractable problems.

It is useful to begin by distinguishing between two different causes of intractability allowed by our definition. The first, which is the one we usually have in mind, is that the problem is so difficult that an exponential amount of time is needed to discover a solution. The second is that the solution *itself* is required to be so extensive that it cannot be described with an expression having length bounded by a polynomial function of the input length.

This second cause occurs, for example, in the variant of the traveling salesman problem that includes a number B as an additional parameter and that asks for *all* tours having total length B or less. It is easy to construct instances of this problem in which exponentially many tours are shorter than the given bound, so that no polynomial time algorithm could possibly list them all.

Intractability of this sort is by no means insignificant, and it is important to recognize it when it occurs. However, in most cases its existence is apparent from the problem definition. In fact, this type of intractability can be regarded as a signal that the problem is not defined realistically, because we are asking for more information than we could ever hope to use. Thus, from now on we shall restrict our attention to the first type of intractability. Accordingly, only problems for which the solution length is bounded by a polynomial function of the input length will be considered.

The earliest intractability results for such problems are the classical undecidability results of Alan Turing. Over 40 years ago, Turing demonstrated that certain problems are so hard that they are "undecidable," in the sense that no algorithm at all can be given for solving them. He proved, for example, that it is impossible to specify *any* algorithm which, given an arbitrary computer program and an arbitrary input to that program, can decide whether or not the program will eventually halt when applied to that input [Turing, 1936]. A variety of other problems are now known to be undecidable, including the triviality problem for finitely presented groups [Rabin, 1958], Hilbert's tenth problem (solvability of polynomial equations in integers) [Matijasevic, 1970], and several problems of "tiling the plane" [Berger, 1966]. Since these undecidable problems cannot be solved by *any* algorithm, much less a polynomial time algorithm, they indeed are intractable in an especially strong sense.

The first examples of intractable "decidable" problems were obtained in the early 1960's, as part of work on complexity "hierarchies" by Hartmanis and Stearns [1965]. However, these results involved only "artificial" problems, specifically constructed to have the appropriate properties. It was not until the early 1970's that Meyer and Stockmeyer [1972], Fischer and Rabin [1974], and others finally succeeded in proving some "natural" decidable problems to be intractable. These include a variety of previously studied problems from automata theory, formal language theory, and mathematical logic. In fact, the proofs show that these problems cannot be solved in polynomial time using even a "nondeterministic" computer model, which has the ability to pursue an unbounded number of independent computational sequences in parallel. We shall see that this "unreasonable" computer model plays an important role in the theory of NP-completeness, and its capabilities will be specified more fully in Chapter 2.

All the provably intractable problems known to date fall into the two categories we have just mentioned. They are either undecidable or "nondeterministically" intractable. However, most of the apparently intractable problems encountered in practice *are* decidable and *can* be solved in polynomial time with the aid of a nondeterministic computer. Thus, none of the proof techniques developed so far is powerful enough to verify the apparent intractability of these problems.

1.5 NP-Complete Problems

As theoreticians continue to seek more powerful methods for proving problems intractable, parallel efforts focus on learning more about the ways in which various problems are interrelated with respect to their difficulty. As we suggested earlier, the discovery of such relationships between problems often can provide information useful to algorithm designers.

The principal technique used for demonstrating that two problems are related is that of "reducing" one to the other, by giving a constructive transformation that maps any instance of the first problem into an equivalent instance of the second. Such a transformation provides the means for converting any algorithm that solves the second problem into a corresponding algorithm for solving the first problem.

Many simple examples of such reductions have been known for some time. For example, Dantzig [1960] reduced a number of combinatorial optimization problems to the general zero-one integer linear programming problem. Edmonds [1962] reduced the graph theoretic problems of "covering all edges with a minimum number of vertices" and "finding a maximum independent set of vertices" to the general "set covering problem." Gimpel [1965] reduced the general set covering problem to the "prime implicant covering problem" of logic design. Dantzig, Blattner, and Rao [1966] described a "well-known" reduction from the traveling salesman problem to the "shortest path problem" with negative edge lengths allowed.

These early reductions, although rather isolated and limited in scope, foreshadow the kind of results proved in the theory of NP-completeness.

The foundations for the theory of NP-completeness were laid in a paper of Stephen Cook, presented in 1971, entitled "The Complexity of Theorem Proving Procedures" [Cook, 1971a]. In this brief but elegant paper Cook did several important things.

First, he emphasized the significance of "polynomial time reducibility," that is, reductions for which the required transformation can be executed by a polynomial time algorithm. If we have a polynomial time reduction from one problem to another, this ensures that any polynomial time algorithm for the second problem can be converted into a corresponding polynomial time algorithm for the first problem.

Second, he focused attention on the class NP of decision problems that can be solved in polynomial time by a nondeterministic computer. (A decision problem is one whose solution is either "yes" or "no".) Most of the apparently intractable problems encountered in practice, when phrased as decision problems, belong to this class.

Third, he proved that one particular problem in NP, called the "satisfiability" problem, has the property that every other problem in NP can be polynomially reduced to it. If the satisfiability problem can be solved with a polynomial time algorithm, then so can every problem in NP, and if any problem in NP is intractable, then the satisfiability problem also must be intractable. Thus, in a sense, the satisfiability problem is the "hardest" problem in NP.

Finally, Cook suggested that other problems in NP might share with the satisfiability problem this property of being the "hardest" member of NP. He showed this to be the case for the problem "Does a given graph G contain a complete subgraph on a given number k of vertices?"

Subsequently, Richard Karp presented a collection of results [Karp, 1972] proving that indeed the decision problem versions of many well known combinatorial problems, including the traveling salesman problem, are just as "hard" as the satisfiability problem. Since then a wide variety of other problems have been proved equivalent in difficulty to these problems, and this equivalence class, consisting of the "hardest" problems in NP, has been given a name: the class of *NP-complete problems.*

Cook's original ideas have turned out to be remarkably powerful. They have provided the means for combining many individual complexity questions into the single question: Are the NP-complete problems intractable? The lists included in the Appendix of this book contain literally hundreds of different problems now known to be NP-complete. As more and more problems of independent interest are shown to belong to this equivalence class, its importance is continually reinforced.

The question of whether or not the NP-complete problems are intractable is now considered to be one of the foremost open questions of contemporary mathematics and computer science. Despite the willingness of most researchers to conjecture that the NP-complete problems are all intractable, little progress has yet been made toward establishing either a proof or a disproof of this far-reaching conjecture. However, even without a proof that NP-completeness implies intractability, the knowledge that a problem is NP-complete suggests, at the very least, that a major breakthrough will be needed to solve it with a polynomial time algorithm.

1.6 An Outline of the Book

Although this book is intended mainly as a primer on how to determine whether or not any particular problem is NP-complete (either by looking it up in the lists we present or by proving it yourself), we shall also discuss some of the options available for dealing with a problem that is known to be NP-complete. A brief outline of subsequent chapters follows.

In Chapter 2, we present the formal underpinnings of NP-completeness and prove Cook's theorem. The central definitions involve certain theoretical concepts, such as "languages" and "Turing machines," which we develop in a straightforward manner, relating them to the notions of problems and computer models already discussed. This chapter should give the reader a good understanding of the technical meaning of NP-completeness.

Chapter 3 is devoted to methods for proving a problem NP-complete. A number of examples are presented to illustrate the usual structure of such proofs, and to indicate how one goes about generating one. In essence, one proves a new problem to be NP-complete by polynomially reducing a known NP-complete problem to it. We survey the known NP-complete problems that have been most useful for this purpose and demonstrate their use.

In Chapter 4, we examine the ways in which the theory of NP-completeness can be used for conducting a detailed analysis of the complexity of a problem, seeking to determine the "boundary" between those cases of the problem that are polynomially solvable and those that are NP-complete.

In Chapter 5, we show how the techniques used for proving NP-completeness can be generalized so that problems other than just decision problems can be proved to be "as hard as" the NP-complete problems. As an aid to reading the published literature on the theory of NP-completeness, we also provide a brief historical survey of the development of the main ideas and the varying terminology that has been used for discussing them.

In Chapter 6, we discuss several approaches for dealing with intractable problems, especially that of finding near-optimal solutions using fast algorithms. Examples of the successes and failures of each approach are described, and we illustrate how the theory of NP-completeness can be applied even here.

Chapter 7 is intended to acquaint the reader with some of the theoretical issues and ideas that have arisen in parallel with the theory of NP-completeness. Among other topics we discuss the polynomial hierarchy, #P-completeness, polynomial space completeness, and the "relativization" of the question of the intractability of the NP-complete problems.

The last third of the book consists of the Appendix, an extensive and annotated list of problems known to be NP-complete or harder. The list is divided into sections, each devoted to problems from a particular subject area, such as graph theory, scheduling, algebra and number theory, covering and partitioning, mathematical programming, program optimization, automata and language theory, and, of course, miscellaneous topics. The list includes references to related problems known to be solvable in polynomial time and to problems whose status remains open in that neither polynomial time algorithms nor NP-completeness proofs are known for them.

2

The Theory of NP-Completeness

In this chapter we present the formal details of the theory of NP-completeness. So that the theory can be defined in a mathematically rigorous way, it will be necessary to introduce formal counterparts for many of our informal notions, such as "problems" and "algorithms." Indeed, one of the main goals of this chapter is to make explicit the connection between the formal terminology and the more intuitive, informal shorthand that is commonly used in its place. Once we have this connection well in hand, it will be possible for us to pursue our discussions primarily at the informal level in later chapters, reverting to the formal level only when necessary for clarity and rigor.

The chapter begins by discussing decision problems and their representation as "languages," equating "solving" a decision problem with "recognizing" the corresponding language. The one-tape Turing machine is introduced as our basic model for computation and is used to define the class P of all languages recognizable deterministically in polynomial time. This model is then augmented with a hypothetical "guessing" ability, and the augmented model is used to define the class NP of all languages recognizable "nondeterministically" in polynomial time. After discussing the relationship between P and NP, we define the notion of a polynomial transformation from one language to another and use it to define what will be our most important class, the class of NP-complete problems. The chapter concludes with the statement and proof of Cook's fundamental theorem, which provides us with our first bona fide NP-complete problem.

2.1 Decision Problems, Languages, and Encoding Schemes

As a matter of convenience, the theory of NP-completeness is designed to be applied only to *decision problems*. Such problems, as mentioned in Chapter 1, have only two possible solutions, either the answer "yes" or the answer "no." Abstractly, a decision problem Π consists simply of a set D_Π of *instances* and a subset $Y_\Pi \subseteq D_\Pi$ of *yes-instances*. However, most decision problems of interest possess a considerable amount of additional structure, and we will describe them in a way that emphasizes this structure. The standard format we will use for specifying problems consists of two parts, the first part specifying a *generic instance* of the problem in terms of various components, which are sets, graphs, functions, numbers, etc., and the second part stating a yes-no *question* asked in terms of the generic instance. The way in which this specifies D_Π and Y_Π should be apparent. An instance belongs to D_Π if and only if it can be obtained from the generic instance by substituting particular objects of the specified types for all the generic components, and the instance belongs to Y_Π if and only if the answer for the stated question, when particularized to that instance, is "yes."

For example, the following describes a well-known decision problem from graph theory:

SUBGRAPH ISOMORPHISM
INSTANCE: Two graphs, $G_1=(V_1,E_1)$ and $G_2=(V_2,E_2)$.
QUESTION: Does G_1 contain a subgraph isomorphic to G_2, that is, a subset $V' \subseteq V_1$ and a subset $E' \subseteq E_1$ such that $|V'|=|V_2|$, $|E'|=|E_2|$, and there exists a one-to-one function $f: V_2 \to V'$ satisfying $\{u,v\} \in E_2$ if and only if $\{f(u),f(v)\} \in E'$?

A decision problem related to the traveling salesman problem can be described as follows:

TRAVELING SALESMAN
INSTANCE: A finite set $C=\{c_1,c_2, \ldots, c_m\}$ of "cities," a "distance" $d(c_i,c_j) \in Z^+$ for each pair of cities $c_i,c_j \in C$, and a bound $B \in Z^+$ (where Z^+ denotes the positive integers).
QUESTION: Is there a "tour" of all the cities in C having total length no more than B, that is, an ordering $<c_{\pi(1)}, c_{\pi(2)}, \ldots, c_{\pi(m)}>$ of C such that

$$\left(\sum_{i=1}^{m-1} d(c_{\pi(i)},c_{\pi(i+1)})\right) + d(c_{\pi(m)},c_{\pi(1)}) \leqslant B \ ?$$

The reader will find many more examples of the use of this format throughout the book, but these two should suffice for now to convey the basic idea. The second example also serves to illustrate an important point about how a decision problem can be derived from an optimization problem. If the optimization problem asks for a structure of a certain type that has minimum "cost" among all such structures (for example, a tour that has minimum length among all tours), we can associate with that problem the decision problem that includes a numerical bound B as an additional parameter and that asks whether there exists a structure of the required type having cost *no more than* B (for example, a tour of length no more than B). Decision problems can be derived from maximization problems in an analogous way, simply by replacing "no more than" by "at least."

The key point to observe about this correspondence is that, so long as the cost function is relatively easy to evaluate, the decision problem can be no harder than the corresponding optimization problem. Clearly, if we could find a minimum length tour for the traveling salesman problem in polynomial time, then we could also solve the associated decision problem in polynomial time. All we need do is find the minimum length tour, compute its length, and compare that length to the given bound B. Thus, if we could demonstrate that TRAVELING SALESMAN is NP-complete (as indeed it is), we would know that the traveling salesman optimization problem is at least as hard. In this way, even though the theory of NP-completeness restricts attention to only decision problems, we can extend the implications of the theory to optimization problems as well. (We shall see in Chapter 5 that decision problems and optimization problems are often even more closely tied: Many decision problems, including TRAVELING SALESMAN, can also be shown to be "no easier" than their corresponding optimization problems.)

The reason for the restriction to decision problems is that they have a very natural, formal counterpart, which is a suitable object to study in a mathematically precise theory of computation. This counterpart is called a "language" and is defined in the following way.

For any finite set Σ of symbols, we denote by Σ^* the set of all finite strings of symbols from Σ. For example, if $\Sigma=\{0,1\}$, then Σ^* consists of the empty string "ϵ," the strings $0,1,00,01,10,11,000,001$, and all other finite strings of 0's and 1's. If L is a subset of Σ^*, we say that L is a *language* over the alphabet Σ. Thus $\{01,001,111,1101010\}$ is a language over $\{0,1\}$, as is the set of all binary representations of integers that are perfect squares, as is the set $\{0,1\}^*$ itself.

The correspondence between decision problems and languages is brought about by the encoding schemes we use for specifying problem instances whenever we intend to compute with them. Recall that an encoding scheme e for a problem Π provides a way of describing each instance of Π by an appropriate string of symbols over some fixed alphabet Σ. Thus the problem Π and the encoding scheme e for Π partition Σ^* into three classes of strings: those that are not encodings of instances of Π, those that encode instances of Π for which the answer is "no," and those that encode instances of Π for which the answer is "yes." This third class of strings is the language we associate with Π and e, setting

$$L[\Pi,e] = \left\{ x \in \Sigma^*: \begin{array}{l} \Sigma \textit{ is the alphabet used by } e, \textit{ and } x \textit{ is the} \\ \textit{encoding under } e \textit{ of an instance } I \in Y_\Pi \end{array} \right\}$$

Our formal theory is applied to decision problems by saying that, if a result holds for the language $L[\Pi,e]$, then it holds for the problem Π under the encoding scheme e.

In fact, we shall usually follow standard practice and be a bit more informal than this. Each time we introduce a new concept in terms of languages, we will observe that the property is essentially encoding independent, so long as we restrict ourselves to "reasonable" encoding schemes. That is, if e and e' are any two reasonable encoding schemes for Π, then the property holds either for both $L[\Pi,e]$ and $L[\Pi,e']$ or for neither. This will allow us to say, informally, that the property holds (or does not hold) for the problem Π, without actually specifying any encoding scheme. However, whenever we do so, the implicit assertion will be that we could, if requested, specify a particular reasonable encoding scheme e such that the property holds for $L[\Pi,e]$.

Notice that when we operate in this encoding-independent manner, we lose contact with any precise notion of "input length." Since we need some parameter in terms of which time complexity can be expressed, it is convenient to assume that every decision problem Π has an associated, encoding-independent function Length: $D_\Pi \rightarrow Z^+$, which is "polynomially related" to the input lengths we would obtain from a reasonable encoding scheme. By *polynomially related* we mean that, for any reasonable encoding scheme e for Π, there exist two polynomials p and p' such that if $I \in D_\Pi$ and x is a string encoding the instance I under e, then Length$[I] \leqslant p(|x|)$ and $|x| \leqslant p'(\text{Length}[I])$, where $|x|$ denotes the length of the string x. In the SUBGRAPH ISOMORPHISM problem, for example, we might take

$$\text{Length}[I] = |V_1| + |V_2|$$

where $G_1 = (V_1,E_1)$ and $G_2 = (V_2,E_2)$ are the graphs making up an instance. In the TRAVELING SALESMAN decision problem we might take

$$\text{Length}[I] = m + \lceil \log_2 B \rceil + \max\{\lceil \log_2 d(c_i,c_j) \rceil : c_i,c_j \in C\}$$

Since any two reasonable encoding schemes for a problem Π will yield polynomially related input lengths, a wide variety of Length functions are possible for Π, and all our results will carry through for any such function that meets the above conditions.

The usefulness of this informal, encoding-independent approach depends, of course, on there being some agreement as to what constitutes a "reasonable" encoding scheme. The generally accepted meaning of "reasonable" includes both the notion of "conciseness," as captured by the two conditions mentioned in Chapter 1, and the notion of "decodability." The intent of "conciseness" is that instances of a problem should be described with the natural brevity we would use in actually specifying those instances for a computer, without any unnatural "padding" of the input. Such padding could be used, for example, to expand the input length so much that we artificially convert an exponential time algorithm into a polynomial time algorithm. The intent of "decodability" is that, given any particular component of a generic instance, one should be able to specify a polynomial time algorithm that is capable of extracting a description of that component from any given encoded instance.

Of course, these elaborations do not provide a formal definition of "reasonable encoding scheme," and we know of no satisfactory way of making such a definition. Even though most people would agree on whether or not a particular encoding scheme for a given problem is reasonable, the absence of a formal definition can be somewhat discomforting. One way of resolving this difficulty would be to require that generic problem instances always be formed from a fixed collection of basic types of set-theoretic objects. We will not impose such a constraint here, but, as an indication of our intent when we refer to "reasonable encoding schemes," we now give a brief description (which first time readers may wish to skip) of how such a standard encoding scheme could be defined.

Our standard encoding scheme will map instances into "structured strings" over the alphabet $\Psi = \{\mathbf{0}, \mathbf{1}, -, [,], (,), ,\}$. We define *structured strings* recursively, as follows:

(1) The binary representation of an integer k as a string of **0**'s and **1**'s (preceded by a minus sign "$-$" if k is negative) is a structured string representing the integer k.

(2) If x is a structured string representing the integer k, then $[x]$ is a structured string that can be used as a "name" (for example, for a vertex in a graph, a set element, or a city in a traveling salesman instance).

(3) If $x_1, x_2, \ldots, x_m$ are structured strings representing the objects $X_1, X_2, \ldots, X_m$, then $(x_1, x_2, \ldots, x_m)$ is a structured string representing the sequence $<X_1, X_2, \ldots, X_m>$.

To derive an encoding scheme for a particular decision problem specified in our standard format, we first note that, once we have built up a representation for each object in an instance as a structured string, the representation of the entire instance is determined using rule (3) above. Thus we need only specify how the representation for each type of object is constructed. For this we shall restrict ourselves to integers, "unstructured elements'' (vertices, elements, cities, etc.), sequences, sets, graphs, finite functions, and rational numbers.

Rules (1) and (3) already tell us how to represent integers and sequences. To represent each of the unstructured elements in an instance, we merely assign it a distinct "name," as constructed by rule (2), in such a way that if the total number of unstructured elements in an instance is N, then no name with magnitude exceeding N is used. The representations for the four other object types are as follows:

A *set* of objects is represented by ordering its elements as a sequence $<X_1, X_2, \ldots, X_m>$ and taking the structured string corresponding to that sequence.

A *graph* with vertex set V and edge set E is represented by a structured string (x,y), where x is a structured string representing the set V, and y is a structured string representing the set E (the elements of E being the two-element subsets of V that are edges).

A *finite function* $f: \{U_1, U_2, \ldots, U_m\} \rightarrow W$ is represented by a structured string $((x_1,y_1),(x_2,y_2), \ldots, (x_m,y_m))$ where x_i is a structured string representing the object U_i and y_i is a structured string representing the object $f(U_i) \in W, 1 \leqslant i \leqslant m$.

A *rational number* q is represented by a structured string (x,y) where x is a structured string representing an integer a, y is a structured string representing an integer b, $a/b = q$, and the greatest common divisor of a and b is 1.

Although it might be convenient to have a wider collection of object types at our disposal, the ones above will suffice for most purposes and are enough to illustrate our notion of a reasonable encoding scheme. Furthermore, there would be no loss of generality in restricting ourselves to just these types for specifying generic instances, since other types of objects can always be expressed in terms of the ones above.

Note that our prescriptions are not sufficient to generate a *unique* string for encoding each instance but merely for ensuring that each string that does encode an instance obeys certain structural restrictions. A different choice of names for the basic elements or a different choice of order for the description of a set could lead to different strings that encode the same instance. In fact, it makes no difference how many strings encode an instance so long as we can decode each to obtain the essential components of the instance. Moreover, our definitions take this into account; for example, in $L[\Pi,e]$, the set of all strings that encode yes-instances of Π under e, each instance may be represented many times.

Before going on, we remind the reader that our standard encoding scheme is intended solely to illustrate how one might define such a standard scheme, although it also provides a reference point for what we mean by a "reasonable'' encoding scheme. There is no reason why some other general scheme could not be used, or why we could not merely devise an individual encoding scheme for each problem of interest. If the chosen scheme is "equivalent" to ours, in the sense that there exist polynomial time algorithms for converting an encoding of an instance under either scheme to an encoding of that instance under the other scheme, then it, too, will be called "reasonable." If the chosen scheme is *not* equivalent to ours in this sense, then one can still prove results with respect to that scheme, but the encoding-independent terminology should not be used for describing them. Throughout this book we will restrict our attention to reasonable encoding schemes for problems.

2.2 Deterministic Turing Machines and the Class P

In order to formalize the notion of an algorithm, we will need to fix a particular model for computation. The model we choose is the *deterministic one-tape Turing machine* (abbreviated DTM), which is pictured schematically in Figure 2.1. It consists of a *finite state control*, a *read-write head*, and a *tape* made up of a two-way infinite sequence of *tape squares*, labeled $\ldots,-2,-1,0,1,2,3,\ldots$

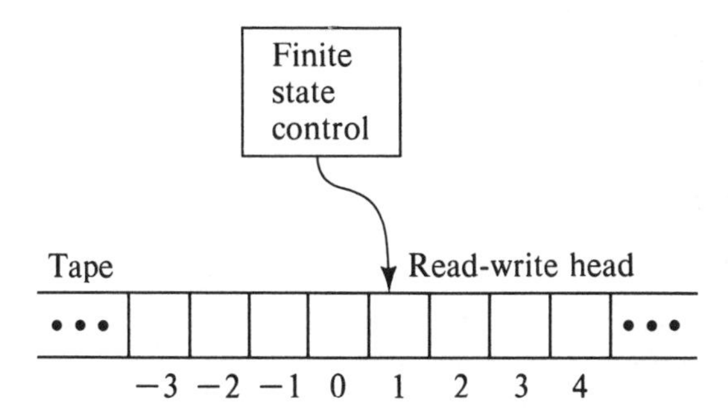

Figure 2.1 Schematic representation of a deterministic one-tape Turing machine (DTM).

A *program* for a DTM specifies the following information:

(1) A finite set Γ of tape *symbols*, including a subset $\Sigma \subset \Gamma$ of *input symbols* and a distinguished *blank symbol* $b \in \Gamma - \Sigma$;

(2) a finite set Q of *states*, including a distinguished *start-state* q_0 and two distinguished *halt-states* q_Y and q_N;

(3) a *transition function* δ: $(Q - \{q_Y, q_N\}) \times \Gamma \rightarrow Q \times \Gamma \times \{-1,+1\}$.

The operation of such a program is straightforward. The *input* to the DTM is a string $x \in \Sigma^*$. The string x is placed in tape squares 1 through $|x|$, one symbol per square. All other squares initially contain the blank symbol. The program starts its operation in state q_0, with the read-write head scanning tape square 1. The computation then proceeds in a step-by-step manner. If the current state q is either q_Y or q_N, then the computation has ended, with the answer being "yes" if $q = q_Y$ and "no" if $q = q_N$. Otherwise the current state q belongs to $Q - \{q_Y, q_N\}$, some symbol $s \in \Gamma$ is in the tape square being scanned, and the value of $\delta(q,s)$ is defined. Suppose $\delta(q,s) = (q', s', \Delta)$. The read-write head then erases s, writes s' in its place, and moves one square to the left if $\Delta = -1$, or one square to the right if $\Delta = +1$. At the same time, the finite state control changes its state from q to q'. This completes one "step" of the computation, and we are ready to proceed to the next step, if there is one.

$$\Gamma = \{0,1,b\},\ \Sigma = \{0,1\}$$

$$Q = \{q_0, q_1, q_2, q_3, q_Y, q_N\}$$

q	0	1	b
q_0	$(q_0,0,+1)$	$(q_0,1,+1)$	$(q_1,b,-1)$
q_1	$(q_2,b,-1)$	$(q_3,b,-1)$	$(q_N,b,-1)$
q_2	$(q_Y,b,-1)$	$(q_N,b,-1)$	$(q_N,b,-1)$
q_3	$(q_N,b,-1)$	$(q_N,b,-1)$	$(q_N,b,-1)$

$$\delta(q,s)$$

Figure 2.2 An example of a DTM program $M = (\Gamma, Q, \delta)$.

An example of a simple DTM program M is shown in Figure 2.2. The transition function δ for M is described in a tabular format, where the entry in row q and column s is the value of $\delta(q,s)$. Figure 2.3 illustrates the computation of M on the input $x = 10100$, giving the state, head position, and contents of the non-blank portion of the tape before and after each step.

Note that this computation halts after eight steps, in state q_Y, so the answer for 10100 is "yes." In general, we say that a DTM program M with input alphabet Σ *accepts* $x \in \Sigma^*$ if and only if M halts in state q_Y when applied to input x. The language L_M *recognized* by the program M is given by

$$L_M = \{x \in \Sigma^*: M \textit{ accepts } x\}$$

It is not hard to see that the DTM program of Figure 2.2 recognizes the language

$$\{x \in \{0,1\}^*: \textit{the rightmost two symbols of } x \textit{ are both } 0\}$$

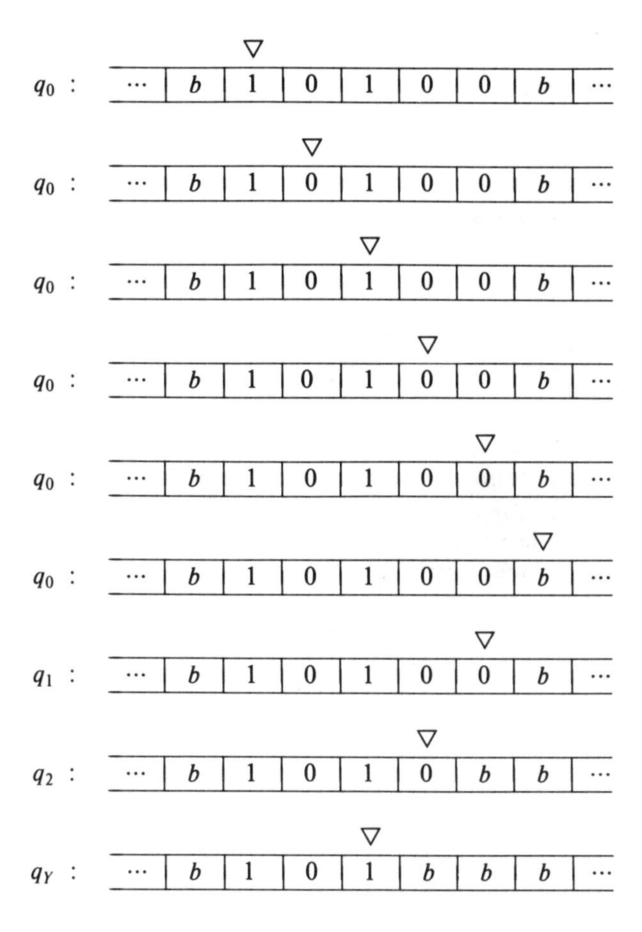

Figure 2.3 The computation of the program M from Figure 2.2 on input 10100.

Observe that this definition of language recognition does not require that M halt for *all* input strings in Σ^*, only for those in L_M. If x belongs to $\Sigma^* - L_M$, then the computation of M on x might halt in state q_N, or it might continue forever without halting. However, for a DTM program to correspond to our notion of an *algorithm*, it must halt on all possible strings over its input alphabet. In this sense, the DTM program of Figure 2.2 is algorithmic, since it will halt for any input string from $\{0,1\}^*$.

The correspondence between "recognizing" languages and "solving" decision problems is straightforward. We say that a DTM program M *solves* the decision problem Π under encoding scheme e if M halts for all input strings over its input alphabet and $L_M = L[\Pi, e]$. The DTM program of Figure 2.2 once more provides an illustration. Consider the following number-theoretic decision problem:

INTEGER DIVISIBILITY BY FOUR
INSTANCE: A positive integer N.
QUESTION: Is there a positive integer m such that $N=4m$?

Under our standard encoding scheme, the integer N is represented by the string of 0's and 1's that is its binary representation. Since a positive integer is divisible by four if and only if the last two digits of its binary representation are 0, this DTM program "solves" the INTEGER DIVISIBILITY BY FOUR problem under our standard encoding scheme.

For future reference, we also point out that a DTM program can be used to compute functions. Suppose M is a DTM program with input alphabet Σ and tape alphabet Γ that halts for all input strings from Σ^*. Then M computes the *function* $f_M: \Sigma^* \rightarrow \Gamma^*$ where, for each $x \in \Sigma^*$, $f_M(x)$ is defined to be the string obtained by running M on input x until it halts and then forming a string from the symbols in tape squares $1, 2, 3,$ etc., in sequence, up to and including the rightmost non-blank tape square. The program M of Figure 2.2 computes the function $f_M: \{0,1\}^* \rightarrow \{0,1,b\}^*$ that maps each string $x \in \{0,1\}^*$ to the string $f_M(x)$ obtained by deleting the last two symbols of x (with $f_M(x)$ equal to the empty string if $|x|<2$).

It is well known that DTM programs are capable of performing much more complicated tasks than those illustrated by our simple example. Even though a DTM has only a single sequential tape and can perform only a very limited amount of work in a single step, a DTM program can be designed to perform any computation that can be performed on an ordinary computer, albeit more slowly. For the reader interested in how this is done, there are a number of excellent references, for example [Minsky, 1967] or [Hopcroft and Ullman, 1969]. For the reader who is *not* interested in how this is done, there is the welcome assurance that no expertise at programming DTMs will be required in this book. The reason for our introduction of the DTM model is to provide us with a formal counterpart of an algorithm upon which to base our definitions.

A formal definition of "time complexity" is now possible. The *time* used in the computation of a DTM program M on an input x is the number of steps occurring in that computation up until a halt state is entered. For a DTM program M that halts for all inputs $x \in \Sigma^*$, its *time complexity function* $T_M: Z^+ \rightarrow Z^+$ is given by

$$T_M(n) = \max \left\{ m : \begin{array}{l} \textit{there is an } x \in \Sigma^*, \textit{ with } |x|=n, \textit{ such that the} \\ \textit{computation of } M \textit{ on input } x \textit{ takes time } m \end{array} \right\}$$

Such a program M is called a *polynomial time DTM program* if there exists a polynomial p such that, for all $n \in Z^+$, $T_M(n) \leqslant p(n)$.

We are now ready to give the formal definition of the first important class of languages that we will be considering, the class P. It is defined as follows:

$$P = \{L: \textit{there is a polynomial time DTM program } M \textit{ for which } L = L_M\}$$

We will say that a decision problem Π belongs to P under the encoding scheme e if $L[\Pi,e] \in P$, that is, if there is a polynomial time DTM program that "solves" Π under encoding scheme e. In light of the previously mentioned equivalence between reasonable encoding schemes, we will usually omit the specification of a particular reasonable encoding scheme, simply saying that the decision problem Π belongs to P.

We also will be informal in our use of the term "polynomial time algorithm." Our formal counterpart for a polynomial time algorithm is the polynomial time DTM program. However, because of the equivalence between "realistic" computer models with respect to polynomial time pointed out in Chapter 1, the formal definition of P could have been rephrased in terms of programs for any such model and the same class of languages would have resulted. Thus we need not tie ourselves to the details of the DTM model when informally demonstrating that certain tasks can be performed by polynomial time algorithms. In fact, we will follow standard practice and discuss algorithms in an almost model-independent manner, speaking of them as operating directly on the components of an instance (the sets, graphs, numbers, etc.) rather than on their encoded descriptions. Here our implicit assertion is that one could, if one desired and had the patience, design a polynomial time DTM program corresponding to each polynomial time algorithm we discuss. Our informal demonstrations should be taken as indicating how this would be done and should be convincing to any reader familiar with the kinds of basic tasks that can be performed in polynomial time on an ordinary computer.

2.3 Nondeterministic Computation and the Class NP

In this section we introduce our second important class of languages/decision problems, the class NP. Before we proceed to the formal definitions in terms of languages and Turing machines, however, it will be useful to provide an intuitive idea of the informal notion this class is intended to capture.

Consider the TRAVELING SALESMAN problem described at the beginning of this chapter: Given a set of cities, the distances between them, and a bound B, does there exist a tour of all the cities having total length B or less? There is no known polynomial time algorithm for solving this problem. However, suppose someone claimed, for a particular instance of this problem, that the answer for that instance is "yes." If we were skeptical, we could demand that they "prove" their claim by providing us with a tour having the required properties. It would then be a simple matter for us to verify the truth or falsity of their claim merely by checking that what they provided us with is actually a tour and, if so, computing its length and comparing that quantity to the given bound B. Furthermore, we could specify our "verification procedure" as a general algorithm that has time complexity polynomial in Length$[I]$.

Another example of a problem with this property is the SUBGRAPH ISOMORPHISM problem of Section 2.1. Given an arbitrary instance I of this problem, consisting of two graphs $G_1=(V_1,E_1)$ and $G_2=(V_2,E_2)$, if the answer for I is "yes," then this fact can be "proved" by giving the required subsets $V' \subseteq V_1$ and $E' \subseteq E_1$ and the required one-to-one function $f: V_2 \rightarrow V'$. Again the validity of the claim can be verified easily in time polynomial in Length$[I]$, merely by checking that V', E', and f satisfy all the stated requirements.

It is this notion of polynomial time "verifiability" that the class NP is intended to isolate. Notice that polynomial time verifiability does not imply polynomial time solvability. In saying that one can verify a "yes" answer for a TRAVELING SALESMAN instance in polynomial time, we are not counting the time one might have to spend in searching among the exponentially many possible tours for one of the desired form. We merely assert that, given any tour for an instance I, we can verify in polynomial time whether or not that tour "proves" that the answer for I is "yes."

Informally we can define NP in terms of what we shall call a *nondeterministic algorithm.* We view such an algorithm as being composed of two separate stages, the first being a *guessing stage* and the second a *checking stage.* Given a problem instance I, the first stage merely "guesses" some structure S. We then provide both I and S as inputs to the checking stage, which proceeds to compute in a normal deterministic manner, either eventually halting with answer "yes," eventually halting with answer "no," or computing forever without halting (as we shall see, the latter two cases need not be distinguished). A nondeterministic algorithm "solves" a decision problem Π if the following two properties hold for all instances $I \in D_\Pi$:

1. If $I \in Y_\Pi$, then there exists some structure S that, when guessed for input I, will lead the checking stage to respond "yes" for I and S.

2. If $I \notin Y_\Pi$, then there exists *no* structure S that, when guessed for input I, will lead the checking stage to respond "yes" for I and S.

For example, a nondeterministic algorithm for TRAVELING SALESMAN could be constructed using a guessing stage that simply guesses an arbitrary sequence of the given cities and a checking stage that is identical to the aforementioned polynomial time "proof verifier" for TRAVELING SALESMAN. Clearly, for any instance I, there will exist a guess S that leads the checking stage to respond "yes" for I and S if and only if there is a tour of the desired length for I.

A nondeterministic algorithm that solves a decision problem Π is said to operate in "polynomial time" if there exists a polynomial p such that, for every instance $I \in Y_\Pi$, there is some guess S that leads the deterministic checking stage to respond "yes" for I and S within time $p(\text{Length}[I])$. Notice that this has the effect of imposing a polynomial bound on the "size" of the guessed structure S, since only a polynomially bounded amount of time can be spent examining that guess.

The class NP is defined informally to be the class of all decision problems Π that, under reasonable encoding schemes, can be solved by polynomial time nondeterministic algorithms. Our example above indicates that TRAVELING SALESMAN is one member of NP. The reader should have no difficulty in providing a similar demonstration for SUBGRAPH ISOMORPHISM.

The use of the term "solve" in these informal definitions should, of course, be taken with a grain of salt. It should be evident that a "polynomial time nondeterministic algorithm" is basically a definitional device for capturing the notion of polynomial time verifiability, rather than a realistic method for solving decision problems. Instead of having just one possible computation on a given input, it has many different ones, one for each possible guess.

There is another important way in which the "solution" of decision problems by nondeterministic algorithms differs from that for deterministic algorithms: the lack of symmetry between "yes" and "no." If the problem "Given I, is X true for I?" can be solved by a polynomial time (deterministic) algorithm, then so can the complementary problem "Given I, is X false for I?" This is because a deterministic algorithm halts for all inputs, so all we need do is interchange the "yes" and "no" responses (interchange states q_Y and q_N in a DTM program). It is not at all obvious that the same holds true for all problems solvable by polynomial time nondeterministic algorithms. Consider, for example, the complement of the TRAVELING SALESMAN problem: Given a set of cities, the intercity distances, and a bound B, is it true that *no* tour of all the cities has length B or less? There is no known way to verify a "yes" answer to this problem short of examining all possible tours (or a large proportion of them). In other words, no polynomial time nondeterministic algorithm for this complementary problem is known. The same is true of many other problems in NP. Thus, although membership in P for a problem Π implies membership in P for its complement, the analogous implication is not known to hold for NP.

We conclude this section by formalizing our definition in terms of languages and Turing machines. The formal counterpart of a nondeterministic algorithm is a program for a *nondeterministic one-tape Turing machine* (NDTM). For simplicity, we will be using a slightly non-standard NDTM model. (More standard versions are described in [Hopcroft and Ullman, 1969] and [Aho, Hopcroft, and Ullman, 1974]. The reader may find it an interesting exercise to verify the equivalence of our model to these with respect to polynomial time.)

The NDTM model we will be using has exactly the same structure as a DTM, except that it is augmented with a *guessing module* having its own *write-only head*, as illustrated schematically in Figure 2.4. The guessing module provides the means for writing down the "guess" and will be used solely for this purpose.

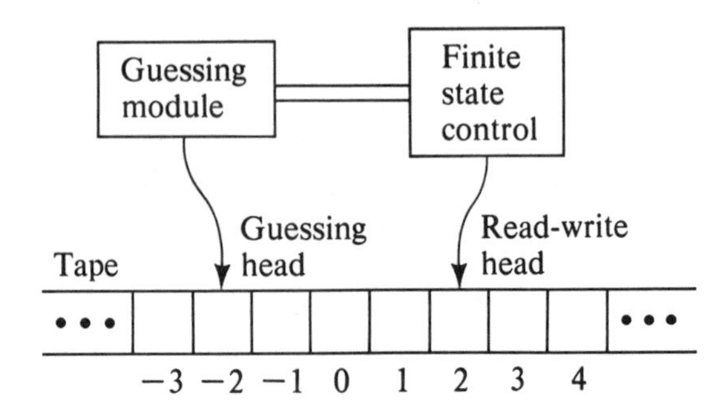

Figure 2.4 Schematic representation of a nondeterministic one-tape Turing machine (NDTM).

An *NDTM program* is specified in exactly the same way as a DTM program, including the tape alphabet Γ, input alphabet Σ, blank symbol b, state set Q, initial state q_0, halt states q_Y and q_N, and transition function $\delta: (Q-\{q_Y,q_N\}) \times \Gamma \rightarrow Q \times \Gamma \times \{-1,+1\}$. The computation of an NDTM program on an input string $x \in \Sigma^*$ differs from that of a DTM in that it takes place in two distinct stages.

The first stage is the "guessing" stage. Initially, the input string x is written in tape squares 1 through $|x|$ (while all other squares are blank), the read-write head is scanning square 1, the write-only head is scanning square -1, and the finite state control is "inactive." The guessing module then directs the write-only head, one step at a time, either to write some symbol from Γ in the tape square being scanned and move one square to the left, or to stop, at which point the guessing module becomes inactive and the finite state control is activated in state q_0. The choice of whether to remain active, and, if so, which symbol from Γ to write, is made by the guessing module in a totally arbitrary manner. Thus the guessing module can write any string from Γ^* before it halts and, indeed, need never halt.

The "checking" stage begins when the finite state control is activated in state q_0. From this point on, the computation proceeds solely under the direction of the NDTM program according to exactly the same rules as for a DTM. The guessing module and its write-only head are no longer involved, having fulfilled their role by writing the guessed string on the tape. Of course, the guessed string can (and usually will) be examined during the checking stage. The computation ceases when and if the finite state control enters one of the two halt states (either q_Y or q_N) and is said to be an *accepting computation* if it halts in state q_Y. All other computations, halting or not, are classed together simply as *non-accepting computations.*

Notice that any NDTM program M will have an infinite number of possible computations for a given input string x, one for each possible guessed string from Γ^*. We say that the NDTM program M *accepts* x if at least one of these is an accepting computation. The language *recognized* by M is

$$L_M = \{x \in \Sigma^* : M \textit{ accepts } x\}$$

The *time* required by an NDTM program M to accept the string $x \in L_M$ is defined to be the minimum, over all accepting computations of M for x, of the number of steps occurring in the guessing and checking stages up until the halt state q_Y is entered. The *time complexity function* $T_M: Z^+ \rightarrow Z^+$ for M is

$$T_M(n) = \max\left\{\{1\} \cup \left\{m: \begin{array}{l}\textit{there is an } x \in L_M \textit{ with } |x|=n \textit{ such} \\ \textit{that the time to accept } x \textit{ by } M \textit{ is } m\end{array}\right\}\right\}$$

Note that the time complexity function for M depends only on the number of steps occurring in *accepting* computations, and that, by convention, $T_M(n)$ is set equal to 1 whenever no inputs of length n are accepted by M.

The NDTM program M is a *polynomial time NDTM program* if there exists a polynomial p such that $T_M(n) \leqslant p(n)$ for all $n \geqslant 1$. Finally, the class NP is formally defined as follows:

$$\text{NP} = \{L : \textit{there is a polynomial time NDTM program } M \text{ for } \textit{which } L_M = L\}$$

It is not hard to see how these formal definitions correspond to the informal definitions that preceded them. The only point deserving special mention is that, whereas we usually envision a nondeterministic algorithm as guessing a structure S that in some way depends on the given instance I, the guessing module of an NDTM entirely disregards the given input. However, since *every* string from Γ^* is a possible guess, we can always design our NDTM program so that the checking stage begins by checking whether or not the guessed string corresponds (under the implicit interpretation our program places on strings) to an appropriate guess for the given input. If not, the program can immediately enter the halt state q_N.

A decision problem Π will be said to belong to NP under encoding scheme e if the language $L[\Pi,e] \in$ NP. As with P, we shall feel free to say that Π is in NP without giving a specific encoding scheme, so long as it is clear that *some* reasonable encoding scheme for Π will yield a language that is in NP.

Furthermore, since any realistic computer model can be augmented with an analogue of our "guessing module with write-only head," we could have rephrased our formal definitions in terms of any of the other standard models of computation. Since all these models are equivalent with respect to deterministic polynomial time, the resulting versions of NP would all be identical. Thus we will be on firm ground when, as already proposed, we identify our formally defined class NP with the class of all decision problems "solvable" by polynomial time nondeterministic algorithms.

In the next section we discuss the relationship between the two classes P and NP as a preliminary to introducing our third and, for this book, most important class, the class of NP-complete problems.

2.1 The Relationship Between P and NP

The relationship between the classes P and NP is fundamental for the theory of NP-completeness. Our first observation, which is implicit in our earlier discussions but which has not been stated explicitly until now, is that P $\subseteq$ NP. Every decision problem solvable by a polynomial time deterministic algorithm is also solvable by a polynomial time nondeterministic algorithm. To see this, one simply needs to observe that any deterministic algorithm can be used as the checking stage of a nondeterministic algorithm. If $\Pi \in$ P, and A is any polynomial time deterministic algorithm for Π, we can obtain a polynomial time nondeterministic algorithm for Π merely by using A as the checking stage and ignoring the guess. Thus $\Pi \in$ P implies $\Pi \in$ NP.

As we also hinted in our discussions, there are many reasons to believe that this inclusion is proper, that is, that P does not equal NP. Polynomial time nondeterministic algorithms certainly appear to be more powerful than polynomial time deterministic ones, and we know of no general methods for converting the former into the latter. In fact, the best general result we can state at present is given by the following:

Theorem 2.1 If $\Pi \in$ NP, then there exists a polynomial p such that Π can be solved by a deterministic algorithm having time complexity $O(2^{p(n)})$.
Proof: Suppose A is a polynomial time nondeterministic algorithm for solv-

ing Π, and let $q(n)$ be a polynomial bound on the time complexity of A. (Without loss of generality, we can assume that q can be evaluated in polynomial time, for example, by taking $q(n) = c_1 n^{c_2}$ for suitably large integer constants c_1 and c_2.) Then we know that, for every accepted input of length n, there must exist some guessed string (over the tape alphabet Γ) of length at most $q(n)$ that leads the checking stage of A to respond "yes" for that input in no more than $q(n)$ steps. Thus the number of possible guesses that need be considered is at most $k^{q(n)}$, where $k=|\Gamma|$, since guesses shorter than $q(n)$ can be regarded as guesses of length exactly $q(n)$ by filling them out with blanks. We can deterministically discover whether A has an accepting computation for a given input of length n by applying the deterministic checking stage of A, until it halts or makes $q(n)$ steps, on each of the $k^{q(n)}$ possible guesses. The simulation responds "yes" if it encounters a guessed string that leads to an accepting computation within the time bound; otherwise it responds "no." This clearly yields a deterministic algorithm for solving Π. Furthermore, its time complexity is essentially $q(n) \cdot k^{q(n)}$, which, although exponential, is $O(2^{p(n)})$ for an appropriately chosen polynomial p. ■

Of course the simulation in the proof of Theorem 2.1 could be speeded up somewhat by using branch-and-bound techniques or backtrack search and by carefully enumerating the guesses so that obviously irrelevant strings are avoided. Nevertheless, despite the considerable savings that might be achieved, there is no known way to perform this simulation in less than exponential time.

Thus the ability of a nondeterministic algorithm to check an exponential number of possibilities in polynomial time might lead one to suspect that polynomial time nondeterministic algorithms are strictly more powerful than polynomial time deterministic algorithms. Indeed, for many individual problems in NP, such as TRAVELING SALESMAN, SUBGRAPH ISOMORPHISM, and a wide variety of others, no polynomial time solution algorithms have been found despite the efforts of many knowledgeable and persistent researchers.

For these reasons, it is not surprising that there is a widespread belief that P$\neq$NP, even though no proof of this conjecture appears on the horizon. Of course, a skeptic might say that our failure to find a proof that P$\neq$NP is just as strong an argument in favor of P$=$NP as our failure to find polynomial time algorithms is an argument for the opposite view. Problems always appear to be intractable until we discover efficient algorithms for solving them. Even a skeptic would be likely to agree, however, that, given our current state of knowledge, it seems more reasonable to operate under the assumption that P$\neq$NP than to devote one's efforts to proving the contrary. In any case, we shall adopt a tentative picture of the world of NP as shown in Figure 2.5, with the expectation (but not the certainty) that the shaded region denoting NP$-$P is not totally uninhabited.

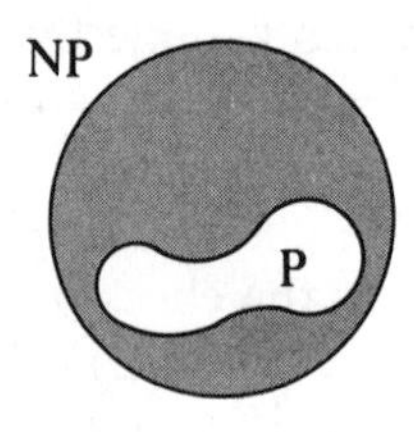

Figure 2.5 A tentative view of the world of NP.

2.5 Polynomial Transformations and NP-Completeness

If P differs from NP, then the distinction between P and NP−P is meaningful and important. All problems in P can be solved with polynomial time algorithms, whereas all problems in NP−P are intractable. Thus, given a decision problem $\Pi \in$ NP, if P$\neq$NP, we would like to know which of these two possibilities holds for Π.

Of course, until we can prove that P$\neq$NP, there is no hope of showing that any particular problem belongs to NP−P. For this reason, the theory of NP-completeness focuses on proving results of the weaker form "if P$\neq$NP, then $\Pi \in$ NP−P." We shall see that, although these conditional results might appear to be almost as difficult to prove as the corresponding unconditional results, there are techniques available that often enable us to prove them in a straightforward way. The extent to which such results should be regarded as evidence for intractability depends on how strongly one believes that P differs from NP.

The key idea used in this conditional approach is that of a polynomial transformation. A *polynomial transformation* from a language $L_1 \subseteq \Sigma_1^*$ to a language $L_2 \subseteq \Sigma_2^*$ is a function $f: \Sigma_1^* \rightarrow \Sigma_2^*$ that satisfies the following two conditions:

1. There is a polynomial time DTM program that computes f.
2. For all $x \in \Sigma_1^*$, $x \in L_1$ if and only if $f(x) \in L_2$.

If there is a polynomial transformation from L_1 to L_2, we write $L_1 \propto L_2$, read "L_1 transforms to L_2" (dropping the modifier "polynomial," which is to be understood).

The significance of polynomial transformations comes from the following lemma:

Lemma 2.1 If $L_1 \propto L_2$, then $L_2 \in$ P implies $L_1 \in$ P (and, equivalently, $L_1 \notin$ P implies $L_2 \notin$ P).

Proof: Let Σ_1 and Σ_2 be the alphabets of L_1 and L_2 respectively, let $f:\Sigma_1^* \rightarrow \Sigma_2^*$ be a polynomial transformation from L_1 to L_2, let M_f denote a polynomial time DTM program that computes f, and let M_2 be a polynomial time DTM program that recognizes L_2. A polynomial time DTM program for recognizing L_1 can be constructed by composing M_f with M_2. For an input $x \in \Sigma_1^*$, we first apply the portion corresponding to program M_f to construct $f(x) \in \Sigma_2^*$. We then apply the portion corresponding to program M_2 to determine if $f(x) \in L_2$. Since $x \in L_1$ if and only if $f(x) \in L_2$, this yields a DTM program that recognizes L_1. That this program operates in polynomial time follows immediately from the fact that M_f and M_2 are polynomial time algorithms. To be specific, if p_f and p_2 are polynomial functions bounding the running times of M_f and M_2, then $|f(x)| \leqslant p_f(|x|)$, and the running time of the constructed program is easily seen to be $O(p_f(|x|)+p_2(p_f(|x|)))$, which is bounded by a polynomial in $|x|$. ■

If Π_1 and Π_2 are decision problems, with associated encoding schemes e_1 and e_2, we shall write $\Pi_1 \propto \Pi_2$ (with respect to the given encoding schemes) whenever there exists a polynomial transformation from $L[\Pi_1,e_1]$ to $L[\Pi_2,e_2]$. As usual, we will omit the reference to specific encoding schemes when we are operating under our standard assumption that only reasonable encoding schemes are used. Thus, at the problem level, we can regard a polynomial transformation from the decision problem Π_1 to the decision problem Π_2 as a function $f: D_{\Pi_1} \rightarrow D_{\Pi_2}$ that satisfies the two conditions:

1. f is computable by a polynomial time algorithm; and
2. for all $I \in D_{\Pi_1}$, $I \in Y_{\Pi_1}$ if and only if $f(I) \in Y_{\Pi_2}$.

Let us obtain a more concrete idea of what this definition means by considering an example. For a graph $G=(V,E)$ with vertex set V and edge set E, a *simple circuit* in G is a sequence $< v_1,v_2, \ldots, v_k >$ of distinct vertices from V such that $\{v_i,v_{i+1}\} \in E$ for $1 \leqslant i < k$ and such that $\{v_k,v_1\} \in E$. A *Hamiltonian circuit* in G is a simple circuit that includes all the vertices of G. The HAMILTONIAN CIRCUIT problem is defined as follows:

HAMILTONIAN CIRCUIT
INSTANCE: A graph $G=(V,E)$.
QUESTION: Does G contain a Hamiltonian circuit?

The reader will no doubt recognize a certain similarity between this problem and the TRAVELING SALESMAN decision problem. We shall show that HAMILTONIAN CIRCUIT (HC) transforms to TRAVELING SALESMAN (TS). This requires that we specify a function f that maps each instance of HC to a corresponding instance of TS and that we prove that this function satisfies the two properties required of a polynomial transformation.

The function f is defined quite simply. Suppose $G=(V,E)$, with $|V|=m$, is a given instance of HC. The corresponding instance of TS has a set C of cities that is identical to V. For any two cities $v_i, v_j \in C$, the intercity distance $d(v_i,v_j)$ is defined to be 1 if $\{v_i,v_j\} \in E$ and 2 otherwise. The bound B on the desired tour length is set equal to m.

It is easy to see (informally) that this transformation f can be computed by a polynomial time algorithm. For each of the $m(m-1)/2$ distances $d(v_i,v_j)$ that must be specified, it is necessary only to examine G to see whether or not $\{v_i,v_j\}$ is an edge in E. Thus the first required property is satisfied. To verify that the second requirement is met, we must show that G contains a Hamiltonian circuit if and only if there is a tour of all the cities in $f(G)$ that has total length no more than B. First, suppose that $<v_1,v_2,\ldots,v_m>$ is a Hamiltonian circuit for G. Then $<v_1,v_2,\ldots,v_m>$ is also a tour in $f(G)$, and this tour has total length $m=B$ because each intercity distance traveled in the tour corresponds to an edge of G and hence has length 1. Conversely, suppose that $<v_1,v_2,\ldots,v_m>$ is a tour in $f(G)$ with total length no more than B. Since any two cities are either distance 1 or distance 2 apart, and since exactly m such distances are summed in computing the tour length, the fact that $B=m$ implies that each pair of successively visited cities must be exactly distance 1 apart. By the definition of $f(G)$, it follows that $\{v_i,v_{i+1}\}$, $1 \leqslant i < m$, and $\{v_m,v_1\}$ are all edges of G, and hence $<v_1,v_2,\ldots,v_m>$ is a Hamiltonian circuit for G.

Thus we have shown that HC $\propto$ TS. Although this proof is much simpler than many we will be describing, it contains all the essential elements of a proof of polynomial transformability and can serve as a model for how such proofs are constructed at the informal level.

The significance of Lemma 2.1 for decision problems now can be illustrated in terms of what it says about HC and TS. In essence, we conclude that if TRAVELING SALESMAN can be solved by a polynomial time algorithm, then so can HAMILTONIAN CIRCUIT, and if HC is intractable, then so is TS. Thus Lemma 2.1 allows us to interpret $\Pi_1 \propto \Pi_2$ as meaning that Π_2 is "at least as hard" as Π_1.

The "polynomial transformability" relation is especially useful because it is transitive, a fact captured by our next lemma.

Lemma 2.2 If $L_1 \propto L_2$ and $L_2 \propto L_3$, then $L_1 \propto L_3$.
Proof: Let Σ_1, Σ_2, and Σ_3 be the alphabets of languages L_1, L_2, and L_3, respectively, let $f_1: \Sigma_1^* \rightarrow \Sigma_2^*$ be a polynomial transformation from L_1 to L_2, and let $f_2: \Sigma_2^* \rightarrow \Sigma_3^*$ be a polynomial transformation from L_2 to L_3. Then the function $f: \Sigma_1^* \rightarrow \Sigma_3^*$ defined by $f(x)=f_2(f_1(x))$ for all $x \in \Sigma_1^*$ is the desired transformation from L_1 to L_3. Clearly, $f(x) \in L_3$ if and only if

$x \in L_1$, and the fact that f can be computed by a polynomial time DTM program follows from an argument analogous to that used in the proof of Lemma 2.1. ■

We can define two languages L_1 and L_2 (two decision problems Π_1 and Π_2) to be *polynomially equivalent* whenever both $L_1 \propto L_2$ and $L_2 \propto L_1$ (both $\Pi_1 \propto \Pi_2$ and $\Pi_2 \propto \Pi_1$). Lemma 2.2 tells us that this is a legitimate equivalence relation and, furthermore, that the relation "$\propto$" imposes a partial order on the resulting equivalence classes of languages (decision problems). In fact, the class P forms the "least" equivalence class under this partial order and hence can be viewed as consisting of the computationally "easiest" languages (decision problems). The class of NP-complete languages (problems) will form another such equivalence class, distinguished by the property that it contains the "hardest" languages (decision problems) in NP.

Formally, a language L is defined to be *NP-complete* if $L \in$ NP and, for all other languages $L' \in$ NP, $L' \propto L$. Informally, a decision problem Π is NP-complete if $\Pi \in$ NP and, for all other decision problems $\Pi' \in$ NP, $\Pi' \propto \Pi$. Lemma 2.1 then leads us to our identification of the NP-complete problems as "the hardest problems in NP." If any single NP-complete problem can be solved in polynomial time, then *all* problems in NP can be so solved. If any problem in NP is intractable, then so are all NP-complete problems. An NP-complete problem Π, therefore, has the property mentioned at the beginning of this section: If P$\neq$NP, then $\Pi \in$ NP$-$P. More precisely, $\Pi \in$ P if and only if P$=$NP.

Assuming that P$\neq$NP, we now can give a more detailed picture of "the world of NP," as shown in Figure 2.6. Notice that NP is not simply partitioned into "the land of P" and "the land of NP-complete." As we shall see in Chapter 7, if P differs from NP, then there must exist problems in NP that are neither solvable in polynomial time nor NP-complete.

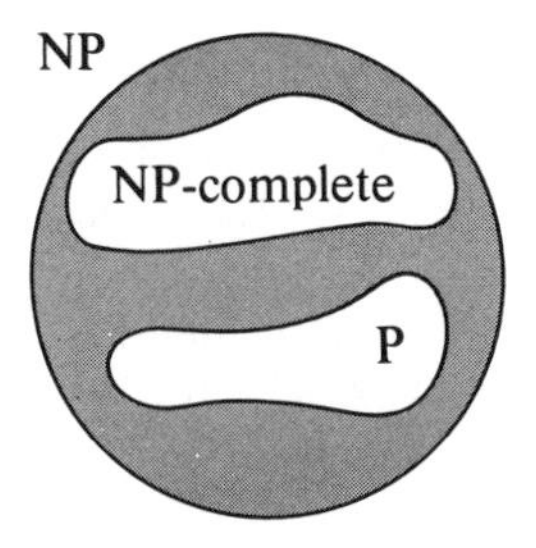

Figure 2.6 The world of NP, revisited.

Our main interest, however, is in the NP-complete problems themselves. Although we suggested at the outset of this section that there are straightforward techniques for proving that a problem is NP-complete, the requirements we have just described would appear to be rather demanding. One must show that *every* problem in NP transforms to our prospective NP-complete problem Π. It is not at all obvious how one might go about doing this. *A priori*, it is not even apparent that any NP-complete problems need exist.

The following lemma, which is an immediate consequence of our definitions and the transitivity of $\propto$, shows that matters would be simplified considerably if we possessed just one problem that we knew to be NP-complete.

Lemma 2.3 If L_1 and L_2 belong to NP, L_1 is NP-complete, and $L_1 \propto L_2$, then L_2 is NP-complete.
Proof: Since $L_2 \in$ NP, all we need to do is show that, for every $L' \in$ NP, $L' \propto L_2$. Consider any $L' \in$ NP. Since L_1 is NP-complete, it must be the case that $L' \propto L_1$. The transitivity of $\propto$ and the fact that $L_1 \propto L_2$ then imply that $L' \propto L_2$. ■

Translated to the decision problem level, this lemma gives us a straightforward approach for proving new problems NP-complete, once we have at least one known NP-complete problem available. To prove that Π is NP-complete, we merely show that

1. $\Pi \in$ NP, and
2. some known NP-complete problem Π' transforms to Π.

Before we can use this approach, however, we still need some first NP-complete problem. Such a problem is provided by Cook's fundamental theorem, which we state and prove in the next section.

2.6 Cook's Theorem

The honor of being the "first" NP-complete problem goes to a decision problem from Boolean logic, which is usually referred to as the SATISFIABILITY problem (SAT, for short). The terms we shall use in describing it are defined as follows:

Let $U = \{u_1, u_2, \ldots, u_m\}$ be a set of Boolean *variables*. A *truth assignment* for U is a function $t: U \rightarrow \{T,F\}$. If $t(u) = T$ we say that u is "true" under t; if $t(u) = F$ we say that u is "false." If u is a variable in U, then u and $\bar{u}$ are *literals* over U. The literal u is true under t if and only if the variable u is true under t; the literal $\bar{u}$ is true if and only if the variable u is false.

A *clause* over U is a set of literals over U, such as $\{u_1, \bar{u}_3, u_8\}$. It represents the disjunction of those literals and is *satisfied* by a truth assignment if and only if at least one of its members is true under that assignment. The clause above will be satisfied by t *unless* $t(u_1) = F$, $t(u_3) = T$, and $t(u_8)=F$. A collection C of clauses over U is *satisfiable* if and only if there exists some truth assignment for U that simultaneously satisfies all the clauses in C. Such a truth assignment is called a *satisfying truth assignment* for C. The SATISFIABILITY problem is specified as follows:

SATISFIABILITY
INSTANCE: A set U of variables and a collection C of clauses over U.
QUESTION: Is there a satisfying truth assignment for C?

For example, $U=\{u_1,u_2\}$ and $C=\{\{u_1,\bar{u}_2\}, \{\bar{u}_1,u_2\}\}$ provide an instance of SAT for which the answer is "yes." A satisfying truth assignment is given by $t(u_1)=t(u_2)=T$. On the other hand, replacing C by $C'=\{\{u_1,u_2\}, \{u_1,\bar{u}_2\}, \{\bar{u}_1\}\}$ yields an instance for which the answer is "no"; C' is not satisfiable.

The seminal theorem of Cook [1971] can now be stated:

Theorem 2.1 (Cook's Theorem) SATISFIABILITY is NP-complete.
Proof: SAT is easily seen to be in NP. A nondeterministic algorithm for it need only guess a truth assignment for the given variables and check to see whether that assignment satisfies all the clauses in the given collection C. This is easy to do in (nondeterministic) polynomial time. Thus the first of the two requirements for NP-completeness is met.

For the second requirement, let us revert to the language level, where SAT is represented by a language $L_{SAT} = L[SAT,e]$ for some reasonable encoding scheme e. We must show that, for all languages $L \in$ NP, $L \propto L_{SAT}$. The languages in NP are a rather diverse lot, and there are infinitely many of them, so we cannot hope to present a separate transformation for each one of them. However, each of the languages in NP can be described in a standard way, simply by giving a polynomial time NDTM program that recognizes it. This allows us to work with a generic polynomial time NDTM program and to derive a generic transformation from the language it recognizes to L_{SAT}. This generic transformation, when specialized to a particular NDTM program M recognizing the language L_M, will give the desired polynomial transformation from L_M to L_{SAT}. Thus, in essence, we will present a simultaneous proof for all $L \in$ NP that $L \propto L_{SAT}$.

To begin, let M denote an arbitrary polynomial time NDTM program, specified by Γ, Σ, b, Q, q_0, q_Y, q_N, and δ, which recognizes the language $L=L_M$. In addition, let $p(n)$ be a polynomial over the integers that bounds the time complexity function $T_M(n)$. (Without loss of generality, we can assume that $p(n) \geqslant n$ for all $n \in Z^+$.) The generic transformation f_L will be derived in terms of M, Γ, Σ, b, Q, q_0, q_Y, q_N, δ, and p.

It will be convenient to describe f_L as if it were a mapping from strings over Σ to instances of SAT, rather than to strings over the alphabet of our encoding scheme for SAT, since the details of the encoding scheme could be filled in easily. Thus f_L will have the property that for all $x \in \Sigma^*$, $x \in L$ if and only if $f_L(x)$ has a satisfying truth assignment. The key to the construction of f_L is to show how a set of clauses can be used to check whether an input x is accepted by the NDTM program M, that is, whether $x \in L$.

If the input $x \in \Sigma^*$ is accepted by M, then we know that there is an accepting computation for M on x such that both the number of steps in the checking stage and the number of symbols in the guessed string are bounded by $p(n)$, where $n = |x|$. Such a computation cannot involve any tape squares except for those numbered $-p(n)$ through $p(n)+1$, since the read-write head begins at square 1 and moves at most one square in any single step. The status of the checking computation at any one time can be specified completely by giving the contents of these squares, the current state, and the position of the read-write head. Furthermore, since there are no more than $p(n)$ steps in the checking computation, there are at most $p(n)+1$ distinct times that must be considered. This will enable us to describe such a computation completely using only a limited number of Boolean variables and a truth assignment to them.

The variable set U that f_L constructs is intended for just this purpose. Label the elements of Q as $q_0, q_1 = q_Y, q_2 = q_N, q_3, \ldots, q_r$, where $r = |Q|-1$, and label the elements of Γ as $s_0 = b, s_1, s_2, \ldots, s_v$, where $v = |\Gamma|-1$. There will be three types of variables, each of which has an intended meaning as specified in Figure 2.7. By the phrase "at time i" we mean "upon completion of the ith step of the checking computation."

Variable	Range	Intended meaning
$Q[i,k]$	$0 \leqslant i \leqslant p(n)$ $0 \leqslant k \leqslant r$	At time i, M is in state q_k.
$H[i,j]$	$0 \leqslant i \leqslant p(n)$ $-p(n) \leqslant j \leqslant p(n)+1$	At time i, the read-write head is scanning tape square j.
$S[i,j,k]$	$0 \leqslant i \leqslant p(n)$ $-p(n) \leqslant j \leqslant p(n)+1$ $0 \leqslant k \leqslant v$	At time i, the contents of tape square j is symbol s_k.

Figure 2.7 Variables in $f_L(x)$ and their intended meanings.

A computation of M induces a truth assignment on these variables in the obvious way, under the convention that, if the program halts before time $p(n)$, the configuration remains static at all later times, maintaining the same halt-state, head position, and tape contents. The tape contents at time 0 consists of the input x, written in squares 1 through n, and the guess w, written in squares -1 through $-|w|$, with all other squares blank.

On the other hand, an arbitrary truth assignment for these variables need not correspond at all to a computation, much less to an accepting computation. According to an arbitrary truth assignment, a given tape square might contain many symbols at one time, the machine might be simultaneously in several different states, and the read-write head could be in any subset of the positions $-p(n)$ through $p(n)+1$. The transformation f_L works by constructing a collection of clauses involving these variables such that a truth assignment is a *satisfying* truth assignment if and only if it is the truth assignment induced by an accepting computation for x whose checking stage takes $p(n)$ or fewer steps and whose guessed string has length at most $p(n)$. We thus will have

$x \in L$	$\Leftrightarrow$	there is an accepting computation of M on x
	$\Leftrightarrow$	there is an accepting computation of M on x with $p(n)$ or fewer steps in its checking stage and with a guessed string w of length exactly $p(n)$
	$\Leftrightarrow$	there is a satisfying truth assignment for the collection of clauses in $f_L(x)$.

This will mean that f_L satisfies one of the two conditions required of a polynomial transformation. The other condition, that f_L can be computed in polynomial time, will be verified easily once we have completed our description of f_L.

The clauses in $f_L(x)$ can be divided into six groups, each imposing a separate type of restriction on any satisfying truth assignment as given in Figure 2.8.

It is straightforward to observe that if all six clause groups perform their intended missions, then a satisfying truth assignment will have to correspond to the desired accepting computation for x. Thus all we need to show is how clause groups performing these missions can be constructed.

Group G_1 consists of the following clauses:

$$\{Q[i,0], Q[i,1], \ldots, Q[i,r]\}, \quad 0 \leqslant i \leqslant p(n)$$

$$\{\overline{Q[i,j]}, \overline{Q[i,j']}\}, \quad 0 \leqslant i \leqslant p(n), \; 0 \leqslant j < j' \leqslant r$$

The first $p(n)+1$ of these clauses can be simultaneously satisfied if and only if, for each time i, M is in *at least* one state. The remaining $(p(n)+1)(r+1)(r/2)$ clauses can be simultaneously satisfied if and only if at no time i is M in *more than one* state. Thus G_1 performs its mission.

Groups G_2 and G_3 are constructed similarly, and groups G_4 and G_5 are both quite simple, each consisting only of one-literal clauses. Figure 2.9 gives a complete specification of the first five groups. Note that the number

Clause group	Restriction imposed
G_1	At each time i, M is in exactly one state.
G_2	At each time i, the read-write head is scanning exactly one tape square.
G_3	At each time i, each tape square contains exactly one symbol from Γ.
G_4	At time 0, the computation is in the initial configuration of its checking stage for input x.
G_5	By time $p(n)$, M has entered state q_Y and hence has accepted x.
G_6	For each time i, $0 \leqslant i < p(n)$, the configuration of M at time $i+1$ follows by a single application of the transition function δ from the configuration at time i.

Figure 2.8 Clause groups in $f_L(x)$ and the restrictions they impose on satisfying truth assignments.

of clauses in these groups, and the maximum number of literals occurring in each clause, are both bounded by a polynomial function of n (since r and v are *constants* determined by M and hence by L).

The final clause group G_6, which ensures that each successive configuration in the computation follows from the previous one by a single step of program M, is a bit more complicated. It consists of two subgroups of clauses.

The first subgroup guarantees that if the read-write head is *not* scanning tape square j at time i, then the symbol in square j does not change between times i and $i+1$. The clauses in this subgroup are as follows:

$$\{\overline{S[i,j,l]}, H[i,j], S[i+1,j,l]\},\ 0 \leqslant i < p(n), -p(n) \leqslant j \leqslant p(n)+1, 0 \leqslant l \leqslant v$$

For any time i, tape square j, and symbol s_l, if the read-write head is not scanning square j at time i, and square j contains s_l at time i but not at time $i+1$, then the above clause based on i, j, and l will fail to be satisfied (otherwise it *will* be satisfied). Thus the $2(p(n)+1)^2(v+1)$ clauses in this subgroup perform their mission.

Clause group	Clauses in group
G_1	$\{Q[i,0],Q[i,1], \ldots, Q[i,r]\}$, $0 \leqslant i \leqslant p(n)$ $\{\overline{Q[i,j]},\overline{Q[i,j']}\}$, $0 \leqslant i \leqslant p(n), 0 \leqslant j < j' \leqslant r$
G_2	$\{H[i,-p(n)],H[i,-p(n)+1], \ldots, H[i,p(n)+1]\}$, $0 \leqslant i \leqslant p(n)$ $\{\overline{H[i,j]},\overline{H[i,j']}\}$, $0 \leqslant i \leqslant p(n), -p(n) \leqslant j < j' \leqslant p(n)+1$
G_3	$\{S[i,j,0],S[i,j,1], \ldots, S[i,j,v]\}$, $0 \leqslant i \leqslant p(n), -p(n) \leqslant j \leqslant p(n)+1$ $\{\overline{S[i,j,k]},\overline{S[i,j,k']}\}$, $0 \leqslant i \leqslant p(n), -p(n) \leqslant j \leqslant p(n)+1, 0 \leqslant k < k' \leqslant v$
G_4	$\{Q[0,0]\},\{H[0,1]\},\{S[0,0,0]\}$, $\{S[0,1,k_1]\},\{S[0,2,k_2]\}, \cdots, \{S[0,n,k_n]\}$, $\{S[0,n+1,0]\},\{S[0,n+2,0]\}, \ldots, \{S[0,p(n)+1,0]\}$, where $x = s_{k_1} s_{k_2} \cdots s_{k_n}$
G_5	$\{Q[p(n),1]\}$

Figure 2.9 The first five clause groups in $f_L(x)$.

The remaining subgroup of G_6 guarantees that the *changes* from one configuration to the next are in accord with the transition function δ for M. For each quadruple (i,j,k,l), $0 \leqslant i < p(n)$, $-p(n) \leqslant j \leqslant p(n)+1$, $0 \leqslant k \leqslant r$, and $0 \leqslant l \leqslant v$, this subgroup contains the following three clauses:

$$\{\overline{H[i,j]}, \overline{Q[i,k]}, \overline{S[i,j,l]}, H[i+1,j+\Delta]\}$$
$$\{\overline{H[i,j]}, \overline{Q[i,k]}, \overline{S[i,j,l]}, Q[i+1,k']\}$$
$$\{\overline{H[i,j]}, \overline{Q[i,k]}, \overline{S[i,j,l]}, S[i+1,j,l']\}$$

where if $q_k \in Q - \{q_Y, q_N\}$, then the values of Δ, k', and l' are such that $\delta(q_k, s_l) = (q_{k'}, s_{l'}, \Delta)$, and if $q_k \in \{q_Y, q_N\}$, then $\Delta = 0$, $k' = k$, and $l' = l$.

Although it may require a few minutes of thought, it is not difficult to see that these $6(p(n))\,(p(n)+1)\,(r+1)\,(v+1)$ clauses impose the desired restriction on satisfying truth assignments.

Thus we have shown how to construct clause groups G_1 through G_6 performing the previously stated missions. If $x \in L$, then there is an accepting computation of M on x of length $p(n)$ or less, and this computation, given the interpretation of the variables, imposes a truth assignment that satisfies all the clauses in $C = G_1 \cup G_2 \cup G_3 \cup G_4 \cup G_5 \cup G_6$.

Conversely, the construction of C is such that any satisfying truth assignment for C must correspond to an accepting computation of M on x. It follows that $f_L(x)$ has a satisfying truth assignment if and only if $x \in L$.

All that remains to be shown is that, for any fixed language L, $f_L(x)$ can be constructed from x in time bounded by a polynomial function of $n=|x|$. Given L, we choose a particular NDTM M that recognizes L in time bounded by a polynomial p (we need not find this NDTM itself in polynomial time, since we are only proving that the desired transformation f_L *exists*). Once we have a specific NDTM M and a specific polynomial p, the construction of the set U of variables and collection C of clauses amounts to little more than filling in the blanks in a standard (though complicated) formula. The polynomial boundedness of this computation will follow immediately once we show that Length $[f_L(x)]$ is bounded above by a polynomial function of n, where Length $[I]$ reflects the length of a string encoding the instance I under a reasonable encoding scheme, as discussed in Section 2.1. Such a "reasonable" Length function for SAT is given, for example, by $|U| \cdot |C|$. No clause can contain more than $2 \cdot |U|$ literals (that's all the literals there are), and the number of symbols required to describe an individual literal need only add an additional $\log |U|$ factor, which can be ignored when all that is at issue is polynomial boundedness. Since r and v are fixed in advance and can contribute only constant factors to $|U|$ and $|C|$, we have $|U| = O(p(n)^2)$ and $|C| = O(p(n)^2)$. Hence Length $[f_L(x)] = |U| \cdot |C| = O(p(n)^4)$, and is bounded by a polynomial function of n as desired.

Thus the transformation f_L can be computed by a polynomial time algorithm (although the particular polynomial bound it obeys will depend on L and on our choices for M and p), and we conclude that, for every $L \in \text{NP}$, f_L is a polynomial transformation from L to SAT (technically, of course, from L to L_{SAT}). It follows, as claimed, that SAT is NP-complete. ■

3

Proving NP-Completeness Results

If every NP-completeness proof had to be as complicated as that for SATISFIABILITY, it is doubtful that the class of known NP-complete problems would have grown as fast as it has. However, as discussed in Section 2.4, once we have proved a single problem NP-complete, the procedure for proving additional problems NP-complete is greatly simplified. Given a problem $\Pi \in NP$, all we need do is show that some already known NP-complete problem Π' can be transformed to Π. Thus, from now on, the process of devising an NP-completeness proof for a decision problem Π will consist of the following four steps:

(1) showing that Π is in NP,

(2) selecting a known NP-complete problem Π',

(3) constructing a transformation f from Π' to Π, and

(4) proving that f is a (polynomial) transformation.

In this chapter, we intend not only to acquaint readers with the end results of this process (the finished NP-completeness proofs) but also to prepare them for the task of constructing such proofs on their own. In Section 3.1 we present six problems that are commonly used as the "known NP-complete problem" in proofs of NP-completeness, and we prove that these six are themselves NP-complete. In Section 3.2 we describe three general approaches for transforming one problem to another, and we demonstrate their use by proving a wide variety of problems NP-complete. A concluding section contains some suggested exercises.

3.1 Six Basic NP-Complete Problems

When seasoned practitioners are confronted with a problem Π to be proved NP-complete, they have the advantage of having a wealth of experience to draw upon. They may well have proved a similar problem Π' NP-complete in the past or have seen such a proof. This will suggest that they try to prove Π NP-complete by mimicking the NP-completeness proof for Π' or by transforming Π' itself to Π. In many cases this may lead rather easily to an NP-completeness proof for Π.

All too often, however, no known NP-complete problem similar to Π can be found (even using the extensive lists at the end of this book). In such cases the practitioner may have no direct intuition as to which of the hundreds of known NP-complete problems is best suited to serve as the basis for the desired proof. Nevertheless, experience can still narrow the choices down to a core of basic problems that have been useful in the past. Even though in theory *any* known NP-complete problem can serve just as well as any other for proving a new problem NP-complete, in practice certain problems do seem to be much better suited for this task. The following six problems are among those that have been used most frequently, and we suggest that these six can serve as a "basic core" of known NP-complete problems for the beginner.

3-SATISFIABILITY (3SAT)
INSTANCE: Collection $C = \{c_1, c_2, \ldots, c_m\}$ of clauses on a finite set U of variables such that $|c_i| = 3$ for $1 \leqslant i \leqslant m$.
QUESTION: Is there a truth assignment for U that satisfies all the clauses in C?

3-DIMENSIONAL MATCHING (3DM)
INSTANCE: A set $M \subseteq W \times X \times Y$, where W, X, and Y are disjoint sets having the same number q of elements.
QUESTION: Does M contain a *matching*, that is, a subset $M' \subseteq M$ such that $|M'| = q$ and no two elements of M' agree in any coordinate?

VERTEX COVER (VC)
INSTANCE: A graph $G = (V,E)$ and a positive integer $K \leqslant |V|$.
QUESTION: Is there a *vertex cover* of size K or less for G, that is, a subset $V' \subseteq V$ such that $|V'| \leqslant K$ and, for each edge $\{u,v\} \in E$, at least one of u and v belongs to V'?

CLIQUE

INSTANCE: A graph $G = (V,E)$ and a positive integer $J \leq |V|$.

QUESTION: Does G contain a *clique* of size J or more, that is, a subset $V' \subseteq V$ such that $|V'| \geq J$ and every two vertices in V' are joined by an edge in E?

HAMILTONIAN CIRCUIT (HC)

INSTANCE: A graph $G = (V,E)$.

QUESTION: Does G contain a Hamiltonian circuit, that is, an ordering $<v_1, v_2, \ldots, v_n>$ of the vertices of G, where $n = |V|$, such that $\{v_n, v_1\} \in E$ and $\{v_i, v_{i+1}\} \in E$ for all i, $1 \leq i < n$?

PARTITION

INSTANCE: A finite set A and a "size" $s(a) \in Z^+$ for each $a \in A$.

QUESTION: Is there a subset $A' \subseteq A$ such that

$$\sum_{a \in A'} s(a) = \sum_{a \in A-A'} s(a) \ ?$$

One reason for the popularity of these six problems is that they all appeared in the original list of 21 NP-complete problems presented in [Karp, 1972]. We shall begin our illustration of the techniques for proving NP-completeness by proving that each of these six problems is NP-complete, noting, whenever appropriate, variants of these problems whose NP-completeness follows more or less directly from that of the basic problems.

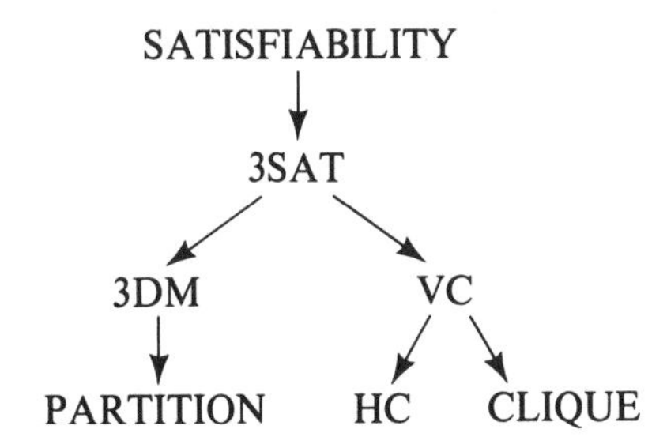

Figure 3.1 Diagram of the sequence of transformations used to prove that the six basic problems are NP-complete.

Our initial transformation will be from SATISFIABILITY, since it is the only "known" NP-complete problem we have so far. However, as we proceed through these six proofs, we will be enlarging our collection of known NP-complete problems, and all problems proved NP-complete before a problem Π will be available for use in proving that Π is NP-complete. The diagram of Figure 3.1 shows which problems we will be transforming to each of our six basic problems, where an arrow is drawn from one problem to another if the first is transformed to the second. This sequence of transformations is not identical to that used by Karp, and, even when his sequence coincides with ours, we have sometimes modified or replaced the original transformation in order to illustrate certain general proof techniques.

3.1.1 3-SATISFIABILITY

The 3-SATISFIABILITY problem is just a restricted version of SATISFIABILITY in which all instances have exactly three literals per clause. Its simple structure makes it one of the most widely used problems for proving other NP-completeness results.

Theorem 3.1. 3-SATISFIABILITY is NP-complete.

Proof: It is easy to see that 3SAT $\in$ NP since a nondeterministic algorithm need only guess a truth assignment for the variables and check in polynomial time whether that truth setting satisfies all the given three-literal clauses.

We transform SAT to 3SAT. Let $U=\{u_1,u_2, \ldots, u_n\}$ be a set of variables and $C=\{c_1,c_2, \ldots, c_m\}$ be a set of clauses making up an arbitrary instance of SAT. We shall construct a collection C' of three-literal clauses on a set U' of variables such that C' is satisfiable if and only if C is satisfiable.

The construction of C' will merely replace each individual clause $c_j \in C$ by an "equivalent" collection C_j' of three-literal clauses, based on the original variables U and some additional variables U_j' whose use will be limited to clauses in C_j'. These will be combined by setting

$$U' = U \cup \left(\bigcup_{j=1}^{m} U_j' \right)$$

and

$$C' = \bigcup_{j=1}^{m} C_j'$$

Thus we only need to show how C_j' and U_j' can be constructed from c_j.

Let c_j be given by $\{z_1, z_2, \ldots, z_k\}$ where the z_i's are all literals derived from the variables in U. The way in which C_j' and U_j' are formed depends on the value of k.

Case 1. $k=1$. $U_j'=\{y_j^1, y_j^2\}$

$C_j' = \{\{z_1,y_j^1,y_j^2\},\{z_1,y_j^1,\bar{y}_j^2\},\{z_1,\bar{y}_j^1,y_j^2\},\{z_1,\bar{y}_j^1,\bar{y}_j^2\}\}$

Case 2. $k=2$. $U_j'=\{y_j^1\}$, $C_j' = \{\{z_1,z_2,y_j^1\},\{z_1,z_2,\bar{y}_j^1\}\}$

Case 3. $k=3$. $U_j'=\phi$, $C_j' = \{\{c_j\}\}$

Case 4. $k>3$. $U_j' = \{y_j^i : 1 \leqslant i \leqslant k-3\}$

$$C_j' = \{\{z_1, z_2, y_j^1\}\} \cup \{\{\bar{y}_j^i, z_{i+2}, y_j^{i+1}\} : 1 \leqslant i \leqslant k-4\}$$
$$\cup \{\{\bar{y}_j^{k-3}, z_{k-1}, z_k\}\}$$

To prove that this is indeed a transformation, we must show that the set C' of clauses is satisfiable if and only if C is. Suppose first that $t: U \rightarrow \{T,F\}$ is a truth assignment satisfying C. We show that t can be extended to a truth assignment $t': U' \rightarrow \{T,F\}$ satisfying C'. Since the variables in $U'-U$ are partitioned into sets U_j' and since the variables in each U_j' occur only in clauses belonging to C_j', we need only show how t can be extended to the sets U_j' one at a time, and in each case we need only verify that all the clauses in the corresponding C_j' are satisfied. We can do this as follows: If U_j' was constructed under either Case 1 or Case 2, then the clauses in C_j' are already satisfied by t, so we can extend t arbitrarily to U_j', say by setting $t'(y)=T$ for all $y \in U_j'$. If U_j' was constructed under Case 3, then U_j' is empty and the single clause in C_j' is already satisfied by t. The only remaining case is Case 4, which corresponds to a clause $\{z_1, z_2, \ldots, z_k\}$ from C with $k>3$. Since t is a satisfying truth assignment for C, there must be a least integer l such that the literal z_l is set true under t. If l is either 1 or 2, then we set $t'(y_j^i)=F$ for $1 \leqslant i \leqslant k-3$. If l is either $k-1$ or k, then we set $t'(y_j^i)=T$ for $1 \leqslant i \leqslant k-3$. Otherwise we set $t'(y_j^i)=T$ for $1 \leqslant i \leqslant l-2$ and $t'(y_j^i)=F$ for $l-1 \leqslant i \leqslant k-3$. It is easy to verify that these choices will insure that all the clauses in C_j' will be satisfied, so all the clauses in C' will be satisfied by t'. Conversely, if t' is a satisfying truth assignment for C', it is easy to verify that the restriction of t' to the variables in U must be a satisfying truth assignment for C. Thus C' is satisfiable if and only if C is.

To see that this transformation can be performed in polynomial time, it suffices to observe that the number of three-literal clauses in C' is bounded by a polynomial in mn. Hence the size of the 3SAT instance is bounded above by a polynomial function of the size of the SAT instance, and, since all details of the construction itself are straightforward, the reader should have no difficulty verifying that this is a polynomial transformation. ■

The restricted structure of 3SAT makes it much more useful than SAT for proving NP-completeness results. Any proof based on SAT (except for the one we have just given) can be converted immediately to one based on 3SAT, without even changing the transformation. In fact, the normalization to clauses having the same size often can simplify the transformations we need to construct and thus make them easier to find. Furthermore, the very smallness of these clauses permits us to use transformations that would not work for instances containing larger clauses. This suggests that it would be still more convenient if we could show that the analogous 2-SATISFIABILITY problem, in which each clause has exactly *two* literals, were NP-complete. However, 2SAT can be solved by "resolution" techniques in time bounded by a polynomial in the product of the number of clauses and the number of variables in the given instance [Cook, 1971] (see also [Even, Itai, and Shamir, 1976]), and hence is in P.

3.1.2 3-DIMENSIONAL MATCHING

The 3-DIMENSIONAL MATCHING problem is a generalization of the classical "marriage problem": Given n unmarried men and n unmarried women, along with a list of all male-female pairs who would be willing to marry one another, is it possible to arrange n marriages so that polygamy is avoided and everyone receives an acceptable spouse? Analogously, in the 3-DIMENSIONAL MATCHING problem, the sets W, X, and Y correspond to *three* different sexes, and each triple in M corresponds to a *3-way marriage* that would be acceptable to all three participants. Traditionalists will be pleased to note that, whereas 3DM is NP-complete, the ordinary marriage problem can be solved in polynomial time (for example, see [Hopcroft and Karp, 1973]).

Theorem 3.2 3-DIMENSIONAL MATCHING is NP-complete.
Proof: It is easy to see that 3DM $\in$ NP, since a nondeterministic algorithm need only guess a subset of $q=|W|=|X|=|Y|$ triples from M and check in polynomial time that no two of the guessed triples agree in any coordinate.

We will transform 3SAT to 3DM. Let $U=\{u_1,u_2, \ldots, u_n\}$ be the set of variables and $C=\{c_1,c_2, \ldots, c_m\}$ be the set of clauses in an arbitrary instance of 3SAT. We must construct disjoint sets W, X, and Y, with $|W|=|X|=|Y|$, and a set $M \subseteq W \times X \times Y$ such that M contains a matching if and only if C is satisfiable.

The set M of ordered triples will be partitioned into three separate classes, grouped according to their intended function: "truth-setting and fan-out," "satisfaction testing," or "garbage collection."

Each truth-setting and fan-out component corresponds to a single variable $u \in U$, and its structure depends on the total number m of clauses in C. This structure is illustrated for the case of $m=4$ in Figure 3.2. In general, the truth-setting and fan-out component for a variable u_i involves "internal" elements $a_i[j] \in X$ and $b_i[j] \in Y$, $1 \leqslant j \leqslant m$, which will not occur in any triples outside of this component, and "external" elements $u_i[j], \bar{u}_i[j] \in W$, $1 \leqslant j \leqslant m$, which *will* occur in other triples. The triples making up this component can be divided into two sets:

$$T_i^t = \{(\bar{u}_i[j], a_i[j], b_i[j]): 1 \leqslant j \leqslant m\}$$
$$T_i^f = \{(u_i[j], a_i[j+1], b_i[j]): 1 \leqslant j < m\} \cup \{(u_i[m], a_i[1], b_i[m])\}$$

Since none of the internal elements $\{a_i[j], b_i[j]: 1 \leqslant j \leqslant m\}$ will appear in any

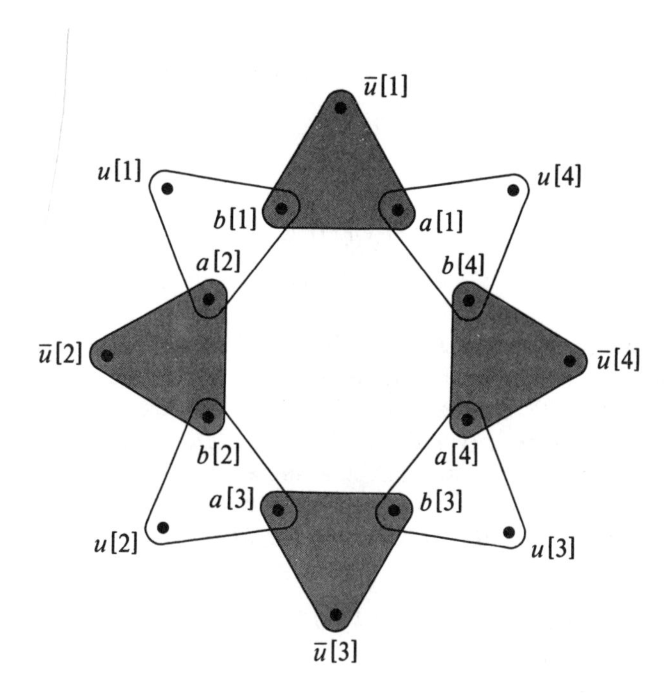

Figure 3.2 Truth setting component T_i when $m=4$ (subscripts have been deleted for simplicity). Either all the sets of T_i^t (the shaded sets) or all the sets of T_i^f (the unshaded sets) must be chosen, leaving uncovered all the $u_i[j]$ or all the $\bar{u}_i[j]$, respectively.

triples outside of $T_i = T_i^t \cup T_i^f$, it is easy to see that any matching M' will have to include exactly m triples from T_i, either all triples in T_i^t or all triples in T_i^f. Hence we can think of the component T_i as forcing a matching to make a choice between setting u_i true and setting u_i false. Thus, in general, a matching $M' \subseteq M$ specifies a truth assignment for U, with the variable u_i being set true if and only if $M' \cap T_i = T_i^t$.

Each satisfaction testing component in M corresponds to a single clause $c_j \in C$. It involves only two "internal" elements, $s_1[j] \in X$ and $s_2[j] \in Y$, and external elements from $\{u_i[j], \bar{u}_i[j]: 1 \leqslant i \leqslant n\}$, determined by which literals occur in clause c_j. The set of triples making up this component is defined as follows:

$$C_j = \{(u_i[j], s_1[j], s_2[j]): u_i \in c_j\} \cup \{(\bar{u}_i[j], s_1[j], s_2[j]): \bar{u}_i \in c_j\}$$

Thus any matching $M' \subseteq M$ will have to contain exactly one triple from C_j. This can only be done, however, if some $u_i[j]$ (or $\bar{u}_i[j]$) for a literal $u_i \in c_j$ ($\bar{u}_i \in c_j$) does not occur in the triples in $T_i \cap M'$, which will be the case if and only if the truth setting determined by M' satisfies clause c_j.

The construction is completed by means of one large "garbage collection" component G, involving internal elements $g_1[k] \in X$ and $g_2[k] \in Y$, $1 \leqslant k \leqslant m(n-1)$, and external elements of the form $u_i[j]$ and $\bar{u}_i[j]$ from W. It consists of the following set of triples:

$$G = \{(u_i[j], g_1[k], g_2[k]), (\bar{u}_i[j], g_1[k], g_2[k]): \\ 1 \leqslant k \leqslant m(n-1), 1 \leqslant i \leqslant n, 1 \leqslant j \leqslant m\}$$

Thus each pair $g_1[k], g_2[k]$ must be matched with a unique $u_i[j]$ or $\bar{u}_i[j]$ that does not occur in any triples of $M' - G$. There are exactly $m(n-1)$ such "uncovered" external elements, and the structure of G insures that they can always be covered by choosing $M' \cap G$ appropriately. Thus G merely guarantees that, whenever a subset of $M - G$ satisfies all the constraints imposed by the truth-setting and fan-out components, then that subset can be extended to a matching for M.

To summarize, we set

$$W = \{u_i[j], \bar{u}_i[j]: 1 \leqslant i \leqslant n, 1 \leqslant j \leqslant m\}$$

$$X = A \cup S_1 \cup G_1$$

where

$$A = \{a_i[j]: 1 \leqslant i \leqslant n, 1 \leqslant j \leqslant m\}$$

$$S_1 = \{s_1[j]: 1 \leqslant j \leqslant m\}$$

$$G_1 = \{g_1[j]: 1 \leqslant j \leqslant m(n-1)\}$$

$$Y = B \cup S_2 \cup G_2$$

where

$$B = \{b_i[j]: 1 \leqslant i \leqslant n, 1 \leqslant j \leqslant m\}$$

$$S_2 = \{s_2[j]: 1 \leqslant j \leqslant m\}$$

$$G_2 = \{g_2[j]: 1 \leqslant j \leqslant m(n-1)\}$$

and

$$M = \left(\bigcup_{i=1}^{n} T_i \right) \cup \left(\bigcup_{j=1}^{m} C_j \right) \cup G$$

Notice that every triple in M is an element of $W \times X \times Y$ as required. Furthermore, since M contains only

$$2mn + 3m + 2m^2 n(n-1)$$

triples and since its definition in terms of the given 3SAT instance is quite direct, it is easy to see that M can be constructed in polynomial time.

From the comments made during the description of M, it follows immediately that M cannot contain a matching unless C is satisfiable. We now must show that the existence of a satisfying truth assignment for C implies that M contains a matching.

Let $t: U \rightarrow \{T,F\}$ be any satisfying truth assignment for C. We construct a matching $M' \subseteq M$ as follows: For each clause $c_j \in C$, let $z_j \in \{u_i, \bar{u}_i : 1 \leqslant i \leqslant n\} \cap c_j$ be a literal that is set true by t (one must exist since t satisfies c_j). We then set

$$M' = \left[\bigcup_{t(u_i)=T} T_i^t \right] \cup \left[\bigcup_{t(u_i)=F} T_i^f \right] \cup \left[\bigcup_{j=1}^{m} \{(z_j[j], s_1[j], s_2[j])\} \right] \cup G'$$

where G' is an appropriately chosen subcollection of G that includes all the $g_1[k], g_2[k]$, and remaining $u_i[j]$ and $\bar{u}_i[j]$. It is easy to verify that such a G' can always be chosen and that the resulting set M' is a matching. ■

In proving NP-completeness results, the following slightly simpler and more general version of 3DM can often be used in its place:

EXACT COVER BY 3-SETS (X3C)
INSTANCE: A finite set X with $|X| = 3q$ and a collection C of 3-element subsets of X.
QUESTION: Does C contain an *exact cover* for X, that is, a subcollection $C' \subseteq C$ such that every element of X occurs in exactly one member of C'?

Note that every instance of 3DM can be viewed as an instance of X3C, simply by regarding it as an unordered subset of $W \cup X \cup Y$, and the matchings for that 3DM instance will be in one-to-one correspondence with the exact covers for the X3C instance. Thus 3DM is just a restricted version of X3C, and the NP-completeness of X3C follows by a trivial transformation from 3DM.

3.1.3 VERTEX COVER and CLIQUE

Despite the fact that VERTEX COVER and CLIQUE are independently useful for proving NP-completeness results, they are really just different ways of looking at the same problem. To see this, it is convenient to consider them in conjunction with a third problem, called INDEPENDENT SET.

An *independent set* in a graph $G = (V,E)$ is a subset $V' \subseteq V$ such that, for all $u,v \in V'$, the edge $\{u,v\}$ is *not* in E. The INDEPENDENT SET problem asks, for a given graph $G = (V,E)$ and a positive integer $J \leqslant |V|$, whether G contains an independent set V' having $|V'| \geqslant J$. The following relationships between independent sets, cliques, and vertex covers are easy to verify.

Lemma 3.1 For any graph $G=(V,E)$ and subset $V' \subseteq V$, the following statements are equivalent:

(a) V' is a vertex cover for G.
(b) $V-V'$ is an independent set for G.
(c) $V-V'$ is a clique in the *complement* G^c of G, where $G^c=(V,E^c)$ with $E^c = \{\{u,v\}: u,v \in V \text{ and } \{u,v\} \notin E\}$.

Thus we see that, in a rather strong sense, these three problems might be regarded simply as "different versions" of one another. Furthermore, the relationships displayed in the lemma make it a trivial matter to transform any one of the problems to either of the others.

For example, to transform VERTEX COVER to CLIQUE, let $G=(V,E)$ and $K \leqslant |V|$ constitute any instance of VC. The corresponding instance of CLIQUE is provided simply by the graph G^c and the integer $J=|V|-K$.

This implies that the NP-completeness of all three problems will follow as an immediate consequence of proving that any one of them is NP-complete. We choose to prove this for VERTEX COVER.

Theorem 3.3 VERTEX COVER is NP-complete.
Proof: It is easy to see that VC $\in$ NP since a nondeterministic algorithm need only guess a subset of vertices and check in polynomial time whether that subset contains at least one endpoint of every edge and has the appropriate size.

We transform 3SAT to VERTEX COVER. Let $U=\{u_1,u_2, \ldots, u_n\}$ and $C=\{c_1,c_2, \ldots, c_m\}$ be any instance of 3SAT. We must construct a graph $G=(V,E)$ and a positive integer $K \leqslant |V|$ such that G has a vertex cover of size K or less if and only if C is satisfiable.

As in the previous proof, the construction will be made up of several components. In this case, however, we will have only truth-setting components and satisfaction testing components, augmented by some additional edges for communicating between the various components.

For each variable $u_i \in U$, there is a truth-setting component $T_i=(V_i,E_i)$, with $V_i=\{u_i,\overline{u}_i\}$ and $E_i=\{\{u_i,\overline{u}_i\}\}$, that is, two vertices joined by a single edge. Note that any vertex cover will have to contain at least one of u_i and $\overline{u}_i$ in order to cover the single edge in E_i.

For each clause $c_j \in C$, there is a satisfaction testing component $S_j=(V'_j,E'_j)$, consisting of three vertices and three edges joining them to form a triangle:

$$V'_j = \{a_1[j],a_2[j],a_3[j]\}$$
$$E'_j = \{\{a_1[j],a_2[j]\},\{a_1[j],a_3[j]\},\{a_2[j],a_3[j]\}\}$$

Note that any vertex cover will have to contain at least two vertices from V_j' in order to cover the edges in E_j'.

The only part of the construction that depends on which literals occur in which clauses is the collection of communication edges. These are best viewed from the vantage point of the satisfaction testing components. For each clause $c_j \in C$, let the three literals in c_j be denoted by x_j, y_j, and z_j. Then the communication edges emanating from S_j are given by:

$$E_j'' = \{\{a_1[j],x_j\},\{a_2[j],y_j\},\{a_3[j],z_j\}\}$$

The construction of our instance of VC is completed by setting $K = n+2m$ and $G=(V,E)$, where

$$V = (\bigcup_{i=1}^{n} V_i) \cup (\bigcup_{j=1}^{m} V_j')$$

and

$$E = (\bigcup_{i=1}^{n} E_i) \cup (\bigcup_{j=1}^{m} E_j') \cup (\bigcup_{j=1}^{m} E_j'')$$

Figure 3.3 shows an example of the graph obtained when $U=\{u_1,u_2,u_3,u_4\}$ and $C=\{\{u_1,\bar{u}_3,\bar{u}_4\},\{\bar{u}_1,u_2,\bar{u}_4\}\}$.

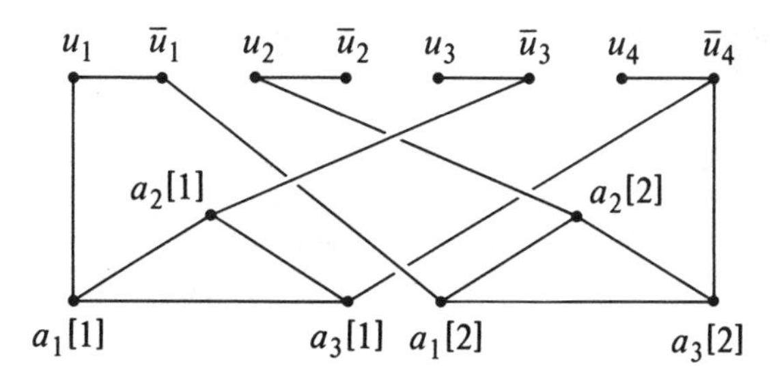

Figure 3.3 VERTEX COVER instance resulting from 3SAT instance in which $U=\{u_1,u_2,u_3,u_4\}$, $C=\{\{u_1,\bar{u}_3,\bar{u}_4\},\{\bar{u}_1,u_2,\bar{u}_4\}\}$. Here $K=n+2m=8$.

It is easy to see how the construction can be accomplished in polynomial time. All that remains to be shown is that C is satisfiable if and only if G has a vertex cover of size K or less.

First, suppose that $V' \subseteq V$ is a vertex cover for G with $|V'| \leqslant K$. By our previous remarks, V' must contain at least one vertex from each T_i and at least two vertices from each S_j. Since this gives a total of at least $n+2m=K$ vertices, V' must in fact contain *exactly* one vertex from each T_i and *exactly* two vertices from each S_j. Thus we can use the way in which V' intersects each truth-setting component to obtain a truth assignment $t: U \rightarrow \{T,F\}$. We merely set $t(u_i)=T$ if $u_i \in V'$ and $t(u_i)=F$ if

$\bar{u}_i \in V'$. To see that this truth assignment satisfies each of the clauses $c_j \in C$, consider the three edges in E_j''. Only two of those edges can be covered by vertices from $V_j' \cap V'$, so one of them must be covered by a vertex from some V_i that belongs to V'. But that implies that the corresponding literal, either u_i or $\bar{u}_i$, from clause c_j is true under the truth assignment t, and hence clause c_j is satisfied by t. Because this holds for every $c_j \in C$, it follows that t is a satisfying truth assignment for C.

Conversely, suppose that $t: U \rightarrow \{T,F\}$ is a satisfying truth assignment for C. The corresponding vertex cover V' includes one vertex from each T_i and two vertices from each S_j. The vertex from T_i in V' is u_i if $t(u_i) = T$ and is $\bar{u}_i$ if $t(u_i) = F$. This ensures that at least one of the three edges from each set E_j'' is covered, because t satisfies each clause c_j. Therefore we need only include in V' the endpoints from S_j of the other two edges in E_j'' (which may or may not also be covered by vertices from truth-setting components), and this gives the desired vertex cover. ■

3.1.4 HAMILTONIAN CIRCUIT

In Chapter 2, we saw that the HAMILTONIAN CIRCUIT problem can be transformed to the TRAVELING SALESMAN decision problem, so the NP-completeness of the latter problem will follow immediately once HC has been proved NP-complete. At the end of the proof we note several variants of HC whose NP-completeness also follows more or less directly from that of HC.

For convenience in what follows, whenever $<v_1, v_2, \ldots, v_n>$ is a Hamiltonian circuit, we shall refer to $\{v_i, v_{i+1}\}$, $1 \leqslant i < n$, and $\{v_n, v_1\}$ as the edges "in" that circuit. Our transformation is a combination of two transformations from [Karp, 1972], also described in [Liu and Geldmacher, 1978].

Theorem 3.4 HAMILTONIAN CIRCUIT is NP-complete
Proof: It is easy to see that HC $\in$ NP, because a nondeterministic algorithm need only guess an ordering of the vertices and check in polynomial time that all the required edges belong to the edge set of the given graph.

We transform VERTEX COVER to HC. Let an arbitrary instance of VC be given by the graph $G = (V,E)$ and the positive integer $K \leqslant |V|$. We must construct a graph $G' = (V',E')$ such that G' has a Hamiltonian circuit if and only if G has a vertex cover of size K or less.

Once more our construction can be viewed in terms of components connected together by communication links. First, the graph G' has K "selector" vertices $a_1, a_2, \ldots, a_K$, which will be used to select K vertices from the vertex set V for G. Second, for each edge in E, G' contains a "cover-testing" component that will be used to ensure that at least one endpoint of that edge is among the selected K vertices. The component for

$e = \{u, v\} \in E$ is illustrated in Figure 3.4. It has 12 vertices,

$$V'_e = \{(u,e,i),(v,e,i): 1 \leqslant i \leqslant 6\}$$

and 14 edges,

$$\begin{aligned} E'_e = \ & \{\{(u,e,i),(u,e,i+1)\}, \{(v,e,i),(v,e,i+1)\}: 1 \leqslant i \leqslant 5\} \\ & \cup \{\{(u,e,3),(v,e,1)\}, \{(v,e,3),(u,e,1)\}\} \\ & \cup \{\{(u,e,6),(v,e,4)\}, \{(v,e,6),(u,e,4)\}\} \end{aligned}$$

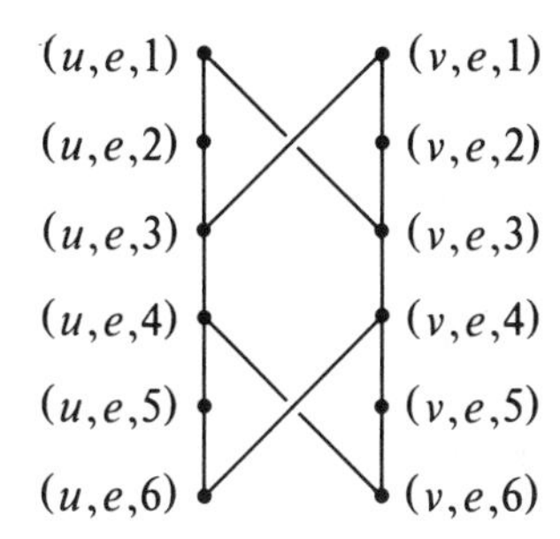

Figure 3.4 Cover-testing component for edge $e = \{u, v\}$ used in transforming VERTEX COVER to HAMILTONIAN CIRCUIT.

In the completed construction, the only vertices from this cover-testing component that will be involved in any additional edges are $(u,e,1)$, $(v,e,1)$, $(u,e,6)$, and $(v,e,6)$. This will imply, as the reader may readily verify, that any Hamiltonian circuit of G' will have to meet the edges in E'_e in exactly one of the three configurations shown in Figure 3.5. Thus, for example, if the circuit "enters" this component at $(u,e,1)$, it will have to "exit" at $(u,e,6)$ and visit either all 12 vertices in the component or just the 6 vertices (u, e, i), $1 \leqslant i \leqslant 6$.

Additional edges in our overall construction will serve to join pairs of cover-testing components or to join a cover-testing component to a selector vertex. For each vertex $v \in V$, let the edges incident on v be ordered (arbitrarily) as $e_{v[1]}, e_{v[2]}, \ldots, e_{v[deg(v)]}$, where $deg(v)$ denotes the *degree* of v in G, that is, the number of edges incident on v. All the cover-testing components corresponding to these edges (having v as endpoint) are joined together by the following connecting edges:

$$E'_v = \{\{(v,e_{v[i]},6),(v,e_{v[i+1]},1)\}: 1 \leqslant i < deg(v)\}$$

As shown in Figure 3.6, this creates a single path in G' that includes exactly those vertices (x,y,z) having $x = v$.

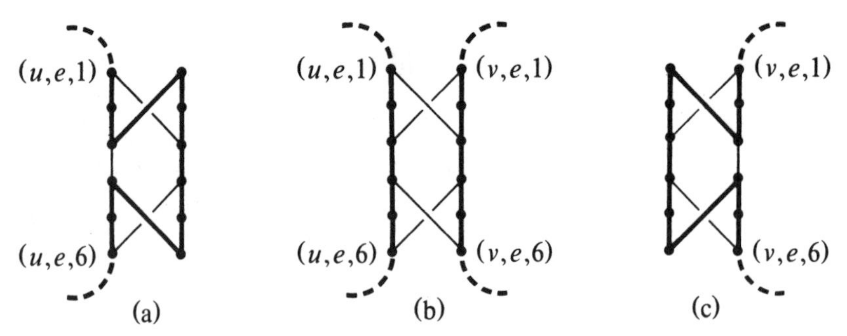

Figure 3.5 The three possible configurations of a Hamiltonian circuit within the cover-testing component for edge $e=\{u,v\}$, corresponding to the cases in which (a) u belongs to the cover but v does not, (b) both u and v belong to the cover, and (c) v belongs to the cover but u does not.

The final connecting edges in G' join the first and last vertices from each of these paths to every one of the selector vertices $a_1,a_2,\ldots,a_K$. These edges are specified as follows:

$$E'' = \{\{a_i,(v,e_{v[1]},1)\},\{a_i,(v,e_{v[deg(v)]},6)\}: 1 \leqslant i \leqslant K,\ v \in V\}$$

The completed graph $G' = (V',E')$ has

$$V' = \{a_i: 1 \leqslant i \leqslant K\} \cup (\bigcup_{e \in E} V'_e)$$

and

$$E' = (\bigcup_{e \in E} E'_e) \cup (\bigcup_{v \in V} E'_v) \cup E''$$

It is not hard to see that G' can be constructed from G and K in polynomial time.

We claim that G' has a Hamiltonian circuit if and only if G has a vertex cover of size K or less. Suppose $<v_1,v_2,\ldots,v_n>$, where $n = |V'|$, is a Hamiltonian circuit for G'. Consider any portion of this circuit that begins at a vertex in the set $\{a_1,a_2,\ldots,a_K\}$, ends at a vertex in $\{a_1,a_2,\ldots,a_K\}$, and that encounters no such vertex internally. Because of the previously mentioned restrictions on the way in which a Hamiltonian circuit can pass through a cover-testing component, this portion of the circuit must pass through a set of cover-testing components corresponding to exactly those edges from E that are incident on some one particular vertex $v \in V$. Each of the cover-testing components is traversed in one of the modes (a), (b), or (c) of Figure 3.5, and no vertex from any other cover-testing component is encountered. Thus the K vertices from $\{a_1,a_2,\ldots,a_K\}$ divide the Hamiltonian circuit into K paths, each path

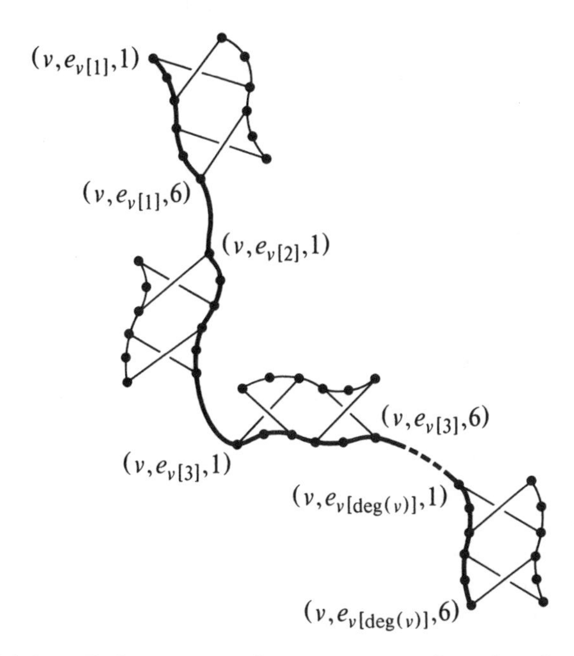

Figure 3.6 Path joining all the cover-testing components for edges from E having vertex v as an endpoint.

corresponding to a distinct vertex $v \in V$. Since the Hamiltonian circuit must include all vertices from every one of the cover-testing components, and since vertices from the cover-testing component for edge $e \in E$ can be traversed only by a path corresponding to an endpoint of e, every edge in E must have at least one endpoint among those K selected vertices. Therefore, this set of K vertices forms the desired vertex cover for G.

Conversely, suppose $V^* \subseteq V$ is a vertex cover for G with $|V^*| \leqslant K$. We can assume that $|V^*| = K$ since additional vertices from V can always be added and we will still have a vertex cover. Let the elements of V^* be labeled as $v_1, v_2, \ldots, v_K$. The following edges are chosen to be "in" the Hamiltonian circuit for G'. From the cover-testing component representing each edge $e = \{u,v\} \in E$, choose the edges specified in Figure 3.5(a), (b), or (c) depending on whether $\{u,v\} \cap V^*$ equals, respectively, $\{u\}$, $\{u,v\}$, or $\{v\}$. One of these three possibilities must hold since V^* is a vertex cover for G. Next, choose all the edges in E'_{v_i} for $1 \leqslant i \leqslant K$. Finally, choose the edges

$$\{a_i, (v_i, e_{v_i[1]}, 1)\}, 1 \leqslant i \leqslant K$$

$$\{a_{i+1},(v_i, e_{v_i[deg(v_i)]}, 6)\}, 1 \leqslant i < K$$

and

$$\{a_1,(v_K, e_{v_K[deg(v_K)]}, 6)\}$$

We leave to the reader the task of verifying that this set of edges actually corresponds to a Hamiltonian circuit for G'. ∎

Several variants of HAMILTONIAN CIRCUIT are also of interest. The HAMILTONIAN PATH problem is the same as HC except that we drop the requirement that the first and last vertices in the sequence be joined by an edge. HAMILTONIAN PATH BETWEEN TWO POINTS is the same as HAMILTONIAN PATH, except that two vertices u and v are specified as part of each instance, and we are asked whether G contains a Hamiltonian path beginning with u and ending with v. Both of these problems can be proved NP-complete using the following simple modification of the transformation just used for HC. We simply modify the graph G' obtained at the end of the construction as follows: add three new vertices, a_0, a_{K+1}, and a_{K+2}, add the two edges $\{a_0,a_1\}$ and $\{a_{K+1},a_{K+2}\}$, and replace each edge of the form $\{a_1,(v, e_{v[deg(v)]}, 6)\}$ by $\{a_{K+1},(v, e_{v[deg(v)]}, 6)\}$. The two specified vertices for the latter variation of HC are a_0 and a_{K+2}.

All three Hamiltonian problems mentioned so far also remain NP-complete if we replace the undirected graph G by a directed graph and replace the undirected Hamiltonian circuit or path by a directed Hamiltonian circuit or path. Recall that a directed graph $G=(V,A)$ consists of a vertex set V and a set of *ordered* pairs of vertices called *arcs.* A Hamiltonian path in a directed graph $G=(V,A)$ is an ordering of V as $<v_1,v_2, \ldots, v_n>$, where $n=|V|$, such that $(v_i,v_{i+1}) \in A$ for $1 \leqslant i < n$. A Hamiltonian circuit has the additional requirement that $(v_n,v_1) \in A$. Each of the three undirected Hamiltonian problems can be transformed to its directed counterpart simply by replacing each edge $\{u,v\}$ in the given undirected graph by the two arcs (u,v) and (v,u). In essence, the undirected versions are merely special cases of their directed counterparts.

3.1.5 PARTITION

In this section we consider the last of our six basic NP-complete problems, the PARTITION problem. It is particularly useful for proving NP-completeness results for problems involving numerical parameters, such as lengths, weights, costs, capacities, etc.

Theorem 3.5 PARTITION is NP-complete
Proof: It is easy to see that PARTITION $\in$ NP, since a nondeterministic algorithm need only guess a subset A' of A and check in polynomial time

that the sum of the sizes of the elements in A' is the same as that for the elements in $A-A'$.

We transform 3DM to PARTITION. Let the sets W, X, Y, with $|W|=|X|=|Y|=q$, and $M \subseteq W \times X \times Y$ be an arbitrary instance of 3DM. Let the elements of these sets be denoted by

$$W = \{w_1, w_2, \ldots, w_q\}$$
$$X = \{x_1, x_2, \ldots, x_q\}$$
$$Y = \{y_1, y_2, \ldots, y_q\}$$

and

$$M = \{m_1, m_2, \ldots, m_k\}$$

where $k=|M|$. We must construct a set A, and a size $s(a) \in Z^+$ for each $a \in A$, such that A contains a subset A' satisfying

$$\sum_{a \in A'} s(a) = \sum_{a \in A-A'} s(a)$$

if and only if M contains a matching.

The set A will contain a total of $k+2$ elements and will be constructed in two steps. The first k elements of A are $\{a_i: 1 \leqslant i \leqslant k\}$, where the element a_i is associated with the triple $m_i \in M$. The size $s(a_i)$ of a_i will be specified by giving its binary representation, in terms of a string of 0's and 1's divided into $3q$ "zones" of $p = \lceil \log_2(k+1) \rceil$ bits each. Each of these zones is labeled by an element of $W \cup X \cup Y$, as shown in Figure 3.7.

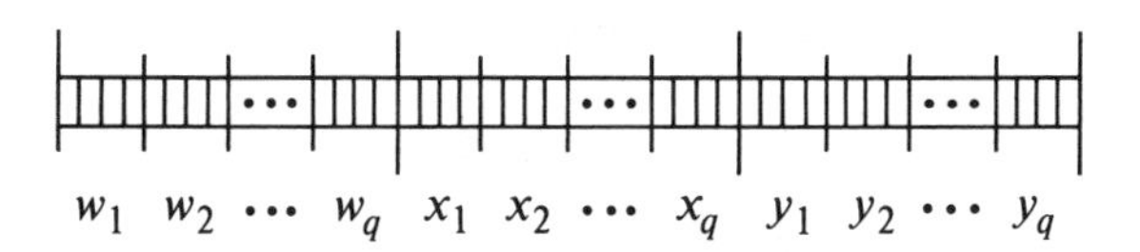

Figure 3.7 Labeling of the $3q$ "zones," each containing $p = \lceil \log_2(k+1) \rceil$ bits of the binary representation for $s(a)$, used in transforming 3DM to PARTITION.

The representation for $s(a_i)$ depends on the corresponding triple $m_i = (w_{f(i)}, x_{g(i)}, y_{h(i)}) \in M$ (where f, g, and h are just the functions that give the subscripts of the first, second, and third components for each m_i). It has a 1 in the rightmost bit position of the zones labeled by $w_{f(i)}$, $x_{g(i)}$, and $y_{h(i)}$ and 0's everywhere else. Alternatively, we can write

$$s(a_i) = 2^{p(3q-f(i))} + 2^{p(2q-g(i))} + 2^{p(q-h(i))}$$

Since each $s(a_i)$ can be expressed in binary with no more than $3pq$ bits, it is clear that $s(a_i)$ can be constructed from the given 3DM instance in polynomial time.

The important thing to observe about this part of the construction is that, if we sum up all the entries in any zone, over all elements of $\{a_i: 1 \leqslant i \leqslant k\}$, the total can never exceed $k = 2^p - 1$. Hence, in adding up $\sum_{a \in A'} s(a)$ for any subset $A' \subseteq \{a_i: 1 \leqslant i \leqslant k\}$, there will never be any "carries" from one zone to the next. It follows that if we let

$$B = \sum_{j=0}^{3q-1} 2^{pj}$$

(which is the number whose binary representation has a 1 in the rightmost position of every zone), then any subset $A' \subseteq \{a_i: 1 \leqslant i \leqslant k\}$ will satisfy

$$\sum_{a \in A'} s(a) = B$$

if and only if $M' = \{m_i: a_i \in A'\}$ is a matching for M.

The final step of the construction specifies the last two elements of A. These are denoted by b_1 and b_2 and have sizes defined by

$$s(b_1) = 2\left(\sum_{i=1}^{k} s(a_i)\right) - B$$

and

$$s(b_2) = \left(\sum_{i=1}^{k} s(a_i)\right) + B$$

Both of these can be specified in binary with no more than $(3pq+1)$ bits and thus can be constructed in time polynomial in the size of the given 3DM instance.

Now suppose we have a subset $A' \subseteq A$ such that

$$\sum_{a \in A'} s(a) = \sum_{a \in A - A'} s(a)$$

Then both of these sums must be equal to $2\sum_{i=1}^{k} s(a_i)$, and one of the two sets, A' or $A - A'$, contains b_1 but not b_2. It follows that the remaining elements of that set form a subset of $\{a_i: 1 \leqslant i \leqslant k\}$ whose sizes sum to B, and hence, by our previous comments, that subset corresponds to a matching M' in M. Conversely, if $M' \subseteq M$ is a matching, then the set $\{b_1\} \cup \{a_i: m_i \in M'\}$ forms the desired set A' for the PARTITION instance. Therefore, 3DM $\propto$ PARTITION, and the theorem is proved. ■

3.2 Some Techniques for Proving NP-Completeness

The techniques used for proving NP-completeness results vary almost as widely as the NP-complete problems themselves, and we cannot hope to illustrate them all here. However, there are several general types of proofs that occur frequently and that can provide a suggestive framework for deciding how to go about proving a new problem NP-complete. We call these (a) restriction, (b) local replacement, and (c) component design.

In this section we shall indicate what we mean by each of these proof types, primarily by giving examples. It would be sheer folly to attempt to define them explicitly. Many proofs can be interpreted in ways that would place them arbitrarily in any one of the three categories. Other proofs depend on decidedly problem-specific methods, so that no such limited set of categories could possibly include them in a natural way. Thus, we caution the reader *not* to interpret this as a way to classify all NP-completeness proofs. Rather, our sole intent is to illustrate several ways of thinking about NP-completeness proofs that the authors (and others) have found to be both intuitively appealing and constructive.

For brevity in what follows, we shall be omitting from all our proofs the verification that the given problem is in NP. Each of the problems we consider is easily seen to be solvable in polynomial time by a nondeterministic algorithm, and the reader should have no difficulty supplying such an algorithm whenever required.

3.2.1 Restriction

Proof by restriction is the simplest, and perhaps the most frequently applicable, of our three proof types. An NP-completeness proof by restriction for a given problem $\Pi \in$ NP consists simply of showing that Π contains a known NP-complete problem Π' as a special case. The heart of such a proof lies in the specification of the additional restrictions to be placed on the instances of Π so that the resulting restricted problem will be identical to Π'. We do not require that the restricted problem and the known NP-complete problem be *exact* duplicates of one another, but rather that there be an "obvious" one-to-one correspondence between their instances that preserves "yes" and "no" answers. This one-to-one correspondence, which provides the required transformation from Π' to Π, is usually so apparent that it need not even be given explicitly.

We have already seen several examples of this type of proof. In Section 3.1.2, the problem EXACT COVER BY 3-SETS was shown to be NP-complete by restricting its instances to 3-sets that contain one element from a set W, one from a set X, and one from a set Y, where W, X, and Y are disjoint sets having the same cardinality, thereby obtaining a problem identical to the 3DM problem. In Section 3.1.4, DIRECTED HAMILTONIAN

CIRCUIT was shown to be NP-complete by restricting its instances to directed graphs in which each arc (u,v) occurs only in conjunction with the oppositely directed arc (v,u), thereby obtaining a problem identical to the undirected HAMILTONIAN CIRCUIT problem.

Thus proofs by restriction can be seen to embody a different way of looking at things than the standard NP-completeness proofs. Instead of trying to discover a way of transforming a known NP-complete problem to our target problem, we focus on the target problem itself and attempt to restrict away its "inessential" aspects until a known NP-complete problem appears.

We now give a number of additional examples of problems proved NP-complete by restriction, stating each proof with the brevity it deserves.

(1) **MINIMUM COVER**
INSTANCE: Collection C of subsets of a set S, positive integer K.
QUESTION: Does C contain a *cover* for S of size K or less, that is, a subset $C' \subseteq C$ with $|C'| \leqslant K$ and such that $\bigcup_{c \in C'} c = S$?
Proof: Restrict to X3C by allowing only instances having $|c|=3$ for all $c \in C$ and having $K = |S|/3$.

(2) **HITTING SET**
INSTANCE: Collection C of subsets of a set S, positive integer K.
QUESTION: Does S contain a *hitting set* for C of size K or less, that is, a subset $S' \subseteq S$ with $|S'| \leqslant K$ and such that S' contains at least one element from each subset in C?
Proof: Restrict to VC by allowing only instances having $|c|=2$ for all $c \in C$.

(3) **SUBGRAPH ISOMORPHISM**
INSTANCE: Two graphs, $G=(V_1,E_1)$ and $H=(V_2,E_2)$.
QUESTION: Does G contain a subgraph *isomorphic* to H, that is, a subset $V \subseteq V_1$ and a subset $E \subseteq E_1$ such that $|V|=|V_2|, |E|=|E_2|$, and there exists a one-to-one function $f: V_2 \rightarrow V$ satisfying $\{u,v\} \in E_2$ if and only if $\{f(u),f(v)\} \in E$?
Proof: Restrict to CLIQUE by allowing only instances for which H is a complete graph, that is, E_2 contains all possible edges joining two members of V_2.

(4) **BOUNDED DEGREE SPANNING TREE**
INSTANCE: A graph $G=(V,E)$ and a positive integer $K \leqslant |V|-1$.
QUESTION: Is there a *spanning tree* for G in which no vertex has degree exceeding K, that is, a subset $E' \subseteq E$ such that $|E'|=|V|-1$, the graph $G'=(V,E')$ is connected, and no vertex in V is included in more than K edges from E'?
Proof: Restrict to HAMILTONIAN PATH by allowing only instances in which $K=2$.

(5) **MINIMUM EQUIVALENT DIGRAPH**
INSTANCE: A directed graph $G=(V,A)$ and a positive integer $K \leqslant |A|$.
QUESTION: Is there a directed graph $G'=(V,A')$ such that $A' \subseteq A$, $|A'| \leqslant K$, and such that, for every pair of vertices u and v in V, G' contains a directed path from u to v if and only if G contains a directed path from u to v.
Proof: Restrict to DIRECTED HAMILTONIAN CIRCUIT by allowing only instances in which G is strongly connected, that is, contains a path from every vertex u to every vertex v, and $K=|V|$. Note that this is actually a restriction to DIRECTED HAMILTONIAN CIRCUIT FOR STRONGLY CONNECTED DIGRAPHS, but the NP-completeness of that problem follows immediately from the constructions we gave for HC and DIRECTED HC.

(6) **KNAPSACK**
INSTANCE: A finite set U, a "size" $s(u) \in Z^+$ and a "value" $v(u) \in Z^+$ for each $u \in U$, a size constraint $B \in Z^+$, and a value goal $K \in Z^+$.
QUESTION: Is there a subset $U' \subseteq U$ such that

$$\sum_{u \in U'} s(u) \leqslant B \quad \text{and} \quad \sum_{u \in U'} v(u) \geqslant K$$

Proof: Restrict to PARTITION by allowing only instances in which $s(u)=v(u)$ for all $u \in U$ and $B=K=\frac{1}{2}\sum_{u \in U} s(u)$.

(7) **MULTIPROCESSOR SCHEDULING**
INSTANCE: A finite set A of "tasks," a "length" $l(a) \in Z^+$ for each $a \in A$, a number $m \in Z^+$ of "processors," and a "deadline" $D \in Z^+$.
QUESTION: Is there a partition $A = A_1 \cup A_2 \cup \cdots \cup A_m$ of A into m disjoint sets such that

$$\max \left\{ \sum_{a \in A_i} l(a) : 1 \leqslant i \leqslant m \right\} \leqslant D \ ?$$

Proof: Restrict to PARTITION by allowing only instances in which $m=2$ and $D=\frac{1}{2}\sum_{a \in A} l(a)$.

As a final comment, we observe that, of all the approaches to proving NP-completeness we shall discuss, proof by restriction is the one that would profit most from an extensive knowledge of the class of known NP-complete problems — beyond the basic six and their variants. Many problems that arise in practice are simply more complicated versions of problems that appear on our lists of NP-complete problems, and the ability to recognize this can often lead to a quick NP-completeness proof by restriction.

3.2.2 Local Replacement

In proofs by local replacement, the transformations are sufficiently nontrivial to warrant spelling out in the standard proof format, but they still tend to be relatively uncomplicated. All we do is pick some aspect of the known NP-complete problem instance to make up a collection of *basic units*, and we obtain the corresponding instance of the target problem by replacing each basic unit, in a uniform way, with a different structure. The transformation from SAT to 3SAT in Section 3.1.1 was of this type. In that transformation, the basic units of an instance of SAT were the clauses, and each clause was replaced by a collection of clauses according to the same general rule. The key point to observe is that each replacement constituted only local modification of structure. The replacements were essentially independent of one another, except insofar as they reflected parts of the original instance that were not changed.

Let us flesh these generalities out with some more examples. The following decision problem corresponds to a problem of minimizing the number of multiplications needed to compute a given collection of products of elementary terms, where the multiplication operation is assumed to be associative and commutative:

ENSEMBLE COMPUTATION
INSTANCE: A collection C of subsets of a finite set A and a positive integer J.
QUESTION: Is there a sequence

$$<z_1 = x_1 \cup y_1, z_2 = x_2 \cup y_2, \ldots, z_j = x_j \cup y_j>$$

of $j \leqslant J$ union operations, where each x_i and y_i is either $\{a\}$ for some $a \in A$ or z_k for some $k < i$, such that x_i and y_i are disjoint for $1 \leqslant i \leqslant j$ and such that for every subset $c \in C$ there is some z_i, $1 \leqslant i \leqslant j$, that is identical to c?

Theorem 3.6 ENSEMBLE COMPUTATION is NP-complete.
Proof: We transform VERTEX COVER to ENSEMBLE COMPUTATION. Let the graph $G = (V, E)$ and the positive integer $K \leqslant |V|$ constitute an arbitrary instance of VC.

The basic units of the instance of VC are the edges of G. Let a_0 be some new element not in V. The local replacement just substitutes for each edge $\{u, v\} \in E$ the subset $\{a_0, u, v\} \in C$. The instance of ENSEMBLE COMPUTATION is completely specified by:

$$A = V \cup \{a_0\}$$
$$C = \{\{a_0, u, v\} : \{u, v\} \in E\}$$
$$J = K + |E|$$

It is easy to see that this instance can be constructed in polynomial time. We claim that G has a vertex cover of size K or less if and only if the desired sequence of $j \leqslant J$ operations exists for C.

First, suppose V' is a vertex cover for G of size K or less. Since we can add additional vertices to V' and it will remain a vertex cover, there is no loss of generality in assuming that $|V'| = K$. Label the elements of V' as $v_1, v_2, \ldots, v_K$ and label the edges in E as $e_1, e_2, \ldots, e_m$, where $m = |E|$. Since V' is a vertex cover, each edge e_j contains at least one element from V'. Thus we can write each e_j as $e_j = \{u_j, v_{r[j]}\}$, where $r[j]$ is an integer satisfying $1 \leqslant r[j] \leqslant K$. The following sequence of $K + |E| = J$ operations is easily seen to have all the required properties:

$$< z_1 = \{a_0\} \cup \{v_1\},\ z_2 = \{a_0\} \cup \{v_2\},\ \ldots,\ z_k = \{a_0\} \cup \{v_K\},$$
$$z_{K+1} = \{u_1\} \cup z_{r[1]},\ z_{K+2} = \{u_2\} \cup z_{r[2]},\ \ldots,\ z_J = \{u_m\} \cup z_{r[m]} >$$

Conversely, suppose $S = < z_1 = x_1 \cup y_1, \ldots, z_j = x_j \cup y_j >$ is the desired sequence of $j \leqslant J$ operations for the ENSEMBLE COMPUTATION instance. Furthermore, let us assume that S is the shortest such sequence for this instance and that, among all such minimum sequences, S contains the fewest possible operations of the form $z_i = \{u\} \cup \{v\}$ for $u, v \in V$. Our first claim is that S can contain *no* operations of this latter form. For suppose that $z_i = \{u\} \cup \{v\}$ with $u, v \in V$ is included. Since $\{u, v\}$ is not in C and since S has minimum length, we must have $\{u, v\} \in E$, and $\{a_0, u, v\} = \{a_0\} \cup z_i$ (or $z_i \cup \{a_0\}$) must occur later in S. However, since $\{u, v\}$ is a subset of only one member of C, z_i cannot be used in any other operation in this minimum length sequence. It follows that we can replace the two operations

$$z_i = \{u\} \cup \{v\} \text{ and } \{a_0, u, v\} = \{a_0\} \cup z_i$$

by

$$z_i = \{a_0\} \cup \{u\} \text{ and } \{a_0, u, v\} = \{v\} \cup z_i$$

thereby reducing the number of proscribed operations without lengthening the overall sequence, a contradiction to the choice of S. Hence S consists only of operations having one of the two forms, $z_i = \{a_0\} \cup \{u\}$ for $u \in V$ or $\{a_0, u, v\} = \{v\} \cup z_i$ for $\{u, v\} \in E$ (where we disregard the relative order of the two operands in each case). Because $|C| = |E|$ and because every member of C contains three elements, S must contain exactly $|E|$ operations of the latter form and exactly $j - |E| \leqslant J - |E| = K$ of the former.

Therefore the set

$$V' = \{u \in V\colon z_i = \{a_0\} \cup \{u\} \text{ is an operation in } S\}$$

contains at most K vertices from V and, as can be verified easily from the construction of C, must be a vertex cover for G. ■

Another example of a polynomial time transformation using local replacement, this time from EXACT COVER BY 3-SETS, is the following:

PARTITION INTO TRIANGLES
INSTANCE: A graph $G=(V,E)$, with $|V|=3q$ for a positive integer q.
QUESTION: Is there a partition of V into q disjoint sets $V_1, V_2, \ldots, V_q$ of three vertices each such that, for each $V_i = \{v_{i[1]}, v_{i[2]}, v_{i[3]}\}$, the three edges $\{v_{i[1]}, v_{i[2]}\}$, $\{v_{i[1]}, v_{i[3]}\}$, and $\{v_{i[2]}, v_{i[3]}\}$ all belong to E?

Theorem 3.7 PARTITION INTO TRIANGLES is NP-complete.
Proof: We transform EXACT COVER BY 3-SETS to PARTITION INTO TRIANGLES. Let the set X with $|X|=3q$ and the collection C of 3-element subsets of X be an arbitrary instance of X3C. We shall construct a graph $G=(V,E)$, with $|V|=3q'$, such that the desired partition exists for G if and only if C contains an exact cover.

The basic units of the X3C instance are the 3-element subsets in C. The local replacement substitutes for each such subset $c_i = \{x_i, y_i, z_i\} \in C$ the collection E_i of 18 edges shown in Figure 3.8. Thus $G=(V,E)$ is defined by

$$V = X \cup \bigcup_{i=1}^{|C|} \{a_i[j]\colon 1 \leqslant j \leqslant 9\}$$

$$E = \bigcup_{i=1}^{|C|} E_i$$

Notice that the only vertices that appear in edges belonging to more than a single E_i are those that are in the set X. Notice also that $|V| = |X| + 9|C| = 3q + 9|C|$ so that $q' = q + 3|C|$. It is not hard to see that this instance of PARTITION INTO TRIANGLES can be constructed in polynomial time from the X3C instance.

If $c_1, c_2, \ldots, c_q$ are the 3-element subsets from C in any exact cover for X, then the corresponding partition $V = V_1 \cup V_2 \cup \cdots \cup V_{q'}$ of V is given by taking

$$\{a_i[1], a_i[2], x_i\},\ \{a_i[4], a_i[5], y_i\}$$
$$\{a_i[7], a_i[8], z_i\},\ \{a_i[3], a_i[6], a_i[9]\}$$

from the vertices meeting E_i whenever $c_i = \{x_i, y_i, z_i\}$ is in the exact cover,

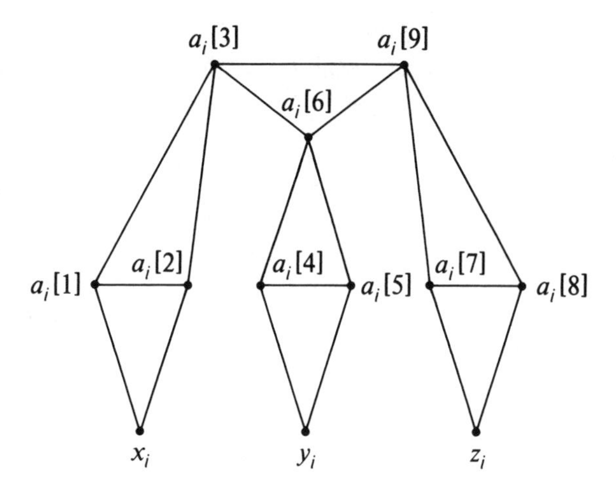

Figure 3.8 Local replacement for $c_i = (x_i, y_i, z_i) \in C$ for transforming X3C to PARTITION INTO TRIANGLES.

and by taking

$$\{a_i[1], a_i[2], a_i[3]\},\ \{a_i[4], a_i[5], a_i[6]\},\ \{a_i[7], a_i[8], a_i[9]\}$$

from the vertices meeting E_i whenever c_i is *not* in the exact cover. This ensures that each element of X is included in exactly one 3-vertex subset in the partition.

Conversely, if $V = V_1 \cup V_2 \cup \cdots \cup V_{q'}$ is any partition of G into triangles, the corresponding exact cover is given by choosing those $c_i \in C$ such that $\{a_i[3], a_i[6], a_i[9]\} = V_j$ for some j, $1 \leqslant j \leqslant q'$. We leave to the reader the straightforward task of verifying that the two partitions we have constructed are as claimed. ■

Both examples we have just seen represent what might be called "pure" local replacement proofs. The structure of the target instance was completely determined by the structure of the given problem instance and the local replacements. It is often advantageous to augment this with a limited amount of additional structure that acts as an "enforcer,"† imposing certain additional restrictions on the ways in which a "yes" answer to the target instance can be obtained. For a target problem having the form "Given an instance I, does there exist an X_I having the desired property?" the enforcer portion of I acts to limit the possible X_I's so that the remaining choices all mirror the choices available in the original problem instance, whereas that portion of I obtained by applying local replacement to the original instance provides the means for making those choices and for ensuring that they have the desired properties. The two elements b_1 and b_2 in the

† A picturesque term suggested by Szymanski [1978].

NP-completeness proof for PARTITION acted as such an enforcer. We give two further examples of local replacement proofs using enforcers, beginning with that for the following scheduling problem:

SEQUENCING WITHIN INTERVALS
INSTANCE: A finite set T of "tasks" and, for each $t \in T$, an integer "release time" $r(t) \geqslant 0$, a "deadline" $d(t) \in Z^+$, and a "length" $l(t) \in Z^+$.
QUESTION: Does there exist a *feasible schedule* for T, that is, a function $\sigma: T \rightarrow Z^+$ such that, for each $t \in T$, $\sigma(t) \geqslant r(t)$, $\sigma(t)+l(t) \leqslant d(t)$, and, if $t' \in T-\{t\}$, then either $\sigma(t')+l(t') \leqslant \sigma(t)$ or $\sigma(t') \geqslant \sigma(t)+l(t)$? (The task t is "executed" from time $\sigma(t)$ to time $\sigma(t)+l(t)$, cannot start executing until time $r(t)$, must be completed by time $d(t)$, and its execution cannot overlap the execution of any other task t'.)

Theorem 3.8 SEQUENCING WITHIN INTERVALS is NP-complete.
Proof: We transform PARTITION to this problem. Let the finite set A and given size $s(a)$ for each $a \in A$ constitute an arbitrary instance of PARTITION, and let $B = \sum_{a \in A} s(a)$.

The basic units of the PARTITION instance are the individual elements $a \in A$. The local replacement for each $a \in A$ is a single task t_a with $r(t_a)=0$, $d(t_a)=B+1$, and $l(t_a)=s(a)$. The "enforcer" is a single task $\bar{t}$ with $r(\bar{t})=\lceil B/2 \rceil$, $d(\bar{t})=\lceil (B+1)/2 \rceil$, and $l(\bar{t})=1$. Clearly, this instance can be constructed in polynomial time from the PARTITION instance.

The restrictions imposed on feasible schedules by the enforcer are two-fold. First, it ensures that a feasible schedule cannot be constructed whenever B is an odd integer (in which case the desired subset for the PARTITION instance cannot exist), because then we would have $r(\bar{t})=d(\bar{t})$, so that $\bar{t}$ could not possibly be scheduled. Thus from now on, let us assume that B is even. In this case the second restriction comes to the forefront. Since B is even, $r(\bar{t})=B/2$ and $d(\bar{t})=r(\bar{t})+1$, so that any feasible schedule must have $\sigma(\bar{t})=B/2$. This divides the time available for scheduling the remaining tasks into two separate blocks, each of total length $B/2$, as illustrated in Figure 3.9. Thus the scheduling problem is turned into a problem of selecting subsets, those that are scheduled before $\bar{t}$ and those that are scheduled after $\bar{t}$. Since the total amount of time available in the two blocks equals the total length B of the remaining tasks, it follows that each block must be filled up exactly. However, this can be done if and only if there is a subset $A' \subseteq A$ such that

$$\sum_{a \in A'} s(a) = B/2 = \sum_{a \in A-A'} s(a)$$

Thus the desired subset A' exists for the instance of PARTITION if and only if a feasible schedule exists for the corresponding instance of SEQUENCING WITHIN INTERVALS. ■

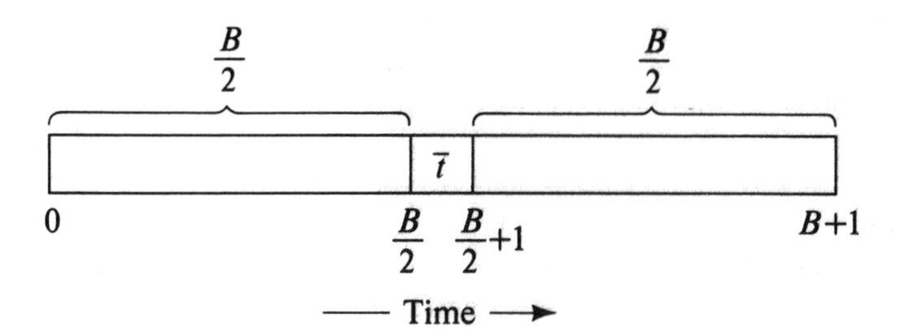

Figure 3.9 Schedule "enforced" by the transformation from PARTITION to SEQUENCING WITHIN INTERVALS.

Our final example of the use of an enforcer in a local replacement proof involves the following problem of diagnostic testing:

MINIMUM TEST COLLECTION
INSTANCE: A finite set A of "possible diagnoses," a collection C of subsets of A, representing binary "tests," and a positive integer $J \leq |C|$.
QUESTION: Is there a subcollection $C' \subseteq C$ with $|C'| \leq J$ such that, for every pair a_i, a_j of possible diagnoses from A, there is some test $c \in C'$ for which $|\{a_i, a_j\} \cap c| = 1$ (that is, a test c that "distinguishes" between a_i and a_j)?

Theorem 3.9 MINIMUM TEST COLLECTION is NP-complete.
Proof: We transform 3DM to this problem. Let the sets W, X, Y, with $|W| = |X| = |Y| = q$, and the collection $M \subseteq W \times X \times Y$ constitute an arbitrary instance of 3DM.

The basic units of the 3DM instance are the ordered triples in M. The local replacement substitutes for each $m = (w, x, y) \in M$ the subset $\{w, x, y\} \in C$. The enforcer is provided by three additional elements, w_0, x_0, and y_0, not belonging to $W \cup X \cup Y$, and two additional tests, $W \cup \{w_0\}$ and $X \cup \{x_0\}$. The complete MINIMUM TEST COLLECTION instance is defined by:

$$A = W \cup X \cup Y \cup \{w_0, x_0, y_0\}$$
$$C = \Big\{\{w, x, y\}: (w, x, y) \in M\Big\} \cup \Big\{ W \cup \{w_0\}, X \cup \{x_0\}\Big\}$$
$$J = q + 2$$

It is easy to see that this instance can be constructed in polynomial time from the given 3DM instance.

Once again the enforcer places certain limitations on the form of the desired entity (in this case, the subcollection C' of tests). First, C' must contain both $W \cup \{w_0\}$ and $X \cup \{x_0\}$, since they are the only tests that distinguish y_0 from w_0 and x_0. Then, since w_0, x_0, and y_0 are not contained in any other tests in C, each element of $W \cup X \cup Y$ must be distinguished from the appropriate one of w_0, x_0, or y_0 by being included in some additional test $c \in C' - \{W \cup \{w_0\}, X \cup \{x_0\}\}$. At most $J-2=q$ such additional tests can be included. Because each of the remaining tests in C contains exactly one member from each of W, X, and Y, and because W, X, and Y are disjoint sets, having q members each, it follows that any such additional q tests in C' must correspond to q triples that form a matching for M. Conversely, given any matching for M, the corresponding q tests from C can be used to complete the desired collection of $J=q+2$ tests. Thus M contains a matching if and only if the required subcollection of tests from C exists. ■

Although the enforcers in both our examples are quite simple, the reader should be placed on notice that this need not always be the case. A particularly complicated enforcing structure is used in the NP-completeness proof for PLANAR DIRECTED HAMILTONIAN PATH in [Garey, Johnson, and Stockmeyer, 1976]. Other relatively complicated enforcers can be found in [Liu and Geldmacher, 1978], [Garey, Johnson, and Sethi, 1976], and [Garey, Graham, Johnson, and Knuth, 1978].

3.2.3 Component Design

Our last type of proof, and the one that tends to be the most complicated, is component design. The NP-completeness proofs given in Section 3.1 for 3-DIMENSIONAL MATCHING, VERTEX COVER, and HAMILTONIAN CIRCUIT are typical examples of this type of proof.

The basic idea is to use the constituents of the target problem instance to design certain "components" that can be combined to "realize" instances of the known NP-complete problem. In these three examples, there are two basic types of components, ones that can be viewed as "making choices" (for example, selecting vertices, choosing truth values for variables) and ones for "testing properties" (for example, checking that each edge is covered, checking that each clause is satisfied). These components are joined together in a target instance in such a way that the choices are communicated to the property testers, and the property testers then check whether the choices made satisfy the required constraints. Interactions between components occur both through direct connections (such as the edges linking the truth setting components to the satisfaction testing components in the transformation from 3SAT to VC) and through global constraints (such as the overall bound K in the transformation from 3SAT to VC, which, together with the structure of the components, ensures that each truth setting component contains exactly one vertex from the cover and that each satisfaction testing component contains exactly two vertices from the cover).

More generally, any proof in which the constructed instance can be viewed as a collection of components, each performing some function in terms of the given instance, can be regarded as a component design proof. The generic transformation used to prove Cook's Theorem in Chapter 2 is a good example of this, with each of the six clause groups being one type of component.

Since component design proofs tend to be rather lengthy and since we have already given a number of examples of such proofs, we shall confine ourselves to a single additional example in this section. (More can be found in [Sethi, 1975], [Even, Itai, and Shamir, 1976], [Garey, Johnson, and Tarjan, 1976] and [Stockmeyer, 1973].) This final example is quite different from the standard ones, and illustrates an approach that has been useful for transforming CLIQUE to several other problems. The target problem is a scheduling problem related to the problem of SEQUENCING WITHIN INTERVALS proved NP-complete in the preceding subsection.

MINIMUM TARDINESS SEQUENCING
INSTANCE: A set T of "tasks," each $t \in T$ having "length" 1 and a "deadline" $d(t) \in Z^+$, a partial order $\lessdot$ on T, and a non-negative integer $K \leqslant |T|$.
QUESTION: Is there a "schedule" $\sigma: T \rightarrow \{0,1, \ldots, |T|-1\}$ such that $\sigma(t) \neq \sigma(t')$ whenever $t \neq t'$, such that $\sigma(t) < \sigma(t')$ whenever $t \lessdot t'$, and such that $|\{t \in T: \sigma(t)+1 > d(t)\}| \leqslant K$?

Theorem 3.10 MINIMUM TARDINESS SEQUENCING is NP-complete.
Proof: Let the graph $G=(V,E)$ and the positive integer $J \leqslant |V|$ constitute an arbitrary instance of CLIQUE. The corresponding instance of MINIMUM TARDINESS SEQUENCING has task set $T = V \cup E$, $K = |E|-(J(J-1)/2)$, and partial order and deadlines defined as follows:

$$t \lessdot t' \iff t \in V, t' \in E, \text{ and vertex } t \text{ is an endpoint of edge } t'$$

$$d(t) = \begin{cases} J(J+1)/2 & \text{if } t \in E \\ |V|+|E| & \text{if } t \in V \end{cases}$$

Thus the "component" corresponding to each vertex is a single task with deadline $|V|+|E|$, and the "component" corresponding to each edge is a single task with deadline $J(J+1)/2$. The task corresponding to an edge is forced by the partial order to occur after the tasks corresponding to its two endpoints in the desired schedule, and only edge tasks are in danger of being tardy (being completed after their deadlines).

It is convenient to view the desired schedule schematically, as shown in Figure 3.10. We can think of the portion of the schedule before the edge task deadline as our "clique selection component." There is room for $J(J+1)/2$ tasks before this deadline. In order to have no more than the specified number of tardy tasks, at least $J(J-1)/2$ of these "early" tasks must be edge tasks. However, if an edge task precedes this deadline, then so must the vertex tasks corresponding to its endpoints. The minimum possible number of vertices that can be involved in $J(J-1)/2$ distinct edges is J (which can happen if and only if those edges form a complete graph on those J vertices). This implies that there must be *at least* J vertex tasks among the "early" tasks. However, there is room for *at most*

$$(J(J+1)/2) - (J(J-1)/2) = J$$

vertex tasks before the edge task deadline. Therefore, any such schedule must have *exactly* J vertex tasks and *exactly* $J(J-1)/2$ edge tasks before this deadline, and these must correspond to a J-vertex clique in G. Conversely, if G contains a complete subgraph of size J, the desired schedule can be constructed as in Figure 3.10. ■

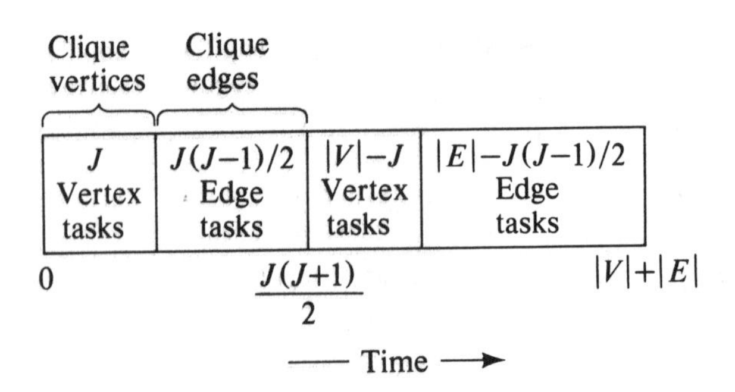

Figure 3.10 Diagram of the desired schedule for an instance of MINIMUM TARDINESS SEQUENCING corresponding to a CLIQUE of size J.

3.3 Some Suggested Exercises

In this section we present the definitions of twelve NP-complete problems and leave to the reader the task of proving that they are NP-complete. None of these problems requires a complicated proof, so we encourage the reader to attempt them all. For the purposes of these exercises, only those "known" NP-complete problems mentioned in Section 3.1 should be used. As a hint for how to proceed, we have grouped the problems according to our own preferred proof technique, but the reader should feel free to ignore these hints whenever an alternative approach seems worthy of pursuit. Those desiring additional (or more difficult) exercises can choose from the lists included in the Appendix, keeping in mind that these lists contain some problems for which only quite elaborate proofs are known.

Restriction

1. **LONGEST PATH**
INSTANCE: Graph $G=(V,E)$, positive integer $K \leqslant |V|$.
QUESTION: Does G contain a simple path (that is, a path encountering no vertex more than once) with K or more edges?

2. **SET PACKING**
INSTANCE: Collection C of finite sets, positive integer $K \leqslant |C|$.
QUESTION: Does C contain K disjoint sets?

3. **PARTITION INTO HAMILTONIAN SUBGRAPHS**
INSTANCE: Graph $G=(V,E)$, positive integer $K \leqslant |V|$.
QUESTION: Can the vertices of G be partitioned into $k \leqslant K$ disjoint sets $V_1, V_2, \ldots, V_k$ such that, for $1 \leqslant i \leqslant k$, the subgraph induced by V_i contains a Hamiltonian circuit?

4. **LARGEST COMMON SUBGRAPH**
INSTANCE: Graphs $G_1=(V_1,E_1)$ and $G_2=(V_2,E_2)$, positive integer K.
QUESTION: Do there exist subsets $E_1' \subseteq E_1$ and $E_2' \subseteq E_2$ such that $|E_1'| = |E_2'| \geqslant K$ and such that the two subgraphs $G_1' = (V_1, E_1')$ and $G_2' = (V_2, E_2')$ are isomorphic?

5. **MINIMUM SUM OF SQUARES**
INSTANCE: Finite set A, "size" $s(a) \in Z^+$ for each $a \in A$, positive integers K and J.
QUESTION: Can the elements of A be partitioned into K disjoint sets $A_1, A_2, \ldots, A_K$ such that $\sum_{i=1}^{K} \left(\sum_{a \in A_i} s(a) \right)^2 \leqslant J$?

Local Replacement

6. **FEEDBACK VERTEX SET**
INSTANCE: Directed graph $G=(V,A)$, positive integer $K \leqslant |V|$.
QUESTION: Is there a subset $V' \subseteq V$ such that $|V'| \leqslant K$ and such that every directed circuit in G includes at least one vertex from V'?

7. **EXACT COVER BY 4-SETS**
INSTANCE: Finite set X with $|X| = 4q$, q an integer, and a collection C of 4-element subsets of X.
QUESTION: Is there a subcollection $C' \subseteq C$ such that every element of X occurs in exactly one member of C'?

8. **DOMINATING SET**
INSTANCE: Graph $G=(V,E)$, positive integer $K \leqslant |V|$.
QUESTION: Is there a subset $V' \subseteq V$ such that $|V'| \leqslant K$ and such that every vertex $v \in V - V'$ is joined to at least one member of V' by an edge in E?

9. **STEINER TREE IN GRAPHS**
INSTANCE: Graph $G=(V,E)$, subset $R \subseteq V$, positive integer $K \leqslant |V|-1$.
QUESTION: Is there a subtree of G that includes all the vertices of R and that contains no more than K edges?

10. **STAR-FREE REGULAR EXPRESSION INEQUIVALENCE**
INSTANCE: Two star-free regular expressions E_1 and E_2 over a finite alphabet Σ, where such expressions are defined by (1) any single symbol $\sigma \in \Sigma$ is a star-free regular expression, and (2) if e_1 and e_2 are star-free regular expressions, then the strings e_1e_2 and $(e_1 \vee e_2)$ are star-free regular expressions.
QUESTION: Do E_1 and E_2 represent different languages over Σ, where the language represented by $\sigma \in \Sigma$ is $\{\sigma\}$, and, if e_1 and e_2 represent the languages L_1 and L_2 respectively, then e_1e_2 represents the language $\{xy: x \in L_1 \text{ and } y \in L_2\}$ and $(e_1 \vee e_2)$ represents the language $L_1 \cup L_2$?

Component Design

11. **SET SPLITTING**
INSTANCE: Collection C of subsets of a finite set S.
QUESTION: Is there a partition of S into two subsets S_1 and S_2 such that no subset in C is entirely contained in either S_1 or S_2?
Hint: Use 3SAT.

12. **PARTITION INTO PATHS OF LENGTH 2**
INSTANCE: Graph $G=(V,E)$, with $|V|=3q$ for a positive integer q.
QUESTION: Is there a partition of V into q disjoint sets $V_1, V_2, \ldots, V_q$ of three vertices each so that, for each $V_i = \{v_{i[1]}, v_{i[2]}, v_{i[3]}\}$, at least two of the three edges $\{v_{i[1]}, v_{i[2]}\}$, $\{v_{i[1]}, v_{i[3]}\}$, and $\{v_{i[2]}, v_{i[3]}\}$ belong to E?
Hint: Use 3DM.

13. **GRAPH GRUNDY NUMBERING**
INSTANCE: Directed graph $G=(V,A)$.
QUESTION: Is there a labeling $L: V \to Z^+$ (where the same label may be assigned to more than one vertex) such that, for each $v \in V$, $L(v)$ is the least non-negative integer not in the set $\{L(u): u \in V, (v,u) \in A\}$?
Hint: Use 3SAT.

14. **GRAPH 3-COLORABILITY**
INSTANCE: Graph $G=(V,E)$.
QUESTION: Is G 3-colorable, that is, does there exist a function $f: V \to \{1,2,3\}$ such that $f(u) \neq f(v)$ whenever $\{u,v\} \in E$?
Hint: Use 3SAT.

4

Using NP-Completeness to Analyze Problems

Now that we have the basic tools of NP-completeness well in hand, we can begin to examine how this theory can be used for analyzing problems.

The discussions in Chapter 1 suggest that, whenever we are confronted with a new problem, a natural first question to ask is: Can it be solved with a polynomial time algorithm? If the answer to this question is obviously "yes," then nothing further can be said about the problem from the standpoint of NP-completeness. We can concentrate our efforts on trying to find as efficient a polynomial time algorithm as possible. However, if no polynomial time algorithm is apparent, an appropriate second question to ask is: "Is the problem NP-complete?"

So that this question is meaningful, let us suppose that we have stated our problem as a decision problem and, further, that we know the decision problem belongs to NP. Just as it might have been obvious that our problem is polynomially solvable, it might now be obvious that it is NP-complete. If so, we have strong evidence that it *cannot* be solved with a polynomial time algorithm.

In most cases, neither of these questions will have an obvious answer. Usually our problem will be neither obviously polynomially solvable nor obviously NP-complete, and some effort will be required to determine which is the case (if indeed either case holds; recall from Section 2.5 that if

$P \neq NP$ there will be problems in NP that are neither NP-complete nor polynomially solvable). How might we proceed to resolve the status of our problem?

If we have a strong suspicion about what the outcome will be, it is rather tempting to concentrate our efforts in that single direction. However, intuition can be a particularly untrustworthy guide in these matters, since many problems that are polynomially solvable differ only slightly from other problems that are NP-complete. For example, we already have seen that 3-SATISFIABILITY and 3-DIMENSIONAL MATCHING are NP-complete, whereas the related 2-SATISFIABILITY and 2-DIMENSIONAL MATCHING problems can be solved in polynomial time. Figure 4.1 lists several other pairs of similar problems for which one belongs to P and the other is NP-complete. Our intuition is based on our knowledge about related problems, and if we let it lead us into investing all our efforts in just one of the possibilities, we run a serious risk of placing all our eggs in the wrong basket.

Thus it is best to proceed with our analysis using a two-sided approach. While we are attempting, on the one hand, to construct an NP-completeness proof, on the other hand, we should be trying to discover a polynomial time algorithm. Which of these two options we choose to emphasize at any one time certainly will depend on the current state of our expectations, but whenever our current line of attack appears to be foundering we must be prepared to reverse direction and try the other. In fact, as we alternate back and forth, the two approaches will often interact with one another. The failure of a proposed NP-completeness proof might lead to an idea for an algorithm; the failure of a proposed algorithm might suggest a way for proving NP-completeness. Any partial results proved along the way, especially those providing "normal forms" for solutions, can be just as useful for constructing an NP-completeness proof as for designing an efficient algorithm.

It is clear that the successful application of such a two-sided approach demands skill both in constructing NP-completeness proofs *and* in designing polynomial time algorithms. We have already said a great deal about techniques for the former in Chapter 3. For the latter, we refer the reader to any of the standard texts on algorithm design, such as [Aho, Hopcroft, and Ullman, 1974] or [Reingold, Nievergelt, and Deo, 1977]. In the remainder of this chapter, we will direct our attention to the use of a similar two-sided approach for continuing our analysis in more depth once we have proved (as so often seems to be the case) that our initial problem is NP-complete.

In Section 4.1 we discuss how one can probe more deeply into the complexity of an NP-complete problem by investigating its subproblems, trying to "map the boundary" between those subproblems that are polynomially solvable and those that are NP-complete. In Section 4.2 we focus on a special type of subproblem that often merits attention for problems in which numbers play a significant role. This leads us to introduce the concepts of

P	NP-complete
SHORTEST PATH BETWEEN TWO VERTICES INSTANCE: Graph $G=(V,E)$, length $l(e) \in Z^+$ for each $e \in E$, specified vertices $a,b \in V$, positive integer B. QUESTION: Is there a simple path from a to b in G having total length B or less?	**LONGEST PATH BETWEEN TWO VERTICES** INSTANCE: Graph $G=(V,E)$, length $l(e) \in Z^+$ for each $e \in E$, specified vertices $a,b \in V$, positive integer B. QUESTION: Is there a simple path from a to b in G having total length B or more?
EDGE COVER INSTANCE: Graph $G=(V,E)$, positive integer K. QUESTION: Is there an $E' \subseteq E$ with $\|E'\| \leqslant K$ such that for each $v \in V$ there is some $e \in E'$ for which $v \in e$?	**VERTEX COVER** INSTANCE: Graph $G=(V,E)$, positive integer K. QUESTION: Is there a $V' \subseteq V$ with $\|V'\| \leqslant K$ such that for each $e \in E$ there is some $v \in V'$ for which $v \in e$?
TRANSITIVE REDUCTION INSTANCE: Directed graph $G=(V,A)$, positive integer K. QUESTION: Is there an $A' \subseteq V \times V$ with $\|A'\| \leqslant K$ such that for all $u,v \in V$ $G'=(V,A')$ contains a path from u to v if and only if G does?	**MINIMUM EQUIVALENT DIGRAPH** INSTANCE: Directed graph $G=(V,A)$, positive integer K. QUESTION: Is there an $A' \subseteq A$ with $\|A'\| \leqslant K$ such that for all $u,v \in V$ $G'=(V,A')$ contains a path from u to v if and only if G does?
INTREE SCHEDULING INSTANCE: Set T of unit length tasks, deadline $d(t) \in Z^+$ for each $t \in T$, partial order $<$ on T such that each task has at most one immediate successor, positive integer m. QUESTION: Can T be scheduled on m processors to obey the partial order and meet all the deadlines?	**OUTTREE SCHEDULING** INSTANCE: Set T of unit length tasks, deadline $d(t) \in Z^+$ for each $t \in T$, partial order $<$ on T such that each task has at most one immediate predecessor, positive integer m. QUESTION: Can T be scheduled on m processors to obey the partial order and meet all the deadlines?

Figure 4.1 Pairs of similar problems, one belonging to P and the other NP-complete.

"pseudo-polynomial time algorithm" and "strong NP-completeness," and also to present an additional (seventh) "basic" NP-complete problem. Section 4.3 concludes the chapter with a brief discussion of how analyzing subproblems can be used to study the effect of individual problem parameters (rather than just the conglomerated "input length") on the complexity of a problem.

4.1 Analyzing Subproblems

Suppose we have just succeeded in demonstrating that our initial problem is NP-complete. Even though this effectively answers the two questions with which we began our analysis, there are still many appropriate follow-up questions that should be asked. The problem we have been analyzing is often distilled from a less elegant applied problem, and some of the details that were dropped in the distillation process might alter the problem enough to make it polynomially solvable. If not, there still might be significant special cases that *can* be solved in polynomial time. It might even be the case that the instances for which the problem is hard are relatively rare and possess easily recognizable features that would allow us to identify them beforehand. Such possibilities can be investigated by analyzing subproblems of our original problem.

As we have been describing decision problems, each consists of two parts: a domain D that is the set of all instances of the problem, and a yes-set Y containing all instances from D for which the answer is "yes." By a *subproblem* (or "special case") of a problem $\Pi = (D, Y)$, we mean a problem $\Pi' = (D', Y')$ such that $D' \subseteq D$ and $Y' = Y \cap D'$. In other words, Π' is a subproblem of Π if it asks the same question as Π, but only over a subset of the domain for Π.

Thus a subproblem of a given problem is obtained whenever we place additional restrictions on the allowed instances. For graph theoretic problems, for example, we might restrict the instances to those in which the graphs are planar, or bipartite, or acyclic, or some combination of these. For problems involving sets, we might restrict the sets to be no larger than a certain size, or to be such that no element occurs in more than a specified number of sets. Any values assigned to set elements might be required to come from some fixed, limited set of allowed values. From all these possible subproblems, the ones we choose to analyze in detail usually are determined by the application we have in mind or are in some sense "natural" subproblems that might be expected to arise in some application.

It should be apparent that, even though a problem Π is NP-complete, each of the subproblems of Π might independently be either NP-complete or polynomially solvable. (Of course, if Π belongs to P, then any subproblem of Π whose instances are themselves recognizable in polynomial time must also be in P, and in general we always restrict our attention to such subproblems.) We already have noted two subproblems of SATISFIABILITY that differ in this respect, 3-SATISFIABILITY and 2-SATISFIABILITY. Assuming that $P \neq NP$, we can view the subproblems of any NP-complete problem Π as lying on different sides of an imaginary "boundary" between polynomial time solvability and intractability. Our goal in analyzing the problem is to determine which subproblems lie on each side.

Actually it is perhaps more accurate to think in terms of there being, at any particular time, a "frontier" between those subproblems we *know* to be polynomially solvable and those we *know* to be NP-complete. The frontier consists of the subproblems whose NP-completeness is still an open question. Figure 4.2 gives a schematic representation for one possible "current state of knowledge" about a collection of subproblems of a problem Π. Whenever we determine that a currently open problem is in P or is NP-complete, we narrow the frontier, enlarging that part of the world upon which "civilization" has been imposed. Of course, unless Π is polynomially solvable, we will never be able to narrow the frontier completely, except possibly for the limited set of subproblems that hold immediate interest. Even if we restrict our attention to a fixed, finite collection of subproblems, some of them might belong to that annoying group of problems that are neither NP-complete nor in P (as we have already remarked, such problems must exist if P$\neq$NP). Nevertheless, each time we settle one of the remaining open problems we can picture ourselves as closing in on an imaginary "borderline."

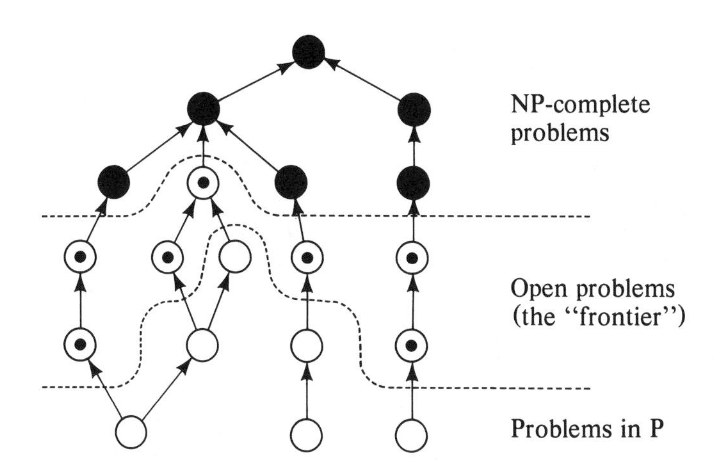

Figure 4.2 One possible state of knowledge about subproblems of an NP-complete problem Π. Problems are represented by circles, filled-in if known to be NP-complete, empty if known to be in P, and dotted if "open." An arrow from Π_1 to Π_2 signifies that Π_1 is a subproblem of Π_2.

To put these ideas in terms of a concrete example, consider the following problem of scheduling equal length tasks subject to precedence constraints (itself a special case of several more general scheduling problems):

PRECEDENCE CONSTRAINED SCHEDULING
INSTANCE: A set T of "tasks" (each assumed to have "length" 1), a partial order $\lessdot$ on T, a number m of "processors," and an overall "deadline" $D \in Z^+$.

QUESTION: Is there a "schedule" $\sigma: T \rightarrow \{0,1,\ldots,D\}$ such that, for each $i \in \{0,1,\ldots,D\}$, $|\{t \in T: \sigma(t)=i\}| \leqslant m$, and such that, whenever $t \lessdot t'$, then $\sigma(t) < \sigma(t')$?

As a hypothetical "application" of this problem, suppose that you are an assistant professor of computer science at State University and have just been assigned the task of helping entering freshmen plan their undergraduate programs. The students provide you with a list of all the courses they intend to take, the number D of semesters in which they expect to graduate, and a maximum number m of courses they are willing to take at one time. State U is sufficiently large that every course is offered each semester, and no two courses selected by your advisees will ever be offered at conflicting times. However, certain courses are required as prerequisites for certain other courses and hence must be taken earlier ($t \lessdot t'$ means that t is a prerequisite for t'). You would like to devise a computer program that takes in all this information and constructs a schedule for each student to follow.

The PRECEDENCE CONSTRAINED SCHEDULING problem is NP-complete [Ullman, 1975], so it is unlikely that you will be able to come up with a general, polynomial time scheduling algorithm. However, there are some natural restrictions that might make the problem easier to solve and that might suffice for most students. For example, it is probably reasonable to place an upper bound on m, such as $m \leqslant 6$, as most students do have a limited capacity for work. The course prerequisites also might satisfy special constraints. Many of your students might select such a varied program of study that none of their chosen courses has any prerequisites, in which case the partial order is empty. Or, in some cases, it might be that each course has only a *single* "explicit" prerequisite, with all other prerequisites for the course also being prerequisites of the explicit one. This gives rise to what is known as a "tree" partial order. Other restrictions are possible, but let us limit our attention to these. Among them they determine the array of subproblems pictured in Figure 4.3, which also displays the current state of knowledge about the complexity of these subproblems.

Note that all the possibilities shown in Figure 4.2 actually occur in Figure 4.3, including the existence of a "frontier." When we have a uniform hierarchy of subproblems like this, it is often possible to specify the same information more concisely by giving the "minimal" NP-complete subproblems and the "maximal" polynomially solvable subproblems. Given a collection C of subproblems of some NP-complete problem, a problem $\Pi \in C$ is a (currently) *minimal* NP-complete subproblem if Π is known to be NP-complete and no subproblem Π' of Π both belongs to C and is known to be NP-complete. A problem $\Pi \in C$ is a (currently) *maximal* polynomially solvable subproblem if Π is known to be in P, and no other problem Π' that contains Π as a subproblem both belongs to C and is known to be polynomially solvable.

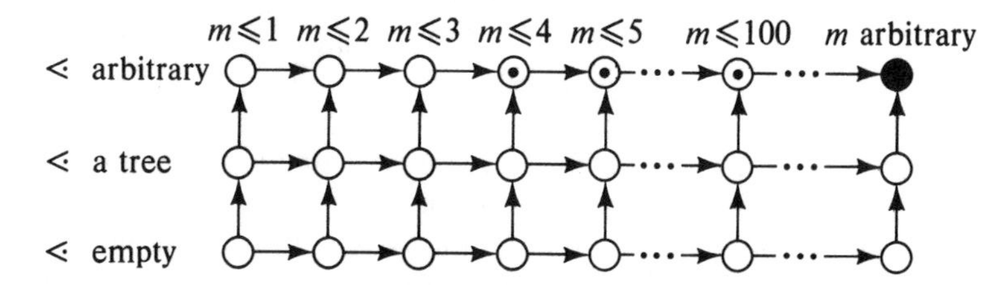

Figure 4.3 Current state of knowledge for a collection of subproblems of PRECEDENCE CONSTRAINED SCHEDULING, using the key given in Figure 4.2.

mially solvable. Minimal and maximal open problems for C can be defined similarly.

For the class of subproblems of PRECEDENCE CONSTRAINED SCHEDULING illustrated in Figure 4.3, the two subproblems specified by "$\lessdot$ arbitrary, $m\leqslant 2$" and "$\lessdot$ a tree, m arbitrary" are (currently) maximal polynomial time solvable subproblems. The general problem itself is the (currently) minimal NP-complete subproblem. The minimal open problem is specified by "$\lessdot$ arbitrary, $m\leqslant 3$," and there is no maximal open problem because "$\lessdot$ arbitrary, $m\leqslant J$" is open for all integers $J\geqslant 3$.

It is natural to investigate the boundary for a particular problem by means of a global version of the two-sided approach discussed earlier. We alternate between analyzing those subproblems that seem most likely to be in P and those that seem most likely to be NP-complete. Using as our starting points the subproblems that are obviously polynomially solvable and the subproblems whose NP-completeness follows trivially from that of the general problem, we gradually enlarge the sets of allowed instances for the former and gradually restrict the sets of allowed instances for the latter. In contrast to analyzing a fixed problem, when we change modes from designing algorithms to proving NP-completeness, we also can change problems, in this case from a more-restricted one to a less-restricted one.

The techniques appropriate for proving a subproblem NP-complete are essentially the same as those described in Chapter 3 for an isolated problem. However, there is one important difference. When we are trying to prove a subproblem NP-complete, we already have an NP-completeness proof for some generalized version of it. This gives us a good candidate for a "known" NP-complete problem to use in our desired proof, and it also gives us an NP-completeness proof that we might be able to modify to obtain a proof for our subproblem. Although this will not always make our task easier, it at least provides us with a head start over trying to construct an NP-completeness proof for an isolated problem.

The advantages of this can be illustrated nicely in terms of some well-known graph theoretic problems. In fact, since problems from graph theory are quite common among the NP-complete problems, and since the restrictions we will be considering are frequently important for such problems, the specific techniques we use for proving NP-completeness results under these restrictions are worth illustrating in their own right.

Although we will be discussing a variety of graph problems, the following "graph coloring" problem (which was an exercise in Chapter 3) will be our primary example:

GRAPH 3-COLORABILITY
INSTANCE: Graph $G=(V,E)$
QUESTION: Is G 3-colorable, that is, does there exist a function $f: V \rightarrow \{1,2,3\}$ such that $f(u) \neq f(v)$ whenever $\{u,v\} \in E$?

This problem is related to the famous Four Color Conjecture (recently proved by Appel and Haken [1977a; 1977b]) and arises in connection with certain scheduling and partitioning problems. It is itself a special case of GRAPH K-COLORABILITY, in which the range of f is $\{1,2, \ldots, K\}$, with K being specified as part of the instance. The NP-completeness of GRAPH 3-COLORABILITY was proved by Stockmeyer [1973] (the proof also appears in [Garey, Johnson, and Stockmeyer, 1976]).

The first restriction we consider is that of bounding the maximum vertex degree (the *degree* of a vertex is the number of edges containing it). Most graph problems can be solved in polynomial time if the maximum vertex degree is restricted to be sufficiently small. For example, if we require that all vertices have degree 2 or less, then HAMILTONIAN CIRCUIT, VERTEX COVER, GRAPH 3-COLORABILITY (and almost any graph problem imaginable) can be trivially solved in polynomial time. The question thus arises: What is the strongest constraint on vertex degree for which the problem remains NP-complete?

The CLIQUE problem is one example for which *no* constant bound on vertex degree preserves its NP-completeness. For if the degree bound is D, then none of our graphs can contain a clique on more than $D+1$ vertices. Thus we can find the largest clique by examining all subsets of $D+1$ or fewer vertices, and the number of such subsets will be polynomially bounded because D is a fixed constant. Note, however, that although this prevents us from proving that the restricted problems are NP-complete (assuming $P \neq NP$), the resulting polynomial time algorithms would not be particularly useful if D were large.

For many other graph problems there *are* degree constrained subproblems that remain NP-complete. Figure 4.4 tabulates some results of this type. Observe that in each case the bound is the best possible (unless $P=NP$), because reducing the bound by only one produces a polynomially solvable subproblem. In fact, each problem then becomes trivial. (For GRAPH 3-COLORABILITY, a theorem of Brooks [1941] asserts that a connected graph with maximum degree 3 is 3-colorable if and only if it is *not* the complete graph on four vertices, an easily verified condition.)

	In P for $D \leqslant$	NP-complete for $D \geqslant$
VERTEX COVER	2	3
HAMILTONIAN CIRCUIT	2	3
GRAPH 3-COLORABILITY	3	4
FEEDBACK VERTEX SET	2	3

Figure 4.4 Classification of subproblems obtained by restricting instances to graphs having no vertex degree larger than D, with respect to polynomial time solvability and NP-completeness.

Each of these degree-limited NP-completeness results can be proved from the general problem using local replacement. The key idea is that of a "vertex substitute," which we illustrate for GRAPH 3-COLORABILITY.

Theorem 4.1. GRAPH 3-COLORABILITY with no vertex degree exceeding 4 is NP-complete.

Proof: Membership in NP for the restricted problem follows immediately from that for the general problem. So suppose $G=(V,E)$ is an arbitrary instance of the general problem. We must construct a corresponding graph $G'=(V',E')$ that has no vertex degree exceeding 4 and that is 3-colorable if and only if G is 3-colorable.

Our vertex substitute is based on the eight vertex graph H_3 shown in Figure 4.5(a), which has three "outlets," labeled by 1, 2, and 3 in the figure. For $k \geqslant 4$, the k-outlet vertex substitute H_k is formed by adjoining to H_{k-1} a copy of H_3 having its first outlet coinciding with outlet $k-1$ of H_{k-1}. The outlet vertices of H_k are the vertices having degree two. The outlets that originally belonged to the H_{k-1} retain the same labels, with the second outlet of the adjoined H_3 becoming outlet $k-1$ and its third outlet becoming outlet k. Figure 4.5(b) shows H_5.

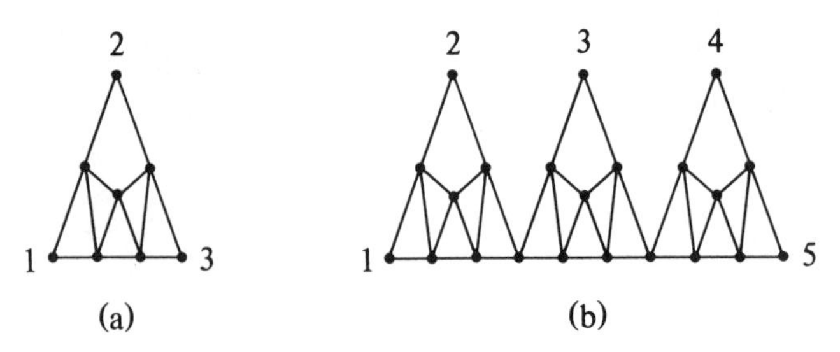

Figure 4.5 The graph H_3 and vertex substitute H_5 (formed from three copies of H_3) used for proving the NP-completeness of degree-restricted GRAPH 3-COLORABILITY.

It is easy to see that, for all $k \geqslant 3$, the following facts hold:

(1) H_k has $7(k-2)+1$ vertices, including k labeled outlets.

(2) No vertex of H_k has degree exceeding 4.

(3) Each outlet of H_k has degree 2.

(4) H_k is 3-colorable, but not 2-colorable, with every way of "3-coloring" H_k assigning the same "color" to all its outlets.

Arbitrarily designate as $v_1, v_2, \ldots, v_r$ the r vertices of the given graph G that have degree exceeding 4. We construct a sequence of graphs.

$$G = G_0, G_1, G_2, \ldots, G_r = G'$$

as follows. Each G_i, $1 \leqslant i \leqslant r$, is constructed from G_{i-1}. Let d be the degree of v_i in G_{i-1} and let $\{u_1, v_i\}, \{u_2, v_i\}, \ldots, \{u_d, v_i\}$ be the edges that include v_i. To form G_i, delete vertex v_i from G_{i-1}, replacing it with a copy of H_d, and replace each edge $\{u_j, v_i\}$ by an edge joining u_j to outlet j of the vertex substitute.

It follows from the construction and previously stated facts that, for $0 \leqslant k \leqslant r$, G_k has $r-k$ vertices of degree exceeding 4 and G_k is 3-colorable if and only if G is 3-colorable. Thus $G' = G_r$ has the desired properties. ■

Different vertex substitutes are required for different problems, and substantial ingenuity may be needed to come up with one that preserves all the necessary properties. However, the frequency with which such degree constraints occur in practice (for example, fan-in, fan-out restrictions on logic circuits) makes it worthwhile to examine their effect on the complexity of any general graph problem.

Another common restriction for graph problems is that to planar graphs. A graph is *planar* if it can be embedded in the plane by identifying each vertex with a unique point and each edge with a line connecting its endpoints, so that no two lines meet except at a common endpoint. Many applications, from map-making to integrated circuit layout, give rise to graphs that are inherently planar, so it is natural to consider the effect of planarity on the complexity of a problem. Once again, CLIQUE is an example of a problem that becomes easy when so restricted, because a planar graph cannot contain a complete subgraph of more than four vertices. A more interesting example is that for the following problem:

MAX CUT
INSTANCE: Graph $G=(V,E)$, "weight" $w(e) \in Z^+$ for each edge $e \in E$, positive integer K.
QUESTION: Can V be partitioned into two disjoint sets V_1 and V_2 such that the sum of the weights of the edges from E that have one endpoint in each set is at least K?

For arbitrary graphs this problem is NP-complete even if we require that all edge weights be equal [Garey, Johnson, and Stockmeyer, 1976]. However, Orlova and Dorfman [1972] and Hadlock [1975] show how matching theory can be used to solve it in polynomial time for planar graphs, with no restrictions on the edge weights.

On the other hand, many graph problems (for example, all those discussed in Chapter 3) remain NP-complete when restricted to planar graphs. There are two common ways to prove such results. The first is to use a planarity preserving transformation from another problem already known to be NP-complete for planar graphs. The second, more basic, technique is to use local replacement applied to the general problem, designing a "crossover" that can be used in place of any edge crossings that occur when a (not necessarily planar) graph is embedded in the plane. We again use GRAPH 3-COLORABILITY to illustrate this technique.

Theorem 4.2. PLANAR GRAPH 3-COLORABILITY is NP-complete.
Proof: Membership in NP follows in the obvious way (planar graphs can be recognized in polynomial time, for example using the linear time algorithm of Hopcroft and Tarjan [1974]). So suppose $G=(V,E)$ is an arbitrary instance of GRAPH 3-COLORABILITY. We must show how to construct a corresponding planar graph $G'=(V',E')$ such that G' is 3-colorable if and only if G is 3-colorable.

The "crossover" used in this proof is the graph H shown in Figure 4.6, and has "outlets" x, x', y, and y' as labeled. This crossover was suggested by M. J. Fischer and is simpler than that used in the original proof of Stockmeyer [1973]. H has 13 vertices and 24 edges and has the following properties (whose verification we leave to the reader):

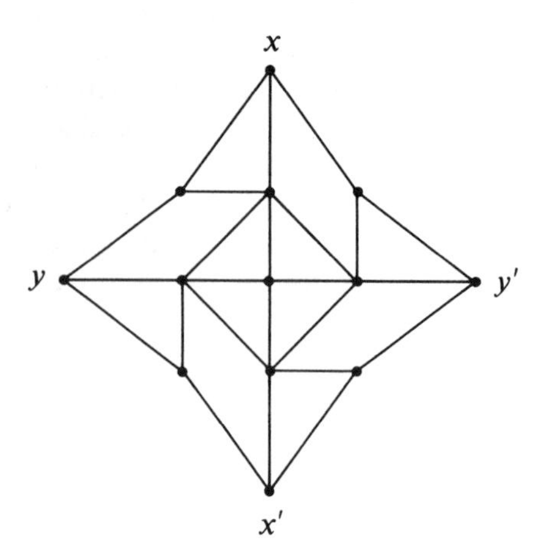

Figure 4.6 Crossover H used in the NP-completeness proof for PLANAR GRAPH 3-COLORABILITY.

1. Any 3-coloring f of H satisfies $f(x)=f(x')$ and $f(y)=f(y')$.
2. There exist 3-colorings f_1 and f_2 for H that satisfy

$$f_1(x)=f_1(x')=f_1(y)=f_1(y')$$

and

$$f_2(x)=f_2(x')\neq f_2(y)=f_2(y')$$

We construct G' from G as follows:

(a) Embed G in the plane, allowing edges to cross one another, but not allowing any edge to touch a vertex other than its own endpoints and not allowing more than two edges to meet at any point other than a vertex. This can be done easily in polynomial time.

(b) For each edge $\{u,v\} \in E$, call its representation in the plane the $\{u,v\}$-*line.* To each such line that is "crossed" by other lines, add new vertices, one between each endpoint and the nearest crossing to it and one between each pair of adjacent crossings. See Figure 4.7(a) and (b).

(c) Replace each crossing in the graph by a copy of H, identifying the outlets x and x' with the nearest two points on one of the lines involved and y and y' with the nearest two points on the other line. See Figure 4.7(c).

(d) For each $\{u,v\} \in E$, choose one endpoint as the *distinguished endpoint* and coalesce it with the nearest *new* point on the $\{u,v\}$-

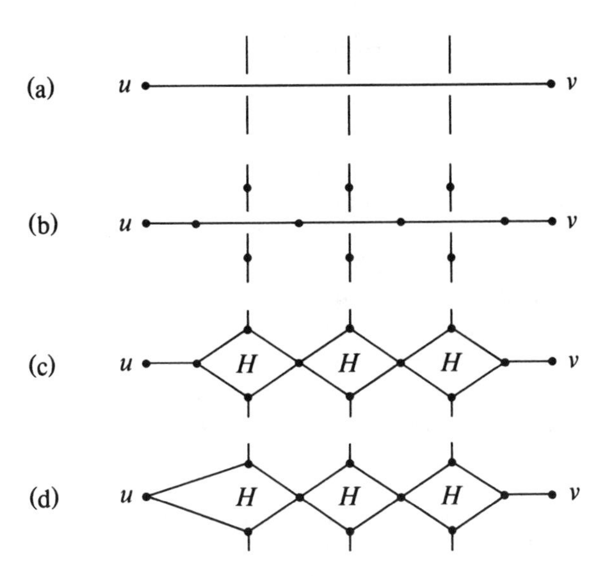

Figure 4.7 The construction of a planar graph G' from a given graph G, using the crossover H of Figure 4.6, so that 3-colorability is preserved.

line. See Figure 4.7(d). The original vertex (from V) retains its identity, and no coalescing occurs unless there *are* new points on the $\{u,v\}$-line. The edge between the other endpoint of the $\{u,v\}$-line and *its* nearest new point on the $\{u,v\}$-line will be called the *operant edge* of the $\{u,v\}$-line.

This completes the description of G'. It is not difficult to see that G' is planar and can be constructed in polynomial time. It remains for us to show that G' is 3-colorable if and only if G is.

Suppose $f\colon V' \rightarrow \{1,2,3\}$ is any 3-coloring for G'. We claim that f restricted to V is a 3-coloring for G. For suppose not. Then there must be a $\{u,v\} \in E$ such that $f(u)=f(v)$. Consider the $\{u,v\}$-line in G', and assume without loss of generality that u is the distinguished endpoint for this line chosen in step (d) of the construction. Then by property 1 of H, all the new points on the $\{u,v\}$-line must be assigned the same "color" as u. Therefore both endpoints of the operant edge for that line have the same color, and this contradicts the assumption that f was a 3-coloring for G'.

Conversely, suppose $f\colon V \rightarrow \{1,2,3\}$ is any 3-coloring for G. It can be extended to a 3-coloring for G' as follows: For each $\{u,v\} \in E$, color each new point on the $\{u,v\}$-line with color $f(u)$, where u is the distinguished vertex for that line. This ensures that both endpoints of every operant edge are colored differently (because $f(u) \neq f(v)$). By property 2 of H, this partial coloring of G' can be extended to a 3-coloring of G' by an appropriate 3-coloring of the interior vertices of each of the crossovers, and the desired result follows. ■

A perusal of the lists contained in the Appendix will provide many other examples of restrictions that have been analyzed for graph theoretic problems. We have attempted throughout these lists to provide as much information as possible about the complexity of various subproblems of each problem. Thus the lists can be used as one source of suggestions for restrictions that might be analyzed for a given problem. Other restrictions will be suggested by the context in which the problem arises. Instances arising in a particular application will often satisfy special constraints that could affect the complexity of the problem, even though these constraints might not be apparent at first. In the next section we discuss a special type of restriction that is often of interest for problems having numerical parameters.

4.2 Number Problems and Strong NP-Completeness

Nowhere does the need for analyzing subproblems of an NP-complete problem have more import than in the case of problems involving numbers. The reasons for this can be illustrated by considering the following "dynamic programming" approach to solving the PARTITION problem.

Let the set $A = \{a_1, a_2, \ldots, a_n\}$ and the sizes $s(a_1), s(a_2), \ldots, s(a_n)$ in Z^+ constitute an arbitrary given instance of PARTITION. Define B to be equal to $\sum_{a \in A} s(a)$. If B is not evenly divisible by 2, then we know that no subset $A' \subseteq A$ can possibly satisfy

$$\sum_{a \in A'} s(a) = \sum_{a \in A - A'} s(a)$$

so we can immediately respond "no" for this instance. Otherwise, for integers $1 \leqslant i \leqslant n, 0 \leqslant j \leqslant B/2$, let $t(i,j)$ denote the truth value of the statement: "there is a subset of $\{a_1, a_2, \ldots, a_i\}$ for which the sum of the item sizes is exactly j." The values of all the $t(i,j)$ can be viewed as being arranged in a table, as shown in Figure 4.8.

The crux of the approach lies in the very simple procedure that can be used for filling in the table entries. It proceeds row by row, from top to bottom. For the top row, all we need do is observe that $t(1,j) = T$ if and only if either $j = 0$ or $j = s(a_1)$. Each subsequent row is filled in by using the entries in the previous row. For $1 < i \leqslant n$, $0 \leqslant j \leqslant B/2$, the entry $t(i,j)$ in row i has the value T if and only if either $t(i-1, j) = T$ or $s(a_i) \leqslant j$ and $t(i-1, j - s(a_i)) = T$. Finally, we observe that, once the entire table has been filled in, we have solved the given instance of PARTITION, because the answer is "yes" if and only if $t(n, B/2) = T$.

The reader should have no difficulty in specifying an iterative algorithm for filling in the table entries, in the manner described, in time bounded by a low order polynomial in the number of table entries (that is, polynomial

i \ j	0	1	2	3	4	5	6	7	8	9	10	11	12	13
1	**T**	**T**	F	F	F	F	F	F	F	F	F	F	F	F
2	**T**	**T**	F	F	F	F	F	F	F	**T**	**T**	F	F	F
3	**T**	**T**	F	F	F	**T**	**T**	F	F	**T**	**T**	F	F	F
4	**T**	**T**	F	**T**	**T**	**T**	**T**	F	**T**	**T**	**T**	F	**T**	**T**
5	**T**	**T**	F	**T**	**T**	**T**	**T**	F	**T**	**T**	**T**	**T**	**T**	**T**

Figure 4.8 Table of $t(i,j)$ for the instance of PARTITION for which $A=\{a_1,a_2,a_3,a_4,a_5\}$, $s(a_1)=1$, $s(a_2)=9$, $s(a_3)=5$, $s(a_4)=3$, and $s(a_5)=8$. The answer for this instance is "yes," since $t(5,13)=T$, reflecting the fact that $s(a_1)+s(a_2)+s(a_4)=13=26/2$.

in nB). In fact, at first glance this might even appear to give us a polynomial time algorithm for solving PARTITION, thus proving that P=NP and obviating the need for this book. Of course this is not the case. The reason is that, by the "conciseness" requirement for reasonable encoding schemes, each integer $s(a_i)$ would be described in the input by a string of length only $O(\log s(a_i))$. Therefore the length of the entire PARTITION instance would be only $O(n\log B)$, and nB is *not* bounded by any polynomial function of this quantity. Thus, this is not a polynomial time algorithm for PARTITION.

Nevertheless, in view of this algorithm, it is clear that the NP-completeness of PARTITION (*and* its supposed intractability) depends strongly on the fact that extremely large input numbers are allowed. If any upper bound were imposed in advance on these numbers, even a bound that is polynomial in Length$[I]$, this algorithm would be a polynomial time algorithm for the restricted problem. (In the sequel, we will be defining the term "pseudo-polynomial time algorithm" to refer to algorithms having this property.) One might expect such a bound to be satisfied in many practical applications.

For example, in scheduling problems where the numbers represent task lengths, extremely large numbers would be unlikely to occur because we actually intend to perform those tasks and we could not afford to do so if any one of them required an inordinately large amount of time. In other problems, where numbers represent empirically measured quantities, limits on the precision of measurement have the effect of limiting the range of numbers for which our algorithm must apply.

Furthermore, a pseudo-polynomial time algorithm can be useful even when there is no natural bound on the input numbers we expect. It will display "exponential behavior." only when confronted with instances containing "exponentially large" numbers, and instances of this sort might be rare for the application we are interested in. If so, this type of algorithm might serve our purposes almost as well as a polynomial time algorithm.

Thus, the possibility of finding a pseudo-polynomial time algorithm for an NP-complete problem involving numbers can be well worth investigating. We shall see that not all such problems are like PARTITION in this regard. For some the theory of NP-completeness can be used to show that even a pseudo-polynomial time algorithm cannot exist unless P = NP. Section 4.2.1 introduces some new terminology and lays the groundwork for proving such "strong" NP-completeness results. Section 4.2.2 illustrates the proof techniques and presents our seventh "basic" NP-complete problem.

4.2.1 Some Additional Definitions

Our new definitions will involve subproblems obtained by placing restrictions on the magnitudes of the numbers occurring in a problem instance. These restrictions will be stated in terms of two encoding-independent functions, Length: $D_\Pi \to Z^+$ and Max: $D_\Pi \to Z^+$, which we assume to be associated with any decision problem Π. Although in theory these two functions can be entirely arbitrary (just like encoding schemes), the significance of what we do with them will depend on the extent to which they reflect the following intended meanings. The function Length, as discussed in Section 2.1, is intended to map any instance I to an integer Length$[I]$ that corresponds to the number of symbols used to describe I under some reasonable encoding scheme for Π. The function Max, which has not been discussed previously, is intended to map any instance I to an integer Max$[I]$ that corresponds to the magnitude of the largest number in I.

The types of results we will be proving will be sufficiently general that each will hold for a broad class of "polynomially related" Length and Max functions. Two Length functions, say Length and Length′, for a problem Π are said to be *polynomially related* if there exist polynomials p and p' such that, for all instances $I \in D_\Pi$,

$$\text{Length}[I] \leqslant p'(\text{Length}'[I])$$

and

$$\text{Length}'[I] \leqslant p(\text{Length}[I])$$

We will say that the pair of functions (Length, Max) is *polynomially related* to the pair of functions (Length′, Max′) if Length and Length′ are polynomially related as above and there exist two-variable polynomials q and q' such that, for all $I \in D_\Pi$,

$$\text{Max}[I] \leqslant q'(\text{Max}'[I], \text{Length}'[I])$$

and

$$\text{Max}'[I] \leqslant q(\text{Max}[I],\text{Length}[I])$$

All the results we state will hold for any Length and Max functions that are polynomially related to the ones we are using.

As an example, consider the PARTITION problem, in which an instance I consists of a finite set A and a size $s(a) \in Z^+$ for each $a \in A$. Any of the following would be a suitable Length function for PARTITION:

$$\text{Length}[I] = |A| + \sum_{a \in A} \lceil \log_2 s(a) \rceil$$

$$\text{Length}[I] = |A| + \max\{\lceil \log_2 s(a) \rceil : a \in A\}$$

$$\text{Length}[I] = |A| \cdot \lceil \log_2 \sum_{a \in A} s(a) \rceil$$

Similarly, any of the following would be a suitable Max function for PARTITION:

$$\text{Max}[I] = \max\{s(a) : a \in A\}$$

$$\text{Max}[I] = \sum_{a \in A} s(a)$$

$$\text{Max}[I] = \lceil (\sum_{a \in A} s(a)) / |A| \rceil$$

We leave for the reader to verify that any of the nine pairs of Length and Max functions that can be chosen using these two lists is polynomially related to any of the others.

The flexibility we are allowed in choosing Length and Max functions will enable us to avoid explicitly stating the ones we have in mind for a problem Π, since they can be inferred with sufficient accuracy from our description of a generic problem instance. An appropriate Length function is implied by what we consider to be a reasonable encoding scheme for the problem, and the latter follows from our description of the generic instance using the standard conventions set forth at the end of Section 2.1. An appropriate Max function is implied by our specifying that certain objects in the generic instance are numbers (in distinction to sets, sequences, graphs, named elements, etc.). These numbers usually will be integers, and any more complicated "number" in an instance will be viewed as being a composite of one or more separate integers, as has already been done for rational numbers. By convention, we will take Max$[I]$ to be the magnitude of the largest integer occurring in I, or 0 if no integers occur in I.

One final property will be required of the functions Max and Length. This is that, given any reasonable encoding scheme for Π, there must exist polynomial time DTMs that take as input the encoded representation of any instance $I \in D_\Pi$ and that output the values of Length$[I]$ and Max$[I]$, written in binary notation. We need this property solely because we will be considering restrictions on instances defined in terms of Length$[I]$ and Max$[I]$, and we need to be able to decide whether or not a given string encodes an instance meeting these restrictions. Any natural choices for Length and Max will certainly have this property.

The definitions that follow assume that every decision problem Π has an associated Length function and an associated Max function as discussed above. Formal precision at the language level would also require that an encoding scheme be given for each problem Π. However, it is convenient to state the definitions at the problem level without this proviso, operating under our standard assumptions about the use of reasonable encoding schemes. The reader should have no difficulty in filling in the details needed to make these definitions precise at the language level, and it is more natural and informative to continue our discussions in terms of problems.

An algorithm that solves a problem Π will be called a *pseudo-polynomial time algorithm* for Π if its time complexity function is bounded above by a polynomial function of the two variables Length$[I]$ and Max$[I]$. By definition, any polynomial time algorithm is also a pseudo-polynomial time algorithm, because it runs in time bounded by a polynomial in Length$[I]$ alone. However, we have already seen an example of a pseudo-polynomial time algorithm that is *not* a polynomial time algorithm, that given for PARTITION. This shows that, even though an NP-completeness result for a problem Π rules out the possibility of solving Π with a polynomial time algorithm (unless P$=$NP), it does not rule out the possibility of solving Π with a pseudo-polynomial time algorithm.

To be more precise, an NP-completeness result does not *necessarily* rule out the possibility of solving Π with a pseudo-polynomial time algorithm. Many of the decision problems we have considered so far have the property that Max$[I]$ is itself bounded by a polynomial function of Length$[I]$, and for these problems there is no distinction between polynomial time algorithms and pseudo-polynomial time algorithms. For example, the only number that occurs in an instance of CLIQUE is the bound J, and J is constrained to be no larger than the number of vertices in the given graph. SATISFIABILITY involves no numbers at all, except for the subscripts on variables and literals, and these can be ignored because they actually are "names" rather than "numbers." (Our conventions on encoding schemes ensure that such numerical "names" will always be polynomially bounded in terms of Length$[I]$.) The issues we are concerned with here are not relevant for problems like this, so let us give a name to the type of problem for which these issues *are* relevant. We say that a problem Π is a *number problem* if there exists no polynomial p such that Max$[I] \leqslant p($Length$[I])$ for all $I \in D_\Pi$. The only number problem among our six basic NP-complete problems is PARTITION.

As an immediate consequence of this definition, we can make the following observation:

Observation 4.1 If Π is NP-complete and Π is not a number problem, then Π cannot be solved by a pseudo-polynomial time algorithm unless P = NP.

Thus, assuming that P $\neq$ NP, the only NP-complete problems that are potential candidates for being solved by pseudo-polynomial time algorithms are those that are number problems.

For any decision problem Π and any polynomial p (over the integers), let Π_p denote the subproblem of Π obtained by restricting Π to only those instances I that satisfy $\text{Max}[I] \leqslant p(\text{Length}[I])$. Then Π_p is not a number problem. Furthermore, if Π is solvable by a pseudo-polynomial time algorithm, then Π_p must be solvable by a polynomial time algorithm. Given any input string x, all we need do is check that x encodes an instance I satisfying $\text{Max}[I] \leqslant p(\text{Length}[I])$ and, if so, apply the pseudo-polynomial time algorithm for Π to I. By our assumption that $\text{Max}[I]$ and $\text{Length}[I]$ can be computed in polynomial time, the required inequality can be checked in polynomial time. By the definition of pseudo-polynomial time algorithm, the algorithm for Π will be a polynomial time algorithm for the instances that satisfy this inequality. This motivates us to call a decision problem Π *NP-complete in the strong sense* if Π belongs to NP and there exists a polynomial p over the integers for which Π_p is NP-complete. In particular, if Π is NP-complete and Π is not a number problem, then Π is automatically NP-complete in the strong sense.

We then have the following generalization of Observation 4.1:

Observation 4.2 If Π is NP-complete in the strong sense, then Π cannot be solved by a pseudo-polynomial time algorithm unless P = NP.

This second observation provides the means for applying the theory of NP-completeness to questions about the existence of pseudo-polynomial time algorithms. We know that PARTITION cannot be NP-complete in the strong sense, because it can be solved by a pseudo-polynomial time algorithm. However, we have not yet seen any examples of number problems that *are* NP-complete in the strong sense. This situation will be rectified in the next section, where we illustrate how strong NP-completeness results can be proved.

4.2.2 Proving Strong NP-Completeness Results

The most straightforward way to prove that a number problem Π is NP-complete in the strong sense is simply to prove for some specific polynomial p that Π_p is NP-complete. For example, the TRAVELING SALESMAN problem (TS) defined in Section 2.1 is a number problem because there are no constraints on the values of either the intercity distances $d(i,j)$ or the bound B. We proved TS NP-complete by transforming HAMIL-

TONIAN CIRCUIT to it. Moreover, the instances of TS created by this transformation all have intercity distances equal to 1 or 2 and a bound B equal to the number m of cities. Thus if we take Max$[I]$ to be the larger of B and the longest intercity distance, and we take Length$[I]$ to be $m + \lceil \log_2 B \rceil + \sum_{i,j} \lceil \log_2 d(i,j) \rceil$, then all the instances created by this transformation satisfy the bound

$$\text{Max}[I] \leqslant \text{Length}[I]$$

In other words, this transformation actually shows that the *subproblem* of TS made up of all those instances satisfying the above inequality is itself NP-complete. It follows that TRAVELING SALESMAN is NP-complete in the strong sense.

In contrast, the NP-completeness proofs for KNAPSACK, MULTIPROCESSOR SCHEDULING, and SEQUENCING WITHIN INTERVALS, described in Section 3.2, all leave open the possibility that these problems can be solved by pseudo-polynomial time algorithms. It turns out that KNAPSACK *can* be solved in pseudo-polynomial time, using a dynamic programming approach similar to that we used for PARTITION, as delineated in [Dantzig, 1957]. All pseudo-polynomial time algorithms known to us are based on similar techniques, and we refer the reader to [Horowitz and Sahni, 1976], [Lawler, 1977a], [Lawler and Moore, 1969], and [Sahni, 1976] for illustrations of these techniques.

The problems MULTIPROCESSOR SCHEDULING and SEQUENCING WITHIN INTERVALS, however, do turn out to be NP-complete in the strong sense. In order to show this, it is useful to have a number problem that is NP-complete in the strong sense and that is somewhat "more numeric" than any we have seen so far. Such a problem is provided by our seventh "basic" NP-complete problem, 3-PARTITION, which is defined as follows:

3-PARTITION

INSTANCE: A finite set A of $3m$ elements, a bound $B \in Z^+$, and a "size" $s(a) \in Z^+$ for each $a \in A$, such that each $s(a)$ satisfies $B/4 < s(a) < B/2$ and such that $\sum_{a \in A} s(a) = mB$.

QUESTION: Can A be partitioned into m disjoint sets $S_1, S_2, \ldots, S_m$ such that, for $1 \leqslant i \leqslant m$, $\sum_{a \in S_i} s(a) = B$? (Notice that the above constraints on the item sizes imply that every such S_i must contain *exactly* three elements from A.)

We prove that 3-PARTITION is NP-complete in the strong sense in two steps, first proving that the related 4-PARTITION problem is NP-complete in the strong sense. 4-PARTITION is identical to 3-PARTITION except that the set A contains $4m$ elements and each $s(a)$ must satisfy

$B/5 < s(a) < B/3$. Thus each set in the desired partition will contain exactly *four* elements.

Theorem 4.3 4-PARTITION is NP-complete in the strong sense.

Proof: It is easy to see that 4-PARTITION belongs to NP, since all we need do is verify in polynomial time that a given partition of A has all the stated properties. We shall transform 3-DIMENSIONAL MATCHING to a restricted version of 4-PARTITION in which all the element sizes are bounded by a polynomial function of the total number of elements, and hence by a polynomial function of Length$[I]$. In particular, taking Max$[I]$ to be $\max\{s(a): a \in A\}$ we shall show that 4-PARTITION is NP-complete even when restricted to instances I with Max$[I] \leqslant 2^{16} \cdot |A|^4$.

Let $W = \{w_1, w_2, \ldots, w_q\}$, $X = \{x_1, x_2, \ldots, x_q\}$, $Y = \{y_1, y_2, \ldots, y_q\}$, and $M \subseteq W \times X \times Y$ denote an arbitrary instance of 3DM. We may assume without loss of generality that $|M| \geqslant q$. Our corresponding instance of 4-PARTITION has $|A| = 4|M|$ elements, one for each occurrence of a member of $W \cup X \cup Y$ in a triple in M and one for each triple in M.

The elements corresponding to a particular $z \in W \cup X \cup Y$ will be denoted by $z[1], z[2], \ldots, z[N(z)]$, where $N(z)$ denotes the number of triples from M in which z occurs. We shall regard $z[1]$ as being the "actual" element corresponding to z, and $z[2]$ through $z[N(z)]$ as being the "dummy" elements corresponding to z. The sizes of these elements depend on which one of W, X, or Y contains z and on the index of z within that set. These are defined as follows, where r is chosen equal to $32q$:

$$\begin{aligned}
s(w_i[1]) &= 10r^4 + ir + 1 && 1 \leqslant i \leqslant q \\
s(w_i[l]) &= 11r^4 + ir + 1 && 1 \leqslant i \leqslant q,\ 2 \leqslant l \leqslant N(w_i) \\[6pt]
s(x_j[1]) &= 10r^4 + jr^2 + 2 && 1 \leqslant j \leqslant q \\
s(x_j[l]) &= 11r^4 + jr^2 + 2 && 1 \leqslant j \leqslant q,\ 2 \leqslant l \leqslant N(x_j) \\[6pt]
s(y_k[1]) &= 10r^4 + kr^3 + 4 && 1 \leqslant k \leqslant q \\
s(y_k[l]) &= 8r^4 + kr^3 + 4 && 1 \leqslant k \leqslant q,\ 2 \leqslant l \leqslant N(y_k)
\end{aligned}$$

The single element corresponding to a particular triple $m_l = (w_i, x_j, y_k) \in M$ is denoted by u_l, and its size depends on the indices of its members as follows:

$$s(u_l) = 10r^4 - kr^3 - jr^2 - ir + 8$$

Notice that, if we add to $s(u_l)$ the sizes of three elements that correspond to w_i, x_j, and y_k, respectively, then the total will be equal to $40r^4 + 15$ whenever all three are "actual" elements or whenever all three are "dummy" elements. We choose this number to be our bound B, that is,

$$B = 40r^4 + 15 \cdot |A|^3 + 15$$

The reader should have no difficulty verifying that this is a polynomial transformation, that the size of each element is strictly between $B/3$ and $B/5$, and that the sum of all the element sizes is $|M| \cdot B$, as required. Furthermore, we observe that the size of each element is bounded above by $12r^4 \leqslant 12 \cdot 8^4 \cdot |A|^4 < 2^{16} \cdot |A|^4$. Thus all that remains to be done to prove that 4-PARTITION is NP-complete in the strong sense is to show that the desired 4-partition exists if and only if M contains a matching.

First, suppose that $M' \subseteq M$ is a matching. The corresponding 4-partition is made up of $|M|$ 4-sets, each containing a u_l, a $w_i[\cdot]$, an $x_j[\cdot]$, and a $y_k[\cdot]$, where $(w_i, x_j, y_k) = m_l \in M$. If $m_l \in M'$, we group u_l with $w_i[1]$, $x_j[1]$, and $y_k[1]$. If $m_l \in M - M'$, we group u_l with "dummy" elements corresponding to w_i, x_j, and y_k. It is not hard to see that there are enough dummy elements so that this can be done, and by our previous comments the sizes of the four elements in each set will sum exactly to B. Thus we have our required 4-partition.

Now suppose we are given a 4-partition of the required form. Consider any 4-set in this 4-partition. By successively considering the sum of the element sizes modulo r, r^2, r^3, r^4, and r^5, we shall show that this 4-set must contain one element corresponding to each of the three members of that triple, all three being "actual" elements or all three being "dummy" elements. First, since $r > 4 \cdot 8 = 32$, we know that the sum modulo R of the sizes of the four elements, which must equal $B(\text{mod } r) = 15$, is the same as the sum of the item sizes when each is taken modulo r beforehand. The only way these can sum to 15 is for the 4-set to contain one element that corresponds to a member of W, one that corresponds to a member of X, one that corresponds to a member of Y, and one that corresponds to a triple from M. Let w_i, x_j, and y_k denote the corresponding members of W, X, and Y, and let $m_l = (w_{i'}, x_{j'}, y_{k'})$ denote the corresponding triple from M. Then the sum of the element sizes modulo r^2 must equal $((i-i')r + 15)(\text{mod } r^2)$ and, since $(i-i')r + 15 < r^2$, we must have

$$B(\text{mod } r^2) = 15 = (i-i')r + 15$$

It follows from this that $i = i'$. Similarly, since $(j-j')r^2 + 15 < r^3$, we must have

$$B(\text{mod } r^3) = 15 = (j-j')r^2 + 15$$

and hence $j = j'$, and , since $(k-k')r^3 + 15 < r^4$, we must have

$$B(\text{mod } r^4) = 15 = (k-k')r^3 + 15$$

and hence $k = k'$. Thus w_i, x_j, and y_k are indeed the three members of the triple m_l, and we know that the coefficient of r^4 in the sum of the element sizes is simply the sum of the individual coefficients for r^4. Our choice of these coefficients in the construction then guarantees that the only way for them to sum to 40 is for all three elements to be "actual" elements or for all three to be "dummy" elements.

The total collection of $3q$ "actual" elements, one for each member of $W \cup X \cup Y$, must therefore be contained in q of our given 4-sets, each of these 4-sets consisting of one element corresponding to a triple from M and the three "actual" elements corresponding to the members of that triple. Those q triples from M provide the desired matching. ■

Theorem 4.4 3-PARTITION is NP-complete in the strong sense.
Proof: It is easy to see that 3-PARTITION belongs to NP. We shall transform the subproblem of 4-PARTITION in which all instances satisfy $\max\{s(a): a \in A\} \leqslant 2^{16} \cdot |A|^4$ to 3-PARTITION, maintaining the property that all element sizes are bounded by a polynomial function of the total number of elements.

Let $A = \{a_1, a_2, \ldots, a_{4n}\}$, bound B, and item sizes $s(a)$ satisfying $B/5 < s(a) < B/3$ and $s(a) \leqslant 2^{16} \cdot |A|^4$ be a specification of any such instance of 4-PARTITION. Our corresponding instance of 3-PARTITION will have $24n^2 - 3n$ elements, one for each element from A, two for each *pair* of elements from A, and $8n^2 - 3n$ "filler" elements.

Corresponding to each element $a_i \in A$ is a "regular" element w_i, with size defined by

$$s'(w_i) = 4 \cdot (5B + s(a_i)) + 1$$

where we use $s'(\cdot)$ to denote the size function in our 3-PARTITION instance. Corresponding to each *pair* of elements $a_i, a_j \in A$ we have two "pairing" elements, $u[i,j]$ and $\bar{u}[i,j]$, with sizes defined by

$$s'(u[i,j]) = 4 \cdot (6B - s(a_i) - s(a_j)) + 2$$

$$s'(\bar{u}[i,j]) = 4 \cdot (5B + s(a_i) + s(a_j)) + 2$$

Finally, for $1 \leqslant k \leqslant 8n^2 - 3n$, we have a "filler" element u_k^* with size $s'(u_k^*) = 20B$. The bound B' for our 3-PARTITION instance is $64B + 4$.

Once again the reader should encounter no difficulty in verifying that this is a polynomial transformation, that the size of every element is strictly between $B'/4 = 16B + 1$ and $B'/2 = 32B + 2$, and that the sum of all the element sizes is equal to $(8n^2 - n)B'$. Furthermore, since the elements in A are constrained to have sizes no larger than $2^{16} \cdot |A|^4$, the sizes in the 3-PARTITION instance will also satisfy a polynomial bound in terms of $|A|$, hence in terms of the number of elements in the constructed instance I', hence in Length$[I]$. Thus, to complete our demonstration that 3-PARTITION is NP-complete in the strong sense, we need only show that a 3-partition exists for the constructed instance if and only if a 4-partition exists for the original instance.

First, suppose that we have a 4-partition for the original instance. The corresponding 3-partition is constructed as follows: Arbitrarily divide each 4-set $\{a_i,a_j,a_k,a_l\}$ into two 2-sets, say $\{a_i,a_j\}$ and $\{a_k,a_l\}$. Our 3-partition will then contain the two 3-sets $\{w_i,w_j,u[i,j]\}$ and $\{w_k,w_l,\bar{u}[i,j]\}$. (Notice that we could just as well have used $\bar{u}[k,l]$ instead of $u[i,j]$ and $u[k,l]$ instead of $\bar{u}[i,j]$.) The sizes of the elements in each of these 3-sets sums to B' since $s(a_i)+s(a_j)+s(a_k)+s(a_l)=B$. Doing this for each of the n given 4-sets, we obtain $2n$ 3-sets that contain all of the "regular" elements and n matched pairs of "pairing" elements. This leaves $8n^2-3n$ matched pairs of "pairing" elements and $8n^2-3n$ "filler" elements. Since the sum of the sizes of two matched "pairing" elements is $44B+4 = B'-20B$, each such matched pair can be grouped with a remaining one of the "filler" elements to complete the desired 3-partition.

Now suppose that we are given a 3-partition for the constructed instance. By considering the element sizes modulo 4 we see that no 3-set can contain an odd number of "regular" elements, no 3-set can contain three "pairing" elements, and no 3-set can contain two "regular" elements and a "filler" element. It follows that the given 3-partition is made up of $2n$ 3-sets that each contain two "regular" elements and one "pairing" element, along with $8n^2-3n$ 3-sets that each contain two "pairing" elements and one "filler" element. Consider any one of the latter type of 3-sets, and let $u[i,j]$ (or $\bar{u}[k,l]$) be one of the two "pairing" elements in that set. If the other pairing element in this 3-set is not $\bar{u}[i,j]$ (or $u[k,l]$), then it must have the same size as that matching element and so can be interchanged with it to obtain an equivalent 3-partition. This operation can be repeated until we obtain a 3-partition in which *every* "filler" element occurs together with a matched pair $u[i,j]$, $\bar{u}[i,j]$. Thus, any "pairing" element that occurs with two "regular" elements in this 3-partition is such that its "match" also occurs in such a 3-set. This divides the 3-sets containing "regular" elements into n pairs of 3-sets. Since the two "pairing" elements in each such pair of 3-sets are matched, their sizes sum to $44B+4$, and hence the sizes of the four "regular" elements must sum to $84B+4$. This implies that the corresponding four elements from A form a 4-set of elements whose sizes sum to B. Therefore these n pairs of 3-sets provide the required 4-partition. ■

Notice that this last transformation, if viewed as a transformation from the general 4-PARTITION problem to 3-PARTITION, would not be enough to prove strong NP-completeness for 3-PARTITION. We needed to restrict our attention to an NP-complete subproblem of 4-PARTITION in which $\max\{s(a)\}$ was polynomially bounded. However, it is easy to see that the particular polynomial bound that we chose was not essential. Indeed, it would be convenient if we could operate with transformations like this without needing to go into the details of the subproblems and the particular polynomials involved. This can be done using the following definition and lemma.

Let Π and Π' denote arbitrary decision problems with instance sets D_Π and $D_{\Pi'}$, "yes" sets Y_Π and $Y_{\Pi'}$, and specified functions Max, Length, Max$'$, and Length$'$, respectively. A *pseudo-polynomial transformation* from Π to Π' is a function $f: D_\Pi \to D_{\Pi'}$ such that

(a) for all $I \in D_\Pi$, $I \in Y_\Pi$ if and only if $f(I) \in Y_{\Pi'}$,

(b) f can be computed in time polynomial in the two variables Max$[I]$ and Length $[I]$,

(c) there exists a polynomial q_1 such that, for all $I \in D_\Pi$,

$$q_1(\text{Length}'[f(I)]) \geqslant \text{Length}[I]$$

(d) there exists a two-variable polynomial q_2 such that, for all $I \in D_\Pi$,

$$\text{Max}'[f(I)] \leqslant q_2(\text{Max}[I],\text{Length}[I])$$

Lemma 4.1 If Π is NP-complete in the strong sense, $\Pi' \in$ NP, and there exists a pseudo-polynomial transformation from Π to Π', then Π' is NP-complete in the strong sense.

Proof: Let f be such a pseudo-polynomial transformation, with functions q_1 and q_2 as specified in the definition. We can assume without loss of generality that q_1 and q_2 have only positive integer coefficients, since they can be so modified without decreasing their values. Because Π is NP-complete in the strong sense, there is some polynomial p such that Π_p is NP-complete. Furthermore, we can choose such a p that has only positive integer coefficients, because if p_0 is any polynomial over the integers satisfying $p_0(x) \geqslant p(x)$ for all x, then Π_{p_0} will contain all the instances of Π_p and hence must be NP-complete if Π_p is. Let $\hat{p}$ be the polynomial defined by

$$\hat{p}(x) = q_2(p(q_1(x)),q_1(x))$$

We claim that the function f, when restricted to instances of Π_p, becomes a polynomial transformation from Π_p to $\Pi'_{\hat{p}}$, thus proving that $\Pi'_{\hat{p}}$ is NP-complete. First let us see that every instance I of Π_p is mapped by f to an instance of $\Pi'_{\hat{p}}$. Using the definition of Π_p and the inequalities satisfied by q_1 and q_2, we have, for each instance I of Π_p,

$$\begin{aligned} \text{Max}'[f(I)] &\leqslant q_2(\text{Max}[I],\text{Length}[I]) \\ &\leqslant q_2(p(\text{Length}[I]),\text{Length}[I]) \\ &\leqslant q_2(p(q_1(\text{Length}'[f(I)])),q_1(\text{Length}'[f(I)])) \\ &= \hat{p}(\text{Length}'[f(I)]) \end{aligned}$$

Thus $f(I)$ is an instance of $\Pi'_{\hat{p}}$. Conditions (a) and (b) of the definition of pseudo-polynomial transformation, along with the fact that every instance I of Π_p satisfies $\text{Max}[I] \leqslant p(\text{Length}[I])$, then imply immediately that f meets the remaining requirements to be a polynomial transformation.

Hence $\Pi'_{\hat{p}}$ is NP-complete, and it follows that Π' is NP-complete in the strong sense. ■

This lemma frees us from having to deal with particular subproblems Π_p when proving strong NP-completeness results, a great convenience since we are rarely interested in identifying the specific polynomial involved. However, the complicated definition of pseudo-polynomial transformation might appear to be a rather formidable obstacle to using this approach. In fact, it is not as complicated as it seems. Condition (a) is identical to one of the two requirements that must be met by an ordinary polynomial transformation, and condition (b) is almost identical to the other but allows us a bit more freedom in the complexity of our transformation. Condition (c) will be met by all but the most unusual transformations, since it requires only that the transformation not cause a substantial *decrease* in input length. The heart of the definition lies in condition (d), and it serves the purpose of ensuring that the magnitude of the largest number in the constructed instance does not blow up exponentially in terms of the Max and Length of the given instance.

As a first example, the construction we used to prove Theorem 4.4 can be viewed as a pseudo-polynomial transformation from the general 4-PARTITION problem to 3-PARTITION. The 3-PARTITION problem itself earns its title as our seventh "basic NP-complete problem" because of the ease with which pseudo-polynomial transformations can be constructed from it. For instance, we can use such a transformation to show that the SEQUENCING WITHIN INTERVALS problem, proved NP-complete in Section 3.2.2, is actually NP-complete in the strong sense.

Theorem 4.5 SEQUENCING WITHIN INTERVALS is NP-complete in the strong sense.

Proof: Recall that in this problem we are given a set T of tasks, each task $t \in T$ having a length $l(t) \in Z^+$ and a time interval $[r(t),d(t)]$ within which it is to be executed, and we are asked whether the tasks can be sequenced to obey these constraints, with at most one task ever being executed at a time. In Section 3.2.2 we proved it to be NP-complete, and hence we already know that it belongs to NP. We shall give a pseudo-polynomial transformation from 3-PARTITION to SEQUENCING WITHIN INTERVALS.

Let $A = \{a_1, a_2, \ldots, a_{3m}\}$, $B \in Z^+$, and $s(a_1), s(a_2), \ldots, s(a_{3m})$ constitute an arbitrary instance of 3-PARTITION. The corresponding instance of SEQUENCING WITHIN INTERVALS is given by

$$T = A \cup \{t_i : 1 \leqslant i < m\}$$

$$l(t) = \begin{cases} 1 & \text{if } t = t_i, 1 \leqslant i < m \\ s(a_j) & \text{if } t = a_j \in A \end{cases}$$

$$r(t) = \begin{cases} iB+i-1 & \text{if } t=t_i, 1 \leqslant i < m \\ 0 & \text{if } t=a_j \in A \end{cases}$$

$$d(t) = \begin{cases} iB+i & \text{if } t=t_i, 1 \leqslant i < m \\ mB+m-1 & \text{if } t=a_j \in A \end{cases}$$

This transformation clearly can be performed in time polynomial in the input length alone, and the length of the constructed instance is polynomially related to the length of the given instance, so conditions (b) and (c) of the definition of pseudo-polynomial transformation are met. Furthermore, the largest number in the constructed instance is $mB+m-1$, so condition (d) is met. All that remains to be shown is that condition (a) is met, just as in our usual NP-completeness proofs.

Any sequence that satisfies the specified constraints must execute each task t_i, $1 \leqslant i < m$, from time $iB+i-1$ to time $iB+i$, as shown in Figure 4.9. This leaves m separate blocks of time, each of length exactly B, and since this is just enough time in total to accommodate all the tasks $t \in A$, each block must be completely filled. These blocks therefore play the same role as the sets $S_1, S_2, \ldots, S_m$ in the desired partition of A. It follows that the desired sequence exists if and only if the desired partition exists for the given 3-PARTITION instance.

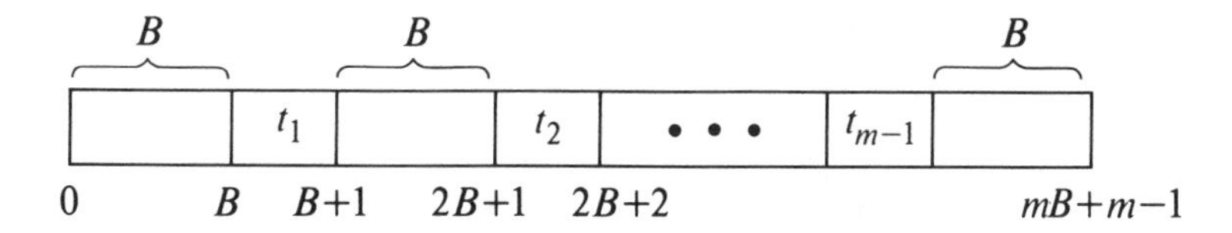

Figure 4.9 The form required of a sequence meeting the constraints of an instance of SEQUENCING WITHIN INTERVALS obtained by transforming an instance of 3-PARTITION in the proof of Theorem 4.5.

Thus condition (a) is met, and we indeed have given a pseudo-polynomial transformation from 3-PARTITION to SEQUENCING WITHIN INTERVALS. By Lemma 4.1, this proves that the latter problem is NP-complete in the strong sense. ■

We suggest as an exercise that the reader try to construct a similar transformation from 3-PARTITION to the MULTIPROCESSOR SCHEDULING problem defined in Section 3.2.1. Our lists of NP-complete problems contain a number of other problems that are proved NP-complete in the strong sense with comparable ease, merely by slightly modifying earlier proofs that used PARTITION to use 3-PARTITION instead. The straightforward nature of these modifications is indicative of the usefulness of 3-PARTITION.

We conclude this section with an example of how a pseudo-polynomial transformation from 3-PARTITION can be useful for proving an ordinary NP-completeness result for a problem that is *not* a number problem. In fact, this problem will involve no numbers at all!

Recall the SUBGRAPH ISOMORPHISM problem defined in Section 2.1: Given two graphs G and H, is H isomorphic to a subgraph of G? We proved this problem NP-complete in Section 3.2.1 simply by noting that it contains CLIQUE as a special case. However, there is one important subproblem of SUBGRAPH ISOMORPHISM that is known to belong to P. This is the problem SUBTREE ISOMORPHISM in which both G and H are required to be *trees* (a tree is a connected graph that contains no cycles). A polynomial time algorithm for this subproblem has been obtained by Edmonds and Matula [1976] (see also [Reyner, 1977]).

Our philosophy of trying to narrow in on the "boundary" between easy and hard subproblems of an NP-complete problem then suggests the following question: What if only *one* of G and H is required to be a tree? In one case the answer is immediate. The version in which only H is required to be a tree contains HAMILTONIAN PATH as a subproblem and hence is NP-complete. The case in which only G is required to be a tree is more interesting. We know that H cannot be a subgraph of such a G unless it is acyclic (contains no cycles), but this does not imply that H must be a tree, since it might be disconnected. In general, an acyclic graph is called a *forest*, with only connected forests being trees (see Figure 4.10).

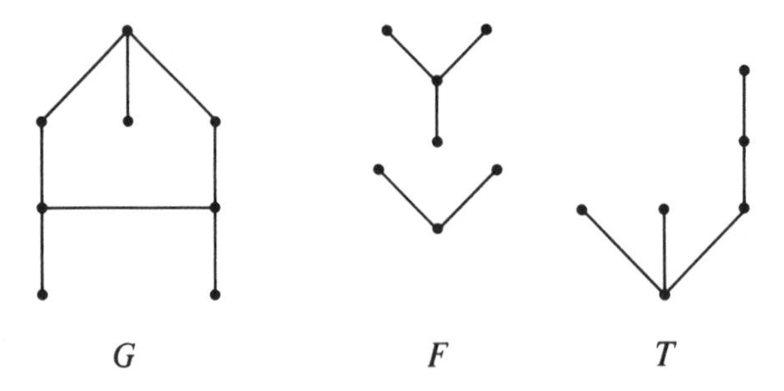

Figure 4.10 Examples of a graph G, forest F, and tree T. G is a graph but not a forest, and F is a forest but not a tree. F is *not* a subforest of T, but each tree in F is a subtree of T.

Let us give the name SUBFOREST ISOMORPHISM to the subproblem of SUBGRAPH ISOMORPHISM in which G is required to be a tree and H is required to be a forest. Despite the similarity of this problem to the polynomially solvable SUBTREE ISOMORPHISM problem, we have the following theorem:

Theorem 4.6 SUBFOREST ISOMORPHISM is NP-complete.

Proof: Membership in NP follows from that for SUBGRAPH ISOMORPHISM. We shall give a pseudo-polynomial transformation from 3-PARTITION to SUBFOREST ISOMORPHISM, and the result will follow by Lemma 4.1.

Let $A = \{a_1, a_2, \ldots, a_m\}$, $B \in Z^+$, and $s(a_1), s(a_2), \ldots, s(a_m)$ in Z^+ constitute an arbitrary instance of 3-PARTITION. The corresponding instance of SUBFOREST ISOMORPHISM is illustrated in Figure 4.11.

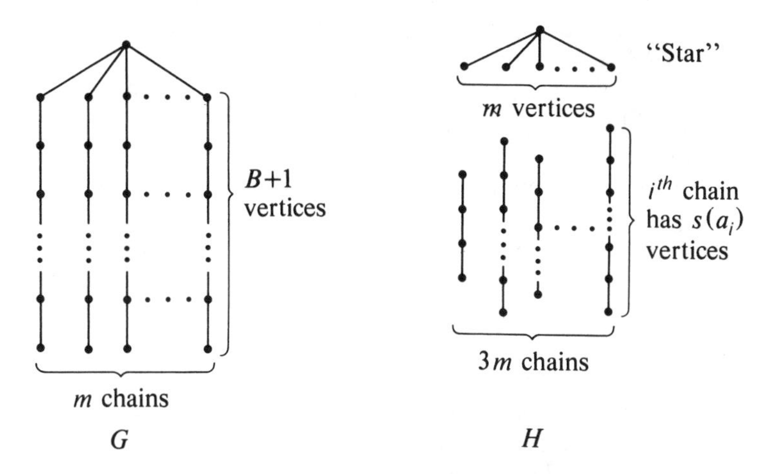

Figure 4.11 The tree G and forest H corresponding to an instance of 3-PARTITION in the proof of Theorem 4.6.

The tree G consists of m chains of $B+1$ vertices each, all attached at one end to an additional common vertex. The forest H consists of $3m+1$ trees, including one "star" on $m+1$ vertices and $3m$ chains, each corresponding to a particular element $a \in A$ and having $s(a)$ vertices.

Any isomorphism from H to a subgraph of G must map the center of the star to the single high-degree vertex of G. The m neighbors of the center of the star in H then must be mapped to the m neighbors of that vertex in G. This leaves m chains, each of B vertices, in G to which the remaining $3m$ chains in H must be mapped by the isomorphism. The mapping of these chains from H to the remainder of G corresponds to a partition of the elements of A into m sets and, by our construction, can be completed if and only if the elements in each set have sizes summing exactly to B. Thus the required isomorphism from H to a subgraph of G will exist if and only if the required 3-partition of A exists.

This confirms condition (a) of a pseudo-polynomial transformation. It is easy to see that this transformation can be performed in time polynomial in m and B, so condition (b) is satisfied. The total number of vertices in G

and H is $2(mB+1)$, so condition (c) is satisfied. Finally, there are no numbers in the constructed instance, so condition (d) holds. Thus by Lemma 4.1, SUBFOREST ISOMORPHISM is NP-complete in the strong sense, which implies that it is NP-complete in the ordinary sense as well. ■

4.3 Time Complexity as a Function of Natural Parameters

So far in this chapter we have motivated the study of subproblems mainly on the basis of the fact that in practice it is often the subproblem, rather than the general problem, that we are called upon to solve. Having mapped the boundary between the NP-complete subproblems and the polynomial time solvable subproblems, one is better prepared to focus the search for algorithms in potentially profitable directions when such a subproblem arises.

Results concerning subproblems also can be used to help guide the search for algorithms that solve the general problem. If the general problem is NP-complete, we know that an exponential time algorithm will be required (unless P = NP), but there are a variety of ways in which the time complexity of an algorithm can be "exponential," some of which might be preferable to others. This is especially evident when, as is customary in practice, we consider time complexity expressed in terms of natural problem parameters instead of the artificially constructed "input length."

For example, consider the MULTIPROCESSOR SCHEDULING problem of Section 3.2.1. Here a collection of natural parameters might consist of the number n of tasks, the number m of processors, and the length L of the longest task. The ordinary NP-completeness result for this problem proved in Section 3.2.1 implies that, unless P = NP, MULTIPROCESSOR SCHEDULING cannot be solved in time polynomial in the three parameters n, m, and $\log L$. However, one can still ask whether it is possible to have an algorithm with time complexity polynomial in m^n and $\log L$, or polynomial in n^m and $\log L$, or polynomial in n, m, and L, or even polynomial in $(nL)^m$.

Our complexity results for subproblems shed some light on these questions. The original NP-completeness result for MULTIPROCESSOR SCHEDULING actually shows that the subproblem in which m is restricted to the value 2 is NP-complete, thus ruling out an algorithm polynomial in n^m and $\log L$ (unless P = NP), since such an algorithm would be a polynomial time algorithm for this subproblem. Our subproblem results do not rule out an algorithm polynomial in m^n and $\log L$, and indeed exhaustive search algorithms having such a time complexity can be designed. Analogously, the strong NP-completeness result for MULTIPROCESSOR SCHEDULING claimed in Section 4.2.2 rules out an algorithm polynomial in n, m, and L (unless P = NP). It leaves open the possibility of an algorithm polynomial in $(nL)^m$ (which would give a pseudo-polynomial time algorithm for each *fixed* value of m), and again such an algorithm can be shown to exist.

Thus by considering the subproblems obtained by placing restrictions on one or more of the natural problem parameters, we obtain useful information about what types of algorithms are possible for the general problem. Care must be taken to ensure that the parameters we choose are sufficiently representative of instance size that Length$[I]$ can be expressed as a polynomial function of them (so that the class of polynomial time algorithms for the problem is identical to the class of algorithms polynomial in the selected parameters), but otherwise we may choose whatever parameters seem most natural and relevant. A general NP-completeness result then will imply that the problem cannot be solved in time polynomial in all the chosen parameters, and information obtained by restricting these parameters can be meaningful with regard to other types of general algorithms.

Although questions concerning strong NP-completeness and pseudo-polynomial time algorithms are especially relevant here, analyses of this type also can be applied fruitfully to problems that are not number problems, since all problems have natural numerical parameters like sizes of sets, values of bounds, etc. Thus, for instance, the NP-completeness of 3-SATISFIABILITY rules out the possibility (unless P=NP) of an algorithm for SATISFIABILITY that runs in time polynomial in $(mn)^M$, where m is the number of clauses, n is the number of literals, and M is the maximum number of literals per clause, whereas for the CLIQUE problem an n^D algorithm is possible, where n is the number of vertices and D is the maximum vertex degree. Thus the theory of NP-completeness can be used to guide our search not only for polynomial time algorithms, but for exponential time algorithms as well.

5

NP-Hardness

This chapter will conclude our coverage of the main concepts and applications of the theory of NP-completeness. It consists of two parts. In the first part, which makes up the major portion of the chapter, we show how the implications of the theory can be extended beyond the class NP by use of a more general type of reducibility between problems. In the second part we briefly survey the historical development of the theory of NP-completeness and mention some of the alternative terminology that has been used in the literature.

5.1 Turing Reducibility and NP-Hard Problems

Although we have restricted our discussions so far mainly to problems that belong to NP, it should be apparent that the techniques used for proving NP-completeness also can be used for proving that problems outside of NP are hard. Any decision problem Π, whether a member of NP or not, to which we can transform an NP-complete problem will have the property that it cannot be solved in polynomial time unless $P = NP$. We might say that such a problem Π is "NP-hard," since it is, in a sense, at least as hard as the NP-complete problems.

Our notion of NP-hardness will be more general than this, however, since it is possible to generalize the notion of a polynomial transformation in such a way that problems other than just decision problems can be proved to be "at least as hard" as the NP-complete problems. As in Chapter 2, all our definitions will be stated both formally in terms of languages and Turing machines and informally in terms of problems and algorithms.

The more general class of problems to which our definitions will apply is the class of "search problems." A *search problem* Π consists of a set D_Π of finite objects called *instances* and, for each instance $I \in D_\Pi$, a set $S_\Pi[I]$ of finite objects called *solutions* for I. An algorithm is said to *solve* a search problem Π if, given as input any instance $I \in D_\Pi$, it returns the answer "no" whenever $S_\Pi[I]$ is empty and otherwise returns some solution s belonging to $S_\Pi[I]$.

For example, the solution set for an instance of the traveling salesman optimization problem consists of *all* tours having the minimum possible length. An algorithm that solves this search problem need only find one such tour for any given instance. An example in which $S_\Pi[I]$ can be empty is provided by the "Hamiltonian circuit construction problem," in which the solution set for a given graph G consists of all Hamiltonian circuits in G. An algorithm for solving this problem must output "no" whenever G does not have a Hamiltonian circuit and otherwise must output one such circuit for G. Notice that any decision problem Π can be formulated as a search problem by defining $S_\Pi[I] = \{\text{"yes"}\}$ if $I \in Y_\Pi$ and $S_\Pi[I] = \phi$ if $I \notin Y_\Pi$. It is convenient to assume that all decision problems have been formulated in this way, so that a decision problem can be considered simply to be a special type of search problem.

The formal counterpart of a search problem is a *string relation*. For a finite alphabet Σ, a string relation over Σ is a binary relation $R \subseteq \Sigma^+ \times \Sigma^+$, where $\Sigma^+ = \Sigma^* - \{\epsilon\}$, the set of all nonempty strings over Σ. A language L over Σ can be identified with the string relation

$$R = \{(x,s): x \in \Sigma^+ \text{ and } x \in L\}$$

where s is any fixed symbol from Σ. (Notice that this ignores whether or not the empty string belongs to L but will not affect the kinds of computational questions in which we are interested.) A function $f:\Sigma^* \rightarrow \Sigma^*$ *realizes* the string relation R if and only if, for each $x \in \Sigma^+$, $f(x) = \epsilon$ whenever there is no $y \in \Sigma^+$ such that $(x,y) \in R$ and otherwise $f(x)$ equals some $y \in \Sigma^+$ for which $(x,y) \in R$. A DTM program M *solves* the string relation R if the function f_M computed by M realizes R.

The correspondence between search problems and string relations is once again accomplished by means of encoding schemes, only now an encoding scheme for Π must give both a string encoding each instance $I \in D_\Pi$ and a string encoding each solution $s \in S_\Pi[I]$. Under the encoding scheme e, the search problem Π corresponds to the string relation $R[\Pi,e]$ defined by:

$$R[\Pi,e] = \left\{ (x,y)\colon \begin{array}{l} x \in \Sigma^+ \textit{ is the encoding under } e \textit{ of an instance} \\ I \in D_\Pi \textit{ and } y \in \Sigma^+ \textit{ is the encoding under } e \\ \textit{of a solution } s \in S_\Pi[I] \end{array} \right\}$$

We say that Π (under encoding scheme e) is solvable by a polynomial time algorithm if there is a polynomial time DTM program that "solves" $R[\Pi,e]$.

The generalization of "polynomial transformation" that we will be using is motivated by the observation that any polynomial transformation from a decision problem Π to a decision problem Π' provides an algorithm A for solving Π by using a hypothetical "subroutine" for solving Π'. Given any instance I of Π, the algorithm first constructs an equivalent instance I' of Π', then applies the subroutine to I', and finally outputs the answer returned by the subroutine, since it is also the correct answer for I. Except for the time required by the subroutine, the algorithm A runs in polynomial time. Thus if the subroutine were itself a polynomial time algorithm for solving Π', then the overall procedure would be a polynomial time algorithm for solving Π.

Notice that this last statement would be true even if A used the subroutine for Π' many times (though no more than a polynomially bounded number of times) and even if the assumed subroutine were for solving a search problem rather than a decision problem. This fact provides the basis for our generalization. A *polynomial time Turing reduction* (or simply Turing reduction) from a search problem Π to a search problem Π' is an algorithm A that solves Π by using a hypothetical subroutine S for solving Π' such that, if S were a polynomial time algorithm for Π', then A would be a polynomial time algorithm for Π.

This notion can be captured formally in terms of what are called oracle machines. For specificity we again use a Turing machine model, although analogous definitions could be made in terms of any standard model of computation. An *oracle Turing machine* (OTM) consists of a standard DTM augmented with an additional *oracle tape*, having tape squares numbered $\ldots,-2,-1,0,1,2,\ldots$, and a read-write *oracle head* for operating with this tape. Such a machine is illustrated schematically in Figure 5.1.

A program for an OTM is similar to that for a DTM and specifies the following:

(1) a finite set Γ of *tape symbols*, including a subset $\Sigma \subset \Gamma$ of *input symbols* and a distinguished *blank symbol* $b \in \Gamma - \Sigma$;

(2) a finite set Q of *states*, including a distinguished *start-state* q_0, a distinguished *halt-state* q_h, a distinguished *oracle-consultation state* q_c, and a distinguished *resume-computation state* q_r;

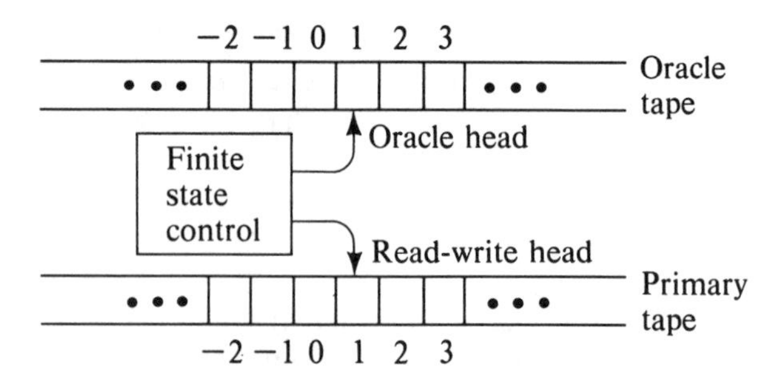

Figure 5.1 Schematic representation of an oracle Turing machine (OTM).

(3) a *transition function*

$$\delta: (Q-\{q_h, q_c\}) \times \Gamma \times \Sigma \rightarrow Q \times \Gamma \times \Sigma \times \{-1,+1\} \times \{-1,+1\}.$$

The computation of an OTM program on an input $x \in \Sigma^*$ is also much like that for a DTM, except that, when the finite state control is in state q_c, what happens in the next step depends on a specified *oracle function* $g:\Sigma^* \rightarrow \Sigma^*$. The computation begins with the symbols of x written in squares 1 through $|x|$ of the primary tape, with the rest of that tape and all of the oracle tape being blank, each tape head scanning square 1 of its tape, and the finite state control in state q_0. The computation proceeds in a step-by-step manner, with one of three possibilities occurring at each step:

a. If the current state is q_h, then the computation has ended and no further steps take place.

b. If the current state is $q \in Q-\{q_h, q_c\}$, then the action taken depends on the symbols being scanned on the two tapes and the transition function δ. Let s_1 be the symbol in the square currently being scanned by the primary tape head, let s_2 be the symbol in the square currently being scanned by the oracle head, and let $(q', s_1', s_2', \Delta_1, \Delta_2)$ be the value of $\delta(q, s_1, s_2)$. The finite state control then changes from state q to state q', the primary tape head writes s_1' in place of s_1 and changes its scanning position by Δ_1 (forward one square if $\Delta_1 = +1$ and backward one square if $\Delta_1 = -1$), and the oracle head writes s_2' in place of s_2 and changes its scanning position by Δ_2. Thus this is just like a step of an ordinary DTM, except that it involves two tapes.

c. If the current state is q_c, then the action taken depends on the contents of the oracle tape and on the oracle function g. Let $y \in \Sigma^*$ be the string appearing in squares 1 through $|y|$ of the oracle tape, where square $|y|+1$ is the first square to the right of square 0 that contains a blank, and let $z \in \Sigma^*$ be the value of $g(y)$. Then in one step the oracle tape is changed to contain the string z in squares 1 through $|z|$, with blanks everywhere else, the oracle head is set to scan square 1, and the finite state control is changed from state q_c to state q_r. Such a step leaves both the contents of the primary tape and the position of its tape head unchanged.

The main difference between a DTM and an OTM is in this third type of step, which provides the means by which an OTM program can "consult" the oracle. If the OTM writes a query string y on the oracle tape and then enters the oracle-consultation state, the answer string $z = g(y)$ will be returned in one step of the computation. Thus this corresponds to calling a hypothetical subroutine for computing the function g. The computation of an OTM program M on an input string x depends both on x and on the associated oracle function g.

Let us use M_g to denote the "relativized" OTM program obtained by combining M with oracle g. If M_g halts for all inputs $x \in \Sigma^*$, then it can be viewed as computing a *function* $f_M^g: \Sigma^* \rightarrow \Gamma^*$, defined in exactly the same way as for a DTM. We shall say that M_g is a *polynomial time OTM program* if there exists a polynomial p such that M_g halts within $p(|x|)$ steps for every input $x \in \Sigma^*$.

Let R and R' be any two string relations over Σ. A *polynomial time Turing reduction* from R to R' is an OTM program M with input alphabet Σ such that, for every function $g: \Sigma^* \rightarrow \Sigma^*$ that realizes R', the relativized program M_g is a polynomial time OTM program and the function f_M^g computed by M_g realizes R. If there is such a reduction from R to R', we shall write $R \propto_T R'$, read "R Turing-reduces to R'." Notice that $\propto_T$, like $\propto$, is transitive.

We are now prepared to define "NP-hard." A string relation R is *NP-hard* if there is some NP-complete language L (itself stated as a string relation, as noted earlier) such that $L \propto_T R$. A search problem Π (under encoding scheme e) is said to be NP-hard if the string relation $R[\Pi,e]$ is NP-hard. Informally, this can be interpreted as saying that a search problem Π is NP-hard if there exists some NP-complete problem Π' that Turing-reduces to Π. It is not difficult to see that if a string relation R (or a search problem Π) is NP-hard, then it cannot be solved in polynomial time unless P = NP.

Note that, by the transitivity of $\propto_T$, if R is any NP-hard string relation and if R Turing-reduces to the string relation R', then R' also must be NP-hard. Furthermore, by our association of languages with string relations and decision problems with search problems, we can immediately say that all NP-complete languages and all NP-complete problems are NP-hard.

We will encounter OTMs once more in Chapter 7, but for now let us continue solely at the problem level, applying the above terminology in the usual informal way, to examine some of the applications of Turing reducibility and NP-hardness.

The first and most trivial application concerns the complements of the NP-complete problems. The *complement* of a decision problem Π is the problem Π^c having domain D_Π and yes-set $D_\Pi - Y_\Pi$. In Chapter 2 we observed that it is not known in general whether $\Pi \in$ NP implies $\Pi^c \in$ NP. Nor is it always apparent that Π can be polynomially transformed to Π^c, because of the reversed roles of "yes" and "no." However, it *is* a trivial matter to give a Turing reduction from Π to Π^c (and vice-versa), so if Π is NP-complete or NP-hard, then Π^c must be NP-hard.

Our second application concerns those search problems, like the traveling salesman optimization problem and the Hamiltonian circuit construction problem, for which the corresponding decision problems are known to be NP-complete. Whenever we show that a polynomial time algorithm for the search problem could be used to solve the corresponding decision problem in polynomial time, we are actually giving a Turing reduction between them, and hence an NP-completeness result for the decision problem can be translated into an NP-hardness result for the search problem. Thus the argument presented in Chapter 2 to show that TRAVELING SALESMAN is no harder than the traveling salesman optimization problem, along with the fact that TRAVELING SALESMAN is NP-complete, constitutes a proof that the optimization problem is NP-hard.

Neither of these examples illustrates the full power of Turing reducibility, however, because each requires only one call of the subroutine (one consultation of the oracle) for any instance. An example that uses Turing reducibility to more advantage involves the following decision problem:

Kth LARGEST SUBSET
INSTANCE: A finite set A, a size $s(a) \in Z^+$ for each $a \in A$, and two non-negative integers $B \leqslant \sum_{a \in A} s(a)$ and $K \leqslant 2^{|A|}$.
QUESTION: Are there at least K distinct subsets $A' \subseteq A$ that satisfy $s(A') \leqslant B$ (where $s(A')$ is defined to be $\sum_{a \in A'} s(a)$)?

It is shown in [Lawler, 1972] that this problem can be solved in pseudo-polynomial time, in fact in time bounded by a polynomial function of $|A| \cdot K \lceil \log s(A) \rceil$. Thus, for any *fixed* value of K, it can be solved in polynomial time. The question then arises, can it be solved *in general* in polynomial time?

Not only does this problem appear not to be in P, it does not appear even to be in NP, since the natural way of solving it nondeterministically involves guessing K subsets of A, and there seems to be no way to write down such a guess using only a polynomial number of symbols in $|A| \cdot \lceil \log K \rceil \cdot \lceil \log s(A) \rceil$. On the other hand, no transformation from an NP-complete problem to this problem is known. However, Johnson and Kashdan [1976] show that the NP-complete PARTITION problem can be Turing-reduced to Kth LARGEST SUBSET.

The argument proceeds as follows: Suppose $S[A,s,B,K]$ is a subroutine for solving the Kth LARGEST SUBSET problem, with parameters $A = \{a_1,a_2, \ldots, a_n\}$, $s: A \rightarrow Z^+$, $B \leqslant s(A)$, and $K \leqslant 2^n$. The corresponding algorithm for solving PARTITION begins by computing $s(A)$, where $A = \{a_1,a_2, \ldots, a_n\}$ and $s: A \rightarrow Z^+$ describe the given instance. If $s(A)$ is not evenly divisible by 2, the algorithm immediately responds "no" for this instance. Otherwise, it sets b equal to $s(A)/2$ and applies the following binary search procedure, using the assumed subroutine, to determine the number L^* of subsets $A' \subseteq A$ satisfying $s(A') \leqslant b$.

Step 1. Set $L_{\text{MIN}} \leftarrow 0$, $L_{\text{MAX}} \leftarrow 2^n$.

Step 2. If $L_{\text{MAX}} - L_{\text{MIN}} = 1$, set $L^* \leftarrow L_{\text{MIN}}$ and halt.

Step 3. Set $L \leftarrow (L_{\text{MAX}} + L_{\text{MIN}})/2$ and call $S[A,s,b,L]$. If the answer is "yes," set $L_{\text{MIN}} \leftarrow L$ and go to Step 2. Otherwise, set $L_{\text{MAX}} \leftarrow L$ and go to Step 2.

This procedure determines L^* using exactly n calls of the subroutine. Only one additional call is needed to determine the answer for our given PARTITION instance, this time to $S[A,s,b-1,L^*]$. If the answer for this call is "yes," then all subsets $A' \subseteq A$ that satisfy $s(A') \leqslant b$ also must satisfy $s(A') \leqslant b-1$, so the answer for the PARTITION instance is "no." If the answer for this call is "no," then there must be some subset $A' \subseteq A$ for which $s(A') = b$, so the answer for the PARTITION instance is "yes."

It is easy to see that this procedure would be a polynomial time algorithm for PARTITION if S were a polynomial time subroutine for Kth LARGEST SUBSET. Thus we have the desired Turing reduction from PARTITION to Kth LARGEST SUBSET. It follows that Kth LARGEST SUBSET is NP-hard and cannot be solved by a polynomial time algorithm unless P = NP.

On the basis of this example it should be evident that we can use the notion of NP-hardness for analyzing the complexity of problems in much the same way as we use NP-completeness. All the types of questions discussed in Chapter 4 are also applicable to NP-hard problems, and we can proceed to consider the complexity of subproblems and such related issues as pseudo-polynomial time algorithms and "strong" NP-hardness (defined analogously to strong NP-completeness, with a search problem being NP-hard in the strong sense if it contains an NP-hard subproblem satisfying a polynomial bound on $\text{Max}[I]$). The only limitation on such results is that, whereas an NP-complete problem can be said to be solvable in polynomial time *if and only if* P = NP, all we can say with certainty about an NP-hard problem is that it cannot be solved in polynomial time *unless* P = NP.

As a final application of Turing reducibility we shall show how even this distinction can often be removed. Recall that, when we pointed out in Chapter 2 that TRAVELING SALESMAN is no harder than the corresponding optimization problem, we also noted that it is in a sense "no easier." What we meant by that statement can now be spelled out: Not only is the decision problem Turing reducible to the optimization problem, but the optimization problem is also Turing reducible to the decision problem. Thus each can be solved in polynomial time if and only if the other can. Since the decision problem is NP-complete, it follows that the optimization problem can be solved in polynomial time if and only if P = NP.

To see this, let us introduce an intermediate problem, defined as follows:

TRAVELING SALESMAN EXTENSION (TSE)

INSTANCE: A finite set $C=\{c_1,c_2,\ldots,c_m\}$ of cities, a distance $d(c_i,c_j)\in Z^+$ for each pair of cities $c_i,c_j\in C$, a bound $B\in Z^+$, and a "partial" tour $\Theta = <c_{\pi(1)},c_{\pi(2)},\ldots,c_{\pi(K)}>$ of K distinct cities from C, $1\leqslant K\leqslant m$.

QUESTION: Can Θ be extended to a full tour

$$<c_{\pi(1)},c_{\pi(2)},\ldots,c_{\pi(K)},c_{\pi(K+1)},\ldots,c_{\pi(m)}>$$

having total length B or less?

It is easy to see that this problem belongs to NP, and hence, by the definition of NP-completeness, TSE $\propto$ TS. Since a transformation is just a special case of a Turing reduction, this in turn implies that TSE $\propto_T$ TS. Thus letting TSO stand for the traveling salesman optimization problem, all we need to show is that TSO $\propto_T$ TSE and by transitivity we will have that TSO is Turing reducible to TS.

So suppose that $S[C,d,\Theta,B]$ is a subroutine for solving TSE, with the parameters standing for the set C of cities, distance function d from pairs of cities to Z^+, partial tour Θ, and bound $B\in Z^+$. Let C and d be any given instance of TSO, and let B^* denote the optimal tour length for this instance, whatever it might be. Since every city must occur in an optimal tour, and since any tour can be cyclically permuted without changing its length, there must be an optimal tour that starts with city c_1. Furthermore, we know that B^* lies between the two values $B_{\text{MIN}} = m$ and $B_{\text{MAX}} = m\cdot(\max\{d(c_i,c_j): c_i,c_j\in C\})$. Thus by using a binary search procedure analogous to that used for K[th] LARGEST SUBSET, we can determine the value of B^* by a sequence of at most $\lceil \log_2 B_{\text{MAX}}\rceil$ calls on the subroutine $S[C,d,<c_1>,B]$, with different values of B.

Once we know the value of B^*, we can proceed to construct an optimal tour using the subroutine S. Let us call a sequence Θ of distinct cities from C an *extendible partial tour* if it can be extended to a complete tour having total length B^*. Clearly, $<c_1>$ is an extendible partial tour. Since $<c_1>$ is extendible, there must exist at least one $c_j\in C-\{c_1\}$ such that $<c_1,c_j>$ is an extendible partial tour. We can find such a c_j by making a sequence of at most $m-2$ calls of the subroutine S, each of the form

$S[C,d,<c_1,c_i>,B^*]$ for a $c_i \in C-\{c_1\}$. (If the first $m-2$ choices fail, then we know that the remaining choice *must* succeed, so we need not apply the subroutine for it.)

In general, if $<c_{\pi(1)}, \ldots, c_{\pi(K)}>$ is an extendible partial tour with $K<m$, then we can find another extendible partial tour $<c_{\pi(1)}, \ldots, c_{\pi(K)}, c_{\pi(K+1)}>$ involving one additional city by a sequence of at most $m-K-1$ calls of the subroutine S. Thus we can build a complete tour for C by using a total of at most $(m-1)(m-2)/2$ calls of S, beyond those used to determine B^*. Since a suitable Length function for traveling salesman instances is $\text{Length}[I] = m + \log_2 B_{\text{MAX}}$, this clearly gives a (polynomial time) Turing reduction from TSO to TSE. Thus we have that TSO $\propto_T$ TSE $\propto_T$ TS, and consequently TSO $\propto_T$ TS as desired.

Notice here that the key relation is TSO $\propto_T$ TSE. Once we know that the traveling salesman optimization problem is Turing reducible to *some* problem in NP, then we know that it can be solved in polynomial time if P = NP and hence can be "no harder" than the NP-complete problems. Thus, let us call a search problem Π *NP-easy* whenever there exists a problem $\Pi' \in$ NP for which $\Pi \propto_T \Pi'$.

Once the reader has absorbed the techniques of the TSO $\propto_T$ TSE procedure, it should not be difficult to see how the same approach can be used to prove that many other search problems are NP-easy, especially those whose decision problem counterparts we have shown to be NP-complete. The problem Π' from NP is usually defined as was TRAVELING SALESMAN EXTENSION, so that it can be used to build up the desired solution sequentially (see for instance [Valiant, 1976a]). For optimization problems, the solution construction procedure is preceded by an initial binary search phase in which the optimum value is determined. As an exercise, we suggest that the reader attempt to prove that the search problems related to the six basic NP-complete problems of Section 3.1 are all NP-easy — namely, finding a satisfying truth assignment, finding a three dimensional matching, finding a minimum cardinality vertex cover, finding a maximum cardinality clique, finding a Hamiltonian circuit, and finding a partition of a set of "sized elements" into two subsets having the same total size.

In fact, we now observe that the restriction of the basic theory to decision problems has caused no substantial loss of generality, since most often the search problems whose decision problem counterparts have been proved to be NP-complete are themselves NP-easy and hence of equivalent complexity. As we have been building our equivalence class of NP-complete problems, we have at the same time been building a much larger class of equivalent search problems: those that are both NP-hard and NP-easy (and which therefore might be called "NP-equivalent"). Although the larger class contains many problems that do not belong to NP, it still retains the familiar property we associate with the NP-complete problems: No problem in this class can be solved in polynomial time unless P = NP, and if P = NP *all* problems in this class can be solved in polynomial time.

5.2 A Terminological History

As we conclude our description of the theory of NP-completeness, it is appropriate to take a brief look backward at the historical development of the main ideas and at the checkered career of the terminology used for discussing them.

We have already mentioned, in Chapter 1, two early papers that discussed the significance of polynomial time complexity. Cobham [1964] noted the wide variety of mathematical functions that can be computed in polynomial time and observed that the class of all such functions remains the same under many different models of computation. Edmonds [1965a] informally identified the term "good algorithm" with the notion of a polynomial time algorithm. A third early paper, [Edmonds, 1965b], also introduced an informal notion analogous to NP. In that paper, it is proposed that a problem be said to have a "good characterization" if for every solution there exists a polynomial time checkable "proof" that it is indeed a solution.

All of these early discussions were in terms of functions or in terms of what we have called search problems. Part of the fundamental contribution of Cook was in seeing the value of restricting discussion to languages and the decision problems they encode. In [Cook, 1971a], the classes of languages we now call P and NP are first identified, and Cook's fundamental theorem is proved. However, this paper diverges from later practice in two significant ways. First, Cook's basic notion of reducibility between languages, which he called "P-reducibility," involved polynomial time Turing reductions rather than polynomial transformations. For this reason, one occasionally sees polynomial time Turing reducibility referred to as "Cook-reducibility." Second, the unnamed equivalence class of problems that Cook built up about SATISFIABILITY was not restricted to what are now called the NP-complete problems, but instead contained all those decision problems we have termed "NP-equivalent."

It was in [Karp, 1972] that the theory of NP-completeness took on its present form, though not all of its current terminology. Karp introduced the terms P and NP, and observed that Cook's Theorem would remain true if the notion of Turing reducibility were replaced by the simpler and more manageable notion of polynomial transformability. This latter type of reducibility is sometimes referred to as "Karp-reducibility." It also is occasionally called "many-one reducibility," because a polynomial transformation is a many-one function. Karp himself simply used the term "reducible." He then presented the current definition of "NP-complete problem," although the term he used was "polynomial complete problem," thus suggesting the analogous "polynomial reduction." (The term "complete" comes from the recursion theoretic notion of a language being "complete" for a class with respect to a given type of reducibility if it belongs to the class and every other member reduces to it.) In light of the distinction between the two classes defined by Cook and Karp, it is amusing to see occasional references in the literature to the "Cook-Karp class," which usually can be taken to mean the Karp class.

Karp's paper, and the talks he gave preceding its publication, had a very stimulating effect. As is perhaps natural, along with the new results that began to appear new terminology was occasionally proposed. In place of "polynomial complete," some grammarians in the field felt constrained to say "polynomially complete," despite its added cumbrousness. More substantially, Sahni [1974] introduced the term "P-complete" as an analogue of "polynomial complete" extended to a more general class of problems (something like our search problems), along with "P-hard" to describe a problem that is at least as hard as the P-complete problems. Meanwhile, L. A. Levin in the Soviet Union independently proved a variation of Cook's Theorem and introduced a term that translates to "universal sequential search problem" [Levin, 1973], and which, like Sahni's "P-complete," applied to a more general class of problems than just decision problems. Our use of the term "search problem" is, in part, borrowed from Levin's paper.

The terminology in common use today (NP-complete, NP-hard, polynomial transformation) is in large part a result of the efforts of Donald Knuth. In 1973, primarily because he wanted a better term for "at least as hard as the polynomial complete problems" to use in Volume 4 of his series *The Art of Computer Programming*, Knuth circulated a private poll to a collection of members of the research community, asking them to rate three proposed alternatives as to their acceptability. He rejected the term "P-hard" for semantic reasons, since if $P \neq NP$, the polynomial complete problems would in fact be *much* harder than those in P. The research community in turn rejected all three of Knuth's alternatives, but responded with a variety of alternative proposals, some not entirely serious. (For instance Shen Lin proposed that the polynomial complete problems be called the "PET problems," not only because they were his personal favorites, but also because PET could be an acronym for "probably exponential time." Moreover, even if the question of whether $P = NP$ were resolved, the name would still have a valid interpretation. If $P \neq NP$, then PET could stand for "provably exponential time." If $P = NP$, then it could stand for "previously exponential time.") An entertaining report on this poll and the discussions it provoked appears in [Knuth, 1974a]. The end result was that the terms "NP-complete" and "NP-hard," both write-in candidates, were declared the victors, with "NP-complete" to replace "polynomial complete" and a decision problem Π to be called "NP-hard" if SATISFIABILITY $\propto \Pi$. The term "NP-complete" was appealing both because it has three fewer syllables than "polynomial complete" and because it is more meaningful in theoretic terms, an NP-complete problem being one that is complete for NP (with respect to polynomial time reducibility). Similarly, "NP-hard" could be interpreted as meaning "as hard as the hardest problem in NP."

The use of these terms has since become nearly universal, although a stray "polynomial complete" still appears every so often. However, although the definition of NP-complete seems to be fairly stable, the definition of "NP-hard" is somewhat less so. In a postscript (which also introduced the term "polynomial transformation" for "polynomial reduction"), Knuth [1974b] suggested that "NP-hard" be redefined to apply to any decision problem to which SATISFIABILITY is *Turing reducible* (rather than just transformable) so that, for instance, the complements of the NP-complete problems could be called "NP-hard." The term "NP-hard" can now be encountered in the literature with both definitions, and neither seems to predominate.

In this book we have gone one step beyond Knuth and presented a still broader definition of "NP-hard," extending it and the notion of Turing reducibility to include search problems, so that these terms will be more generally useful. With a similar philosophy, we have also introduced the terms "NP-easy" and "NP-equivalent," the latter being an attempt to capture something of what Sahni and Levin meant with their "P-complete" and "universal sequential search" problems.

In addition, the "strong NP-completeness" terminology of Section 4.2 was also introduced by the authors, first appearing in [Garey and Johnson, 1978]. An alternative formulation of these concepts is used in [Lageweg, Lenstra, Rinnooy Kan, 1978], where a problem that is NP-complete in the strong sense is called "unary" NP-complete, and a problem that is NP-complete in the ordinary sense is called "binary" NP-complete. This terminology is based on the fact that, if a problem is strongly NP-complete, then the corresponding language is NP-complete even when the encoding scheme is allowed to represent numbers using the (non-concise) "unary" notation (a string of n 1's representing the number n).

Other than the above-mentioned innovations, and a few non-fundamental idiosyncrasies in our machine models, we have attempted to stay as close as possible to accepted practice and terminology, so that this book would be consistent with at least the current literature. The reader who chooses to use this book as a stepping-off point for an exploration of the literature will no doubt encounter references to many peripheral issues that enliven the study of NP-completeness. In the next two chapters we present a sampler of some of these related topics, so that the reader will be aware of what such references are about and can follow up on those that seem most interesting.

6

Coping with NP-Complete Problems

Let us return for a moment to the bandersnatch department of Section 1.1. Recall that we left you there in a somewhat unresolved position. As the chief algorithm designer, you had just neatly sidestepped potential charges of incompetence by proving that the bandersnatch problem is NP-complete. However, the bandersnatch problem had refused to vanish at the sound of those mighty words, and you were still faced with the task of finding some usable algorithm for dealing with it.

In this chapter we will be discussing several of the ways you might approach this task, concentrating primarily on recent work that seeks to obtain provable "performance guarantees" for algorithms. Before narrowing our focus, however, it is appropriate to survey briefly the alternatives that might be considered. These can be divided roughly into two general categories.

The first category consists of those approaches that, while acknowledging the apparent inevitability of exponential time complexity, seek to obtain as much improvement over straightforward exhaustive search as possible. Among the most widely used approaches to reducing the search effort are those based on "branch-and- bound" or "implicit enumeration" techniques (for example, see [Garfinkel and Nemhauser, 1972]). These generate "partial solutions" within a tree-structured search format and utilize powerful bounding methods to recognize partial solutions that cannot possibly be extended to actual solutions, thereby eliminating entire branches of the search in a single step. Other approaches that provide alternative ways of organizing the search, and which sometimes are used in conjunction with branch-and-bound, include dynamic programming (such as used to obtain the pseudo-polynomial time algorithms discussed in Chapter 4), cutting plane methods (for example, see [Hu, 1969], [Garfinkel and Nemhauser, 1972]), and Lagrangian techniques (for example, see [Geoffrion, 1974], [Held and Karp, 1971]). In addition, it is sometimes possible to reduce substantially the worst case time complexity of exhaustive search merely by making a more clever choice of the objects over which the exhaustive search is performed. Some recent examples of this include algorithms for the PARTITION problem [Horowitz and Sahni, 1974], GRAPH K-COLORABILITY [Lawler, 1976b], and INDEPENDENT SET [Tarjan and Trojanowski, 1977].

The second category of approaches pertains solely to optimization problems (not a severe restriction, since so many problems arising in applications are naturally formulated in this way) and involves what might be called a "lowering of our sights." Here we no longer focus on finding an optimal solution, but instead try to find a "good" solution within an acceptable amount of time. Algorithms that do this are loosely termed "heuristic" algorithms, since they frequently are based on sensible "rules of thumb." The methods used for designing such algorithms tend to be rather problem specific, although a few guiding principles have been identified and can provide a useful starting point (see [Lin, 1975] for an excellent discussion of these). The most widely applied technique is that of "neighborhood search," in which a preselected set of local operations is used to repeatedly improve an initial solution, continuing until no further local improvements can be made and a "locally optimum" solution has been obtained. Heuristic algorithms designed via this and other approaches have often proved quite successful in practice, although a considerable amount of fine-tuning is usually required in order to achieve satisfactory performance. As a consequence, it is rarely possible to predict how well such algorithms will perform by formally analyzing them beforehand. Instead, these algorithms are usually evaluated and compared through a combination of empirical studies and common-sense arguments.

Recently, however, a number of results have been obtained that show that heuristic algorithms may not always be so immune to formal analysis. In some cases it is possible to prove that the solutions found by a heuristic algorithm will never differ from optimal by more than some specified percentage. Results like this can be viewed as providing "performance guarantees" for algorithms, and we shall concentrate on them for the remainder of the chapter.

In Section 6.1 we survey the different types of performance guarantees that are possible, and in Section 6.2 we show how the theory of NP-completeness can be used to make inferences about the best possible guarantee for a problem. Finally, in Section 6.3, we discuss some practical considerations concerning the applicability of this type of result, and we mention some recent theoretical work directed toward analyzing the "average performance" of heuristic algorithms.

6.1 Performance Guarantees for Approximation Algorithms

Let us begin by presenting a formal description of what we will mean by an "optimization problem."

A *combinatorial optimization problem* Π is either a *minimization problem* or a *maximization problem* and consists of the following three parts:

(1) a set D_Π of *instances*;

(2) for each instance $I \in D_\Pi$, a finite set $S_\Pi(I)$ of *candidate solutions* for I; and

(3) a function m_Π that assigns to each instance $I \in D_\Pi$ and each candidate solution $\sigma \in S_\Pi(I)$ a positive rational number $m_\Pi(I,\sigma)$, called the *solution value* for σ.

If Π is a minimization [maximization] problem, then an *optimal solution* for an instance $I \in D_\Pi$ is a candidate solution $\sigma^* \in S_\Pi(I)$ such that, for all $\sigma \in S_\Pi(I)$, $m_\Pi(I,\sigma^*) \leqslant m_\Pi(I,\sigma)$ $[m_\Pi(I,\sigma^*) \geqslant m_\Pi(I,\sigma)]$. We will use $\text{OPT}_\Pi(I)$ to denote the value $m_\Pi(I,\sigma^*)$ of an optimal solution for I (usually dropping the subscript Π when the problem is clear from context).

An algorithm A is an *approximation algorithm* for Π if, given any instance $I \in D_\Pi$, it finds a candidate solution $\sigma \in S_\Pi(I)$. The value $m_\Pi(I,\sigma)$ of the candidate solution σ found by A when applied to I will be denoted by $A(I)$. If $A(I) = \text{OPT}(I)$ for all $I \in D_\Pi$, then A is called an *optimization algorithm* for Π.

These definitions can be illustrated by considering our old friend the traveling salesman problem. It is a minimization problem, and the set of instances consists of all finite sets of cities together with their intercity distances. The candidate solutions for a particular instance are all the permutations of the given cities. The solution value for such a permutation is the length of the corresponding tour. Thus an approximation algorithm for this problem need only find some permutation of the given set of cities, whereas an optimization algorithm must always find a permutation that corresponds to a minimum length tour.

If, as in this case, the optimization problem is NP-hard, then we know that a polynomial time optimization algorithm cannot be found unless P=NP. A more reasonable goal is that of finding an approximation algorithm A that runs in low-order polynomial time and that has the property that, for all instances I, $A(I)$ is "close" to $\text{OPT}(I)$. The following example illustrates the type of results we will be interested in.

Consider the "bin packing" problem: Given a finite set $U = \{u_1, u_2, \ldots, u_n\}$ of "items" and a rational "size" $s(u) \in [0,1]$ for each item $u \in U$, find a partition of U into disjoint subsets $U_1, U_2, \ldots, U_k$ such that the sum of the sizes of the items in each U_i is no more than 1 and such that k is as small as possible. We can view each subset U_i as specifying a set of items to be placed in a single unit-capacity "bin," with our objective being to pack the items from U in as few such bins as possible.

This problem is NP-hard in the strong sense (it contains 3-PARTITION as a special case), so there is little hope of finding even a pseudo-polynomial time optimization algorithm for it. However, there are a number of simple approximation algorithms for it that are worth considering.

One of these is known as the "First Fit" algorithm. Imagine that we start with an infinite sequence $B_1, B_2, \ldots$ of unit-capacity bins, all of which are empty. The algorithm then places the items into the bins, one at a time, in order of increasing index. It does so according to the following simple rule: always place the next item u_i into the lowest-indexed bin for which the sum of the sizes of the items already in that bin does not exceed $1 - s(u_i)$. In other words, u_i is always placed into the first bin in which it will fit (without exceeding the bin capacity). Figure 6.1 shows an example, where each item is represented by a rectangle having height proportional to its size.

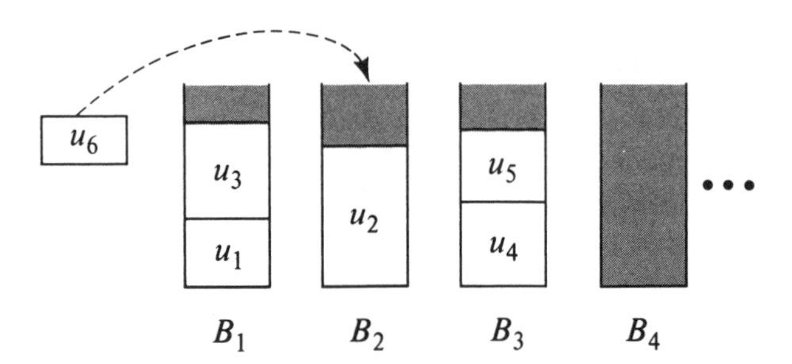

Figure 6.1 An example of a First Fit placement, where u_6 is placed in bin B_2 since that is the lowest indexed bin in which it fits.

Intuitively this seems to be a very natural and reasonable algorithm. It never starts a new bin until all the nonempty bins are too full. What can be proved about its performance?

A first observation relates the number of bins used by First Fit to a natural function of the problem parameters. Let us use "FF" as an abbreviation for First Fit. Then we have that, so long as $\text{FF}(I) > 1$,

$$FF(I) < \lceil 2 \sum_{i=1}^{n} s(u_i) \rceil$$

This is because there can be at most one nonempty bin in the First Fit packing whose contents total ½ or less. (If not, the first item to go in the higher indexed such bin would have fit in the lower indexed such bin and could not have been placed elsewhere by First Fit.) That this bound is essentially the best possible is apparent when we consider instances of the form $U = \{u_1, u_2, \ldots, u_n\}$ where $s(u_i) = 1/2 + \epsilon, 1 \leq i \leq n$. Here no two items will fit in the same bin, so $FF(I) = n$, even though the sum of the item sizes is $(n/2) + n\epsilon$, which can be made as close to $n/2$ as desired by choosing $\epsilon > 0$ suitably small.

This observation also gives us a bound on how bad a First Fit packing can be relative to an optimal packing, since we clearly have

$$OPT(I) \geq \lceil \sum_{i=1}^{n} s(u_i) \rceil$$

We thus conclude that, for all instances I,

$$FF(I) < 2 \cdot OPT(I)$$

However, First Fit actually obeys a better bound of this form, given by the following theorem from [Johnson, Demers, Ullman, Garey, and Graham, 1974]:

Theorem 6.1 For all instances I of the bin packing problem,

$$FF(I) \leq \frac{17}{10} OPT(I) + 2$$

Furthermore, there exist instances I with $OPT(I)$ arbitrarily large such that

$$FF(I) \geq \frac{17}{10} (OPT(I) - 1)$$

Thus Theorem 6.1 characterizes the asymptotic worst-case performance of the First Fit algorithm. First Fit never differs from optimal by significantly more than 70 percent and it can on occasion be essentially this bad. (A slight improvement on the constant term in the upper bound, replacing $(17/10)OPT(I) + 2$ by $\lceil (17/10)OPT(I) \rceil$, is obtained in [Garey, Graham, Johnson, and Yao, 1976]). Although we omit the lengthy proof of Theorem 6.1, we note that worst-case behavior almost as bad as that given by the theorem can be seen from the class of examples described in Figure 6.2, for which $FF(I) = (5/3)OPT(I)$. (The proof of the lower bound in the theorem is a complicated extension of these examples.)

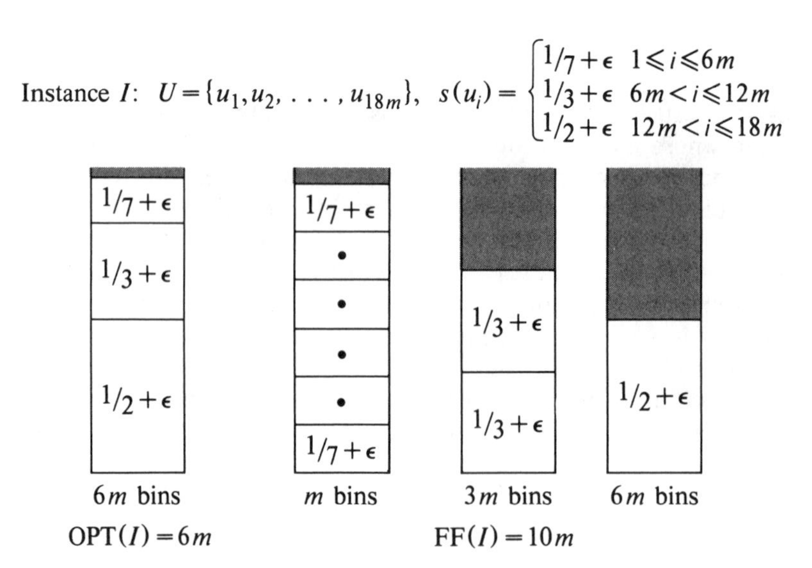

Figure 6.2 Instances I with OPT(I) arbitrarily large such that FF(I) equals (5/3)OPT(I).

These results for First Fit provide a starting point for analyzing approximation algorithms for bin packing. One can now go on to analyze other algorithms that might have better guarantees. An obvious modification to the First Fit algorithm, for example, is that obtained by using the following more sophisticated placement rule: Always place the next item u_i in that bin which has current contents closest to, but not exceeding, $1-s(u_i)$ (choosing the one with lowest index in case of ties). This is known as the "Best Fit" algorithm. Unfortunately, and perhaps surprisingly, Best Fit has essentially the same worst case performance as First Fit [Johnson et al., 1974].

A better approximation algorithm is obtained by observing that the worst performance for First Fit (and Best Fit) seems to occur when the smaller items appear before the larger items in the ordering used by the algorithm. Suppose that, instead of merely taking the items from U in the given order, we first sort them by size and reindex them so that $s(u_1)\geqslant s(u_2)\geqslant \cdots \geqslant s(u_n)$. The algorithm that applies First Fit to such a reordered list is called the "First Fit Decreasing" algorithm (FFD). Its performance is characterized by the following theorem, due to Johnson [1973] (the proof is sketched in [Johnson et al., 1974]):

Theorem 6.2 For all instances I of the bin packing problem,

$$\text{FFD}(I) = \frac{11}{9}\text{OPT}(I)+4$$

Furthermore, there exist instances I with OPT(I) arbitrarily large such that

$$\text{FFD}(I) = \frac{11}{9}\text{OPT}(I)$$

Thus First Fit Decreasing is guaranteed never to be more than about 22 percent worse than optimal, and it can on occasion be this bad. An identical result holds for the analogous "Best Fit Decreasing" algorithm. Figure 6.3 illustrates a class of examples that suffice to prove the lower bound in both cases.

$$\text{Instance } I\text{: } U=\{u_1,u_2,\ldots,u_{30m}\},\quad s(u_i)=\begin{cases} 1/2+\epsilon & 1\leqslant i\leqslant 6m \\ 1/4+2\epsilon & 6m<i\leqslant 12m \\ 1/4+\epsilon & 12m<i\leqslant 18m \\ 1/4-2\epsilon & 18m<i\leqslant 30m \end{cases}$$

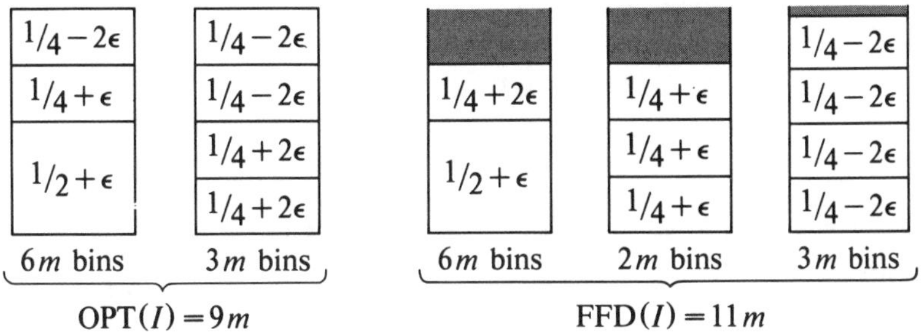

$6m$ bins	$3m$ bins	$6m$ bins	$2m$ bins	$3m$ bins
$1/4-2\epsilon$	$1/4-2\epsilon$	(empty)	(empty)	(empty)
$1/4+\epsilon$	$1/4-2\epsilon$	$1/4+2\epsilon$	$1/4+\epsilon$	$1/4-2\epsilon$
$1/2+\epsilon$	$1/4+2\epsilon$	$1/2+\epsilon$	$1/4+\epsilon$	$1/4-2\epsilon$
	$1/4+2\epsilon$		$1/4+\epsilon$	$1/4-2\epsilon$
				$1/4-2\epsilon$

$\text{OPT}(I)=9m$ $\qquad$ $\text{FFD}(I)=11m$

Figure 6.3 Instances I with $\text{OPT}(I)$ arbitrarily large such that $\text{FFD}(I)$ equals $(11/9)\text{OPT}(I)$.

The proof of the upper bound involves an extremely detailed case analysis, whose recapitulation here would require more pages than we have allotted to this entire chapter. (Although such lengthy proofs appear to be the rule for problems similar to bin packing, we note that results like these for other problems have been obtained without such Herculean effort, and indeed even for bin packing much shorter proofs are obtainable if we are willing to settle for weaker bounds.)

Further modifications of First Fit Decreasing have been suggested ([Johnson, 1973], [Yao, 1978a]) in hopes of obtaining a polynomial time approximation algorithm for bin packing with an even better performance guarantee, but no substantially better bound has yet been proved for any of them.

In summary, our analysis of approximation algorithms for the bin packing problem might be described as follows: We started with a straightforward but apparently sensible algorithm and analyzed its performance, both by proving bounds on what could happen in the worst case and by devising examples to verify that these bounds could not be improved. With this analysis in mind, and especially the insight it provided as to the drawbacks of our initial algorithm, we could then seek alternative algorithms (perhaps just more complicated versions of the original one) and analyze them. We also settled on a general form for our guarantees, in terms of ratios, which was useful for comparison purposes and which seems to express nearness to optimality in a reasonable way. This general approach can serve as a model for our study of other NP-hard optimization problems and indeed has been widely applied (although, of course, on occasion other types of guarantees may be more appropriate or easier to prove, for example, see [Cornuejols, Fisher, and Nemhauser, 1977], [Nemhauser, Wolsey, and Fisher, 1978]).

To formalize this approach, let us make a few more definitions. If Π is a minimization [maximization] problem, and I is any instance in D_Π, we define the ratio $R_A(I)$ by

$$R_A(I) = \frac{A(I)}{\mathrm{OPT}(I)} \qquad \left[R_A(I) = \frac{\mathrm{OPT}(I)}{A(I)} \right]$$

The *absolute performance ratio* R_A for an approximation algorithm A for Π is given by

$$R_A = \inf \{ r \geqslant 1 \colon R_A(I) \leqslant r \textit{ for all instances } I \in D_\Pi \}$$

The *asymptotic performance ratio* R_A^∞ for A is given by

$$R_A^\infty = \inf \left\{ r \geqslant 1 \colon \begin{array}{l} \textit{for some } N \in Z^+,\ R_A(I) \leqslant r \textit{ for all} \\ I \in D_\Pi \textit{ satisfying } \mathrm{OPT}(I) \geqslant N \end{array} \right\}$$

Notice that we have defined these ratios in such a way that the ratio for a minimization problem is the reciprocal of that for a maximization problem. This has been done so that we will have a uniform scale on which to consider approximation algorithms for different types of problems, always having $1 \leqslant R_A \leqslant \infty$ and $1 \leqslant R_A^\infty \leqslant \infty$, with a ratio that is closer to 1 indicating better performance.

Notice also that R_A need not equal R_A^∞. Although Theorem 6.2 shows that $R_{\mathrm{FFD}}^\infty = 11/9$, it is easy to give instances I for which $\mathrm{OPT}(I) = 2$ and $\mathrm{FFD}(I) = 3$, so that $R_{\mathrm{FFD}} \geqslant 3/2$. The asymptotic ratios seem to be the more important ones for bin packing, although for other problems the absolute ratios may be more appropriate, or it may be the case that $R_A = R_A^\infty$ for all the approximation algorithms in which we are interested. At any rate, it will be convenient to have both types of ratios available, and the differences between them are worth keeping in mind when analyzing an approximation algorithm.

As a second example, let us return once more to the traveling salesman problem, only this time with an added restriction. An instance I is still a set C of cities and a specification of the distances between them, but we also require that these distances obey the "triangle inequality"; i.e., for every triple a,b,c of cities from C,

$$d(a,c) \leqslant d(a,b) + d(b,c)$$

This condition is met, for example, whenever the given distances are the actual shortest distances in some standard metric or whenever we allow tours that visit some cities more than once, since in the latter case each of the given distances $d(c_i, c_j)$ can be replaced by the length of the shortest path from c_i to c_j. It is not difficult to see that the problem remains NP-hard under this restriction. Furthermore, all the algorithms that we will be considering will have $R_A = R_A^\infty$, so we can limit our attention to R_A.

Consider the following appealing heuristic, which we shall call the "Nearest Neighbor" algorithm (NN) and which has been proposed, for instance, in [Gavett, 1965]. Let $C = \{c_1, c_2, \ldots, c_m\}$ be the given set of cities. The first city in the tour, $c_{\pi(1)}$, is set to be c_1. In general, if the partial tour built up so far is $<c_{\pi(1)}, c_{\pi(2)}, \ldots, c_{\pi(k)}>$, with $k<m$, the algorithm chooses for $c_{\pi(k+1)}$ that city c that is not yet in the tour and that, among all such cities, is the closest one to $c_{\pi(k)}$, i.e., for which $d(c_{\pi(k)}, c)$ is as small as possible (ties can be broken by choosing the lowest indexed such city). The next theorem, due to Rosenkrantz, Stearns, and Lewis [1977], shows that this approximation algorithm can have much worse behavior than any of those we discussed for bin packing:

Theorem 6.3 For all m-city instances I of the traveling salesman problem with triangle inequality,

$$\text{NN}(I) \leqslant \frac{1}{2}(\lceil \log_2 m \rceil + 1)\,\text{OPT}(I)$$

Furthermore, for arbitrarily large values of m, there exist m-city instances for which

$$\text{NN}(I) > \frac{1}{3}(\log_2(m+1) + \frac{4}{3})\,\text{OPT}(I)$$

The main import of this theorem can be stated quite succinctly: $R_{\text{NN}} = \infty$, a not very promising guarantee. An example of the complicated recursive construction used to prove the lower bound, for $m=15$, is shown in Figure 6.4.

The performance of the Nearest Neighbor algorithm clearly leaves much to be desired. We mention it only as an illustration of how even apparently sensible heuristics can perform very badly, differing from optimal by arbitrarily large multiples.

Fortunately, other approximation algorithms for this problem perform much better. In fact, several approximation algorithms satisfying $R_A = 2$ are described in [Rosenkrantz, Stearns, and Lewis, 1977]. A well-known algorithm that has this behavior is based on the notion of a "minimum spanning tree." Let us view an instance of the traveling salesman problem in terms of a complete graph having the cities as vertices and having each edge

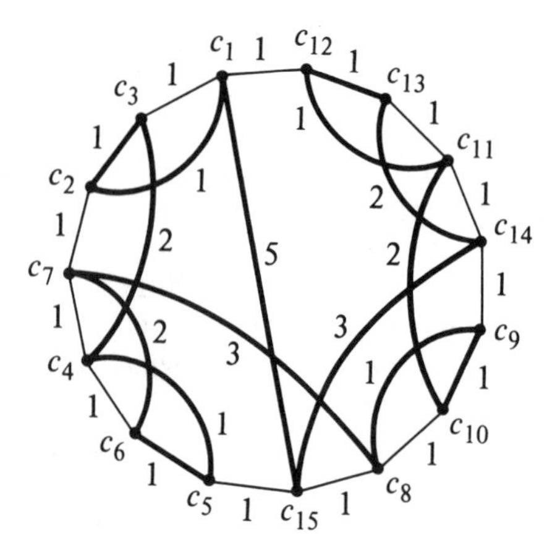

Figure 6.4 Schematic representation of a traveling salesman instance for which the Nearest Neighbor algorithm performs poorly. There are 15 cities, and $d(c_i,c_j)$ is defined to be the length of the shortest path from c_i to c_j using the edges in the figure, which can be determined to obey the triangle inequality. The perimeter gives an optimal tour of length 15, whereas NN would find the darkened tour of length 27. Thus $R_{NN}(I) = 27/15 > 16/9 = (1/3)(\log_2(m+1) + 4/3)$.

$\{c_i,c_j\}$ labeled with the "length" $d(c_i,c_j)$. There is an obvious one-to-one correspondence between the possible tours and the Hamiltonian circuits of this graph. A *spanning tree* is a subgraph that includes all the vertices, is connected, and has no circuits (i.e., for any two vertices, there is *exactly* one path between them). A simple induction proof shows that, if a graph has m vertices, then a spanning tree for that graph must have $m-1$ edges. A *minimum spanning tree* is one for which the sum of the lengths of its edges is as small as possible.

Now, unlike minimum traveling salesman tours, minimum spanning trees are easy to find in low order polynomial time (see [Kruskal, 1956] or [Aho, Hopcroft, and Ullman, 1974]). Furthermore, the length of a minimum spanning tree for the graph obtained from a traveling salesman instance must be less than the length of a minimum tour, because a spanning tree shorter than the tour length can be obtained simply by deleting any single edge from such a tour. This suggests that a reasonably short tour might be obtained by first finding a minimum spanning tree and then visiting all the cities by traversing twice around the tree, as shown in Figure 6.5(b) for the tree of Figure 6.5(a). Although this visits some cities more than once, we can always "shortcut" the tour by proceeding directly to the next unvisited city, as shown in Figure 6.5(c), with the triangle inequality ensuring that this cannot lengthen the tour.

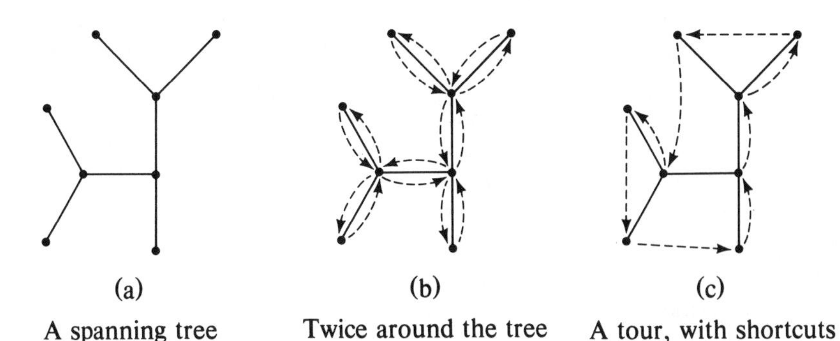

Figure 6.5 How to turn a spanning tree into a traveling salesman tour, using the "twice around the tree" algorithm.

Thus the length of the resulting tour is at most twice the length of the minimum spanning tree, so it is strictly less than twice the length of the optimal tour. Letting "MST" stand for this "twice around the minimum spanning tree" algorithm, we have just proved the following result:

Theorem 6.4 For all instances I of the traveling salesman problem with triangle inequality,

$$\mathrm{MST}(I) < 2 \cdot \mathrm{OPT}(I)$$

The construction of examples to show that this is the best possible bound for the MST algorithm, and hence that $R_{\mathrm{MST}} = 2$, is not difficult, and we leave it as an exercise.

The idea behind the MST algorithm has been extended by Christofides [1976] to devise an even better performing heuristic for this problem. It combines the use of "matching" techniques with the notions of an "Eulerian graph" and an "Eulerian tour." An *Eulerian graph* is simply a graph in which every vertex has even degree. An *Eulerian tour* in a graph is a circuit that traverses every edge exactly once. It can be shown that a necessary and sufficient condition for the existence of an Eulerian tour in a graph G is that G be an Eulerian graph (for example, see [Liu, 1968]), and furthermore it is a simple matter to give a polynomial time algorithm for finding an Eulerian tour in such a graph. Christofides observed that the MST algorithm can be viewed as taking place in four stages: (1) Find a minimum spanning tree, (2) convert the spanning tree into an Eulerian graph by doubling each edge of the tree, (3) find an Eulerian tour of the resulting graph, and (4) convert the Eulerian tour into a traveling salesman tour by using shortcuts. Since the added edges have total length equal to the length of the minimum spanning tree (and thus less than the length of the minimum traveling salesman tour), and since the triangle inequality ensures that the shortcuts will not make the tour longer, we know that the resulting traveling salesman tour must have length less than twice the length of the minimum possible tour. However, Christofides then showed that there is a cheaper way to convert the spanning tree into an Eulerian graph.

This is done by restricting attention to the set $V' = \{a_1, a_2, \ldots, a_{2k}\}$ of vertices that have odd degree in the spanning tree (there must be an even number of such vertices). A *matching* for V' is a partition of V' into k 2-element subsets, and the *weight* of such a matching is the sum of the distances $d(c_i, c_j)$ where $\{c_i, c_j\}$ is a subset in the partition and (to avoid duplicate counting) $i < j$. Any matching for V' provides us with k edges, which, when added to the spanning tree, will convert it into an Eulerian graph, and the total length of those edges is the weight of the matching. A *minimum weight matching* for V' is one that achieves the minimum possible weight. Minimum weight matchings can be found in polynomial time using standard techniques (for example, see [Lawler, 1976a]). The approximation algorithm suggested by Christofides merely replaces stage 2 of the MST algorithm by a stage that finds a minimum weight matching for V' and then converts the spanning tree into an Eulerian graph by adding the corresponding edges.

The key observation for analyzing the performance of this algorithm is that the weight of a minimum weight matching can be at most half the length of a minimum traveling salesman tour. This can be seen by first converting any minimum tour into a tour on just the vertices in V' by skipping over any vertices not in V'. The length of this tour cannot be more than the length of the original tour, by the triangle inequality. Furthermore, this tour on V' provides us with two matchings for V', each formed by taking every other edge, and the shortest of these two matchings cannot have weight exceeding half the tour length. Thus, using "MM" to denote this "minimum matching" algorithm, we have:

Theorem 6.5 For all instances I of the traveling salesman problem with triangle inequality,

$$\mathrm{MM}(I) < \frac{3}{2}\mathrm{OPT}(I)$$

Again, it is not hard to devise examples that show this bound is essentially the best possible one (for example, see [Cornuejols and Nemhauser, 1978]), so $R_{\mathrm{MM}} = 3/2$. No polynomial time approximation algorithm for the traveling salesman problem with triangle inequality is currently known to provide a better guarantee.

These results for the traveling salesman problem, like those for the bin packing problem, are characteristic of what one might expect to achieve with polynomial time approximation algorithms for NP-hard problems. In both cases it was possible to guarantee finding candidate solutions that were fairly close to optimal, although in neither case were the best performance bounds as close to 1 as we might like.

Unfortunately, there are a number of other NP-hard problems for which no polynomial time approximation algorithms that perform even this well have yet been found. One example is the graph coloring problem: Given a graph $G=(V,E)$, find a function $f\colon V \rightarrow \{1,2, \ldots, k\}$ such that $f(u) \neq f(v)$ whenever $\{u,v\} \in E$, and such that k is as small as possible. Here the set of candidate solutions is all functions $f\colon V \rightarrow \{1,2, \ldots, |V|\}$ that satisfy $f(u) \neq f(v)$ whenever $\{u,v\} \in E$, and the value of a candidate solution f is $\max\{f(v)\colon v \in V\}$. In [Johnson, 1974b] it is shown for a large number of polynomial time approximation algorithms A for graph coloring that in each case there exists a positive constant c and infinitely many graphs $G=(V,E)$ such that

$$A(G) > c \cdot |V| \cdot \mathrm{OPT}(G)$$

Furthermore, the best performance bound that is currently known for any polynomial time graph coloring algorithm A [Johnson, 1974b] is

$$A(G) \leqslant \frac{c \cdot |V|}{\log |V|} \mathrm{OPT}(G)$$

Hence no polynomial time approximation algorithm A for graph coloring has yet been found even to come *close* to satisfying $R_A < \infty$.

The graph coloring problem thus appears to be more difficult than either the bin packing problem or the traveling salesman problem with triangle inequality. Not only can we not guarantee an optimal solution in polynomial time unless P=NP, but we do not even know how to guarantee a reasonable approximation to an optimal solution in polynomial time. Two other problems that appear to be equally unmanageable are those of finding a maximum set of independent vertices in a graph and of finding a minimum traveling salesman tour when the triangle inequality is *not* required to hold. (We shall have more to say about these problems in the next section.)

It should be pointed out that this apparent division of problems into the approximable and the nonapproximable does not seem to respect the fact that many of these problems are closely related by polynomial transformations. One might think that such a transformation could be used to convert a good approximation algorithm for one problem into a good approximation algorithm for the other. This is not the case in general, however, as we illustrate with the following example.

Recall from Section 3.1.3 that the problems VERTEX COVER and INDEPENDENT SET are quite closely related. For a graph $G=(V,E)$, the set $V' \subseteq V$ is an independent set for G if and only if the complementary set $V-V'$ is a vertex cover for G. Furthermore, this relationship carries over to the corresponding optimization problems. The set V' is a maximum independent set for G if and only if $V-V'$ is a minimum vertex cover for G (where in both cases the solution value is taken to be the number of vertices in the independent set or vertex cover). However, although we know of no polynomial time approximation algorithm with $R_A < \infty$ for the maximum independent set problem (see [Johnson, 1974a]), there is a straightforward algorithm for the minimum vertex cover problem that has $R_A \leqslant 2$ [Gavril, 1974c].

The algorithm is based on the idea of a "maximal matching" in a graph. A *matching* in a graph $G=(V,E)$ is a set $E' \subseteq E$ such that no two edges in E' share a common endpoint (the relationship between this notion and that used in Christofides' algorithm for the traveling salesman problem should be apparent). A matching E' is *maximal* if every remaining edge in $E-E'$ has an endpoint in common with some member of E'. Maximal matchings can be constructed in polynomial time quite easily, simply by adding edges until no longer possible. Moreover, observe that if E' is a maximal matching for G, then the set of all endpoints of edges in E' must be a vertex cover for G (otherwise E' would not be maximal). This vertex cover has cardinality $2 \cdot |E'|$. The key point to observe is that *all* vertex covers for G must contain at least $|E'|$ vertices, since they must include at least one endpoint from each edge in E'. Thus the vertex cover constructed from E' is never more than twice as large as the minimum possible vertex cover.

The reason for the widely differing behavior of the maximum independent set problem and the minimum vertex cover problem with respect to approximation algorithms is evident when we attempt to translate the above algorithm from one problem to the other, via the transformation between the problems. Suppose we have a graph with 1000 vertices and a minimum vertex cover of size 490. Our approximation algorithm guarantees that we will find a vertex cover of size 980 or less, for a ratio of at most 2, but the corresponding independent sets can have a ratio as large as

$$\frac{1000-490}{1000-980} = \frac{510}{20} = 25.5$$

Although the transformation does preserve optimal solutions, it does not preserve the *ratios* between the values of optimal and suboptimal solutions. There is no reason why a candidate solution that is near-optimal for one problem should map into a candidate solution that is near-optimal for the other. Thus the transformations we use to prove problems NP-complete would need to preserve solution values in a much more uniform way than they do here if they are to be used for converting good approximation algorithms from one problem to another.

The types of behavior of approximation algorithms that we have illustrated so far do not yet exhaust the possibilities. We have seen that some problems appear to be much harder to "approximate" than bin packing. On the other hand, there are some that seem to be much "easier."

Consider, for example, the KNAPSACK problem introduced in Section 3.2.1. As an optimization problem it takes the following form: Given a finite set U of "items," a "size" $s(u) \in Z^+$ for each $u \in U$, a "value"

$v(u) \in Z^+$ for each $u \in U$, and a positive "knapsack capacity" $B \geqslant \max\{s(u): u \in U\}$, find a subset $U' \subseteq U$ such that $\sum_{u \in U'} s(u) \leqslant B$ and such that $\sum_{u \in U'} v(u)$ is as large as possible. That is, we would like to maximize the value of what we place in our "knapsack," subject to the constraint that the total size of all the items not exceed the knapsack capacity.

One simple approximation algorithm for this problem works as follows. Order the set $U = \{u_1, u_2, \ldots, u_n\}$ by "value density," i.e., so that $v(u_1)/s(u_1) \geqslant v(u_2)/s(u_2) \geqslant \cdots \geqslant v(u_n)/s(u_n)$. Starting with U' empty, proceed sequentially through this list, each time adding u_i to U' whenever the sum of the sizes of the items already in U' does not exceed $B - s(u_i)$. Then, compare the value for the solution found by this "greedy procedure" to the value for the solution consisting solely of the maximum value item and take the better of the two. It is not difficult to show that this composite greedy algorithm GA has $R_{\text{GA}} = 2$.

However, one can do significantly better by embedding this algorithm in a more elaborate procedure. Sahni [1975] shows that, for any $k \geqslant 1$, there is a polynomial time approximation algorithm A_k satisfying $R_{A_k} \leqslant 1 + (1/k)$. The basic idea is to try all possible subsets of k or fewer items from U as initial values of U', adding as many items as possible to each of these using the greedy procedure, and then taking the best of the resulting sets as our approximate solution.

Unfortunately, although for each fixed value of k the corresponding algorithm runs in time that is polynomial in n, $\log V$, and $\log S$ (where $V = \max\{v(u): u \in U\}$ and $S = \max\{s(u): u \in U\}$), the polynomials all have k in their exponents. In order to guarantee solutions that are extremely close to optimal, we would have to resort to an algorithm whose time complexity is a rather high degree polynomial and that might well be too expensive to use in practice. Nevertheless, these algorithms do represent at least a theoretical improvement over the type of behavior we saw for bin packing, where no polynomial time approximation algorithm with R_A substantially less than 11/9 is known. Here there are polynomial time approximation algorithms with R_A arbitrarily close to 1.

Moreover, it has recently been shown by Ibarra and Kim [1975a] that the same approximating behavior can be obtained for the knapsack problem without using algorithms that are exponential in k. Their result is based on the fact that the knapsack problem can be solved by a pseudo-polynomial time optimization algorithm. It shows how such an optimization algorithm can be converted into a polynomial time approximation algorithm by rounding and scaling, in such a way that only a limited loss of accuracy is incurred.

To illustrate this, suppose A is a pseudo-polynomial time optimization algorithm for the knapsack problem, say with time complexity $O(n^2 V \log(nVS))$. We can modify each instance of the problem by replacing the value $v(u)$ of every item $u \in U$ by the new value $v'(u) = \lfloor v(u)/K \rfloor$, for some fixed $K>0$. The time complexity for applying A to the resulting instance is

$$O(n^2(V/K)\log(nVS))$$

Furthermore, because an optimal solution cannot contain more than all n items, we have the following relationship between the optimum value $\mathrm{OPT}(I)$ for the original instance and the optimum value $\mathrm{OPT}(I')$ for the corresponding modified instance:

$$\mathrm{OPT}(I) - K\cdot\mathrm{OPT}(I') \leqslant Kn$$

Notice that $K\cdot\mathrm{OPT}(I')$ is less than or equal to the value of the optimum solution for I' when reinterpreted in terms of the original item values. Thus, if we take the optimum set U' for I' as our approximate solution for I, then its value will differ from optimal by at most Kn.

The desired result is obtained by choosing K in a way that depends on the given instance. In particular, we can choose $K = V/(k+1)n$, where k is a fixed positive integer. We then obtain an approximation algorithm A_k having time complexity $O(kn^3\log(nVS))$, which is polynomial in n, $\log V$, $\log S$, *and* k (the time for constructing I' from I is dominated by the time for applying A to I'). Moreover, since

$$A_k(I) \geqslant \mathrm{OPT}(I) - Kn = \mathrm{OPT}(I) - V/(k+1)$$

and since $\mathrm{OPT}(I) \geqslant V$, we have

$$R_{A_k}(I) = \frac{\mathrm{OPT}(I)}{A_k(I)} \leqslant \frac{A_k(I)+(V/(k+1))}{A_k(I)}$$

$$\leqslant 1 + \frac{V/(k+1)}{V-(V/(k+1))} = 1+(1/k)$$

Therefore, this algorithm performs as claimed. (The Ibarra and Kim algorithm is actually a bit more complicated, but this suffices to illustrate the main idea.)

The approximating behavior displayed by these algorithms is not a phenomenon restricted only to knapsack problems but occurs in a number of other situations. Consequently, it is useful to have a general terminology for discussing such results. We define an *approximation scheme* for an optimization problem Π to be an algorithm A that takes as input both an instance $I \in D_\Pi$ and an "accuracy requirement" $\epsilon > 0$, and that then outputs a candidate solution $\sigma \in S_\Pi(I)$ such that

$$R_{A_\epsilon}(I) \leqslant 1+\epsilon$$

The term "scheme" is used here because A actually provides a range of approximation algorithms for Π, one for each fixed value of $\epsilon > 0$.

We say that A is a *polynomial time approximation scheme* if for each fixed $\epsilon > 0$ the derived approximation algorithm A_ϵ is a polynomial time algorithm. We say that A is a *fully polynomial time approximation scheme* if the time complexity of A itself is bounded by a polynomial function of Length$[I]$ and $1/\epsilon$. Thus, the sequence of algorithms given by Sahni [1975] constitutes a polynomial time approximation scheme for the knapsack problem, but it is not a *fully* polynomial time approximation scheme. The sequence of algorithms due to Ibarra and Kim [1975a] constitutes a fully polynomial time approximation scheme for this problem.

Some improvements on the Ibarra and Kim approximation scheme are discussed in [Lawler, 1977b]. Similar approximation schemes for other knapsack-like problems and several scheduling problems can be found in [Sahni, 1976] and [Horowitz and Sahni, 1976]. The key idea in all these cases is like that described above: Starting with a pseudo-polynomial time optimization algorithm, rounding and scaling techniques are used to trade a limited amount of accuracy for greatly improved time complexity.

In many ways, fully polynomial time approximation schemes are the best that one might hope for in solving NP-hard optimization problems. Unless P = NP, we certainly will not be able to find a polynomial time approximation algorithm A with $R_A = 1$. However, since R_A^∞ is an asymptotic ratio, it might be possible to find a polynomial time algorithm with $R_A^\infty = 1$. For example, Lipton and Tarjan [1977] give a polynomial time approximation algorithm A for finding maximum independent sets in planar graphs (a subproblem that remains NP-hard) that guarantees

$$|A(I) - \mathrm{OPT}(I)| \leqslant O\left(1/\sqrt{\log\log \mathrm{OPT}(I)}\right) \cdot \mathrm{OPT}(I)$$

Another way of achieving $R_A^\infty = 1$ is to have $|A(I) - \mathrm{OPT}(I)|$ bounded by a constant, independent of I. For example, one might be able to obtain a polynomial time approximation algorithm for bin packing that satisfies $A(I) \leqslant \mathrm{OPT}(I)+1$ for all instances I. Although we know of no algorithm like this for bin packing, Horowitz and Sahni [1978] show that such an algorithm can be obtained for the NP-hard problem of packing the maximum possible number of items into two bins. Few such "difference results" are known for other NP-hard problems, but several problems that are not known to be solvable in polynomial time *have* been approximated in this way [Stone and Fuller, 1973], [Kaufman, 1974], [Karp, McKellar, and Wong, 1975].

6.2 Applying NP-Completeness to Approximation Problems

So far we have not presented any evidence that the observed differences in "approximability" among NP-hard problems are due to anything other than our own inability to find good approximation algorithms.

In this section, we shall see that some of these differences are, in fact, inherent and that the theory of NP-completeness can be used for delimiting how closely a given problem can be approximated.

Let us define the *best achievable asymptotic performance ratio* for an optimization problem Π to be

$$R_{\text{MIN}}(\Pi) = \inf \left\{ r \geqslant 1: \begin{array}{l} \textit{there exists a polynomial time approximation} \\ \textit{algorithm } A \textit{ for } \Pi \textit{ with } R_A^\infty = r \end{array} \right\}$$

On the basis of the results described in the previous section, we might suspect that there are some NP-hard problems with $R_{\text{MIN}}(\Pi) = \infty$, some with $1 < R_{\text{MIN}}(\Pi) < \infty$, and some with $R_{\text{MIN}}(\Pi) = 1$. With respect to the last case, we have seen several significant subcases: (1) Π can be solved by a polynomial time approximation scheme, (2) Π can be solved by a fully polynomial time approximation scheme, (3) Π can be solved by a polynomial time approximation algorithm A satisfying $R_A^\infty = 1$, and (4) Π can be solved by a polynomial time approximation algorithm A satisfying $|A(I) - \text{OPT}(I)| \leqslant K$, for some fixed constant K. Other types of behavior are, of course, possible, but these are the main types of behavior investigated up to now, and we shall restrict our attention to them in what follows.

In seeking to demonstrate that a certain one of these possibilities cannot be achieved for a particular problem Π, we are as usual confronted with the fact that P has not yet been proved to differ from NP. Thus it remains possible that all NP-equivalent problems can be solved *exactly* in polynomial time (all the optimization problems we have been discussing are NP-easy as well as NP-hard). For this reason, we are once again constrained to proving conditional results, showing that, if $P \neq NP$, then Π cannot be solved by a polynomial time approximation algorithm of the specified type.

Let us begin with the best of the guarantees one might hope for, a difference result of the form "$|A(I) - \text{OPT}(I)| \leqslant K$ for all instances I," where K is a constant. One example of a problem for which such a guarantee can be ruled out is the knapsack problem, the optimization version of which was defined in the preceding section.

Theorem 6.6 If $P \neq NP$, then no polynomial time approximation algorithm A for the knapsack problem can guarantee

$$|A(I) - \text{OPT}(I)| \leqslant K$$

for a fixed constant K.

Proof: Suppose, to the contrary, that A is such an approximation algorithm, where we can assume without loss of generality that K is a positive integer. We will show how A can be used to solve the knapsack problem exactly, in polynomial time, contradicting the assumption that $P \neq NP$. The procedure is quite simple. Given any instance I of the knapsack problem, we merely construct a new instance I' from I by replacing each item value $v(u)$ by $(K+1)v(u)$, and then apply A to I'. This clearly can be done in polynomial time. Furthermore, the candidate solutions for I' are identical to those for I, and the value of a solution for I' is exactly $K+1$ times the value of the corresponding solution for I. Since all solution values for I' are integer multiples of $K+1$, the fact that $|A(I')-\mathrm{OPT}(I')| \leqslant K$ immediately implies that $|A(I')-\mathrm{OPT}(I')|$ must equal 0, and hence

$$|A'(I)-\mathrm{OPT}(I)| = |A(I')-\mathrm{OPT}(I')|/(K+1) = 0$$

where A' denotes our derived algorithm. Thus the candidate solution found by A' is necessarily optimal, so A' is a polynomial time optimization algorithm for the knapsack problem. This is the desired contradiction, and the theorem is proved. ■

The only property used in this proof is that all solution values can be multiplied by an arbitrarily large constant without changing the set of candidate solutions. Since this property holds for many other problems, for example, the traveling salesman problem (even with the triangle inequality), it is easy to see that this type of proof is widely applicable. What may be less obvious is that essentially the same idea can be applied to problems of a very different nature. For example, consider the maximum independent set problem. Here there are no numbers to be multiplied, so instead we "multiply" the entire graph!

Theorem 6.7 If $\mathrm{P} \neq \mathrm{NP}$, then no polynomial time approximation algorithm for the maximum independent set problem can guarantee

$$|A(G)-\mathrm{OPT}(G)| \leqslant K$$

for a fixed constant K.

Proof: Suppose A is such an approximation algorithm, where we again may assume that K is a positive integer. As in the previous proof, we will show how A can be used to derive a polynomial time optimization algorithm, contradicting the assumption that $\mathrm{P} \neq \mathrm{NP}$. Given as an instance any graph G, the algorithm constructs a new graph G' that consists of $K+1$ isomorphic copies of G. It is easy to see that $\mathrm{OPT}(G')=(K+1)\mathrm{OPT}(G)$. Furthermore, we can construct an independent set for G of size at least $\lceil A(G')/(K+1) \rceil$ merely by determining how many vertices A chooses from each copy of G and taking the largest such subset. Since

$$|A(G')-\mathrm{OPT}(G')| = |A(G')-(K+1)\mathrm{OPT}(G)| \leqslant K$$

it follows that this independent set for G will contain exactly $\mathrm{OPT}(G)$ vertices. Thus this procedure is an optimization algorithm for the maximum independent set problem, and it is a polynomial time algorithm because K is a constant, independent of G. ■

As an exercise, the reader might try to use the same approach to prove such a result for the problem of finding a minimum set cover, derived from the MINIMUM COVER problem defined in Section 3.2.1.

Before turning to the discussion of results concerning ratios instead of differences, we remark that an intermediate type of result is possible. By a variation on the techniques used to prove Theorem 6.7, one can show for the maximum independent set problem that, for all constants K and $\epsilon>0$, no polynomial time approximation algorithm can guarantee

$$|A(G)-\mathrm{OPT}(G)| \leqslant K \cdot \mathrm{OPT}(G)^{1-\epsilon}$$

unless P=NP. Similar results for this and other problems are presented in [Nigmatullin, 1975] and [Kucera, 1976]. However, note that even results of this form do not rule out the possibility of a polynomial time approximation algorithm having $R_A^\infty=1$, as illustrated by the result of Lipton and Tarjan [1977] mentioned at the end of the preceding section. The only way we know for showing that no polynomial time approximation algorithm A can have $R_A^\infty=1$ unless P=NP is to show that P$\neq$NP implies $R_{\mathrm{MIN}}(\Pi)>1$, in principle a stronger result. Since such proofs are more conveniently viewed as providing lower bounds for $R_{\mathrm{MIN}}(\Pi)$ in the case that $1 < R_{\mathrm{MIN}}(\Pi) < \infty$, we shall temporarily postpone discussing them, turning now to the question of whether or not Π can be solved with a fully polynomial time approximation scheme.

Recall that we indicated in the preceding section that all known fully polynomial time approximation schemes for NP-hard problems have been derived from pseudo-polynomial time optimization algorithms. Therefore, it is not surprising that issues of strong NP-completeness turn out to be relevant to the existence of such schemes. In fact, the relationship between pseudo-polynomial time algorithms and fully polynomial time approximation schemes goes both ways. Let Π be an optimization problem with associated instance measures Length$[I]$ and Max$[I]$ (as discussed in Section 4.2) and, in addition, suppose that all solution values are positive integers. The following result is proved in [Garey and Johnson, 1978]:

Theorem 6.8 If there exists a two-variable polynomial q such that for all instances $I \in D_\Pi$

$$\mathrm{OPT}(I) < q(\mathrm{Length}[I], \mathrm{Max}[I])$$

then the existence of a fully polynomial time approximation scheme for Π implies the existence of a pseudo-polynomial time optimization algorithm for Π.

Proof: Suppose that A is such a scheme for Π. The corresponding optimization algorithm A' proceeds as follows: Given an instance I, set

$$\epsilon = q(\mathrm{Length}[I], \mathrm{Max}[I])^{-1}$$

and apply A to the instance I with accuracy requirement ϵ. In time polynomial in Length $[I]$ and

$$q(\text{Length}\,[I],\ \text{Max}[I]$$

and hence in pseudo-polynomial time, A finds a candidate solution for I satisfying

$$R_{A_\epsilon}(I) \leqslant 1+\epsilon$$

Let us assume that Π is a maximization problem (an analogous argument applies if it is a minimization problem). Then we have

$$\text{OPT}(I) \leqslant (1+\epsilon)A_\epsilon(I)$$

or

$$\text{OPT}(I) - A_\epsilon(I) \leqslant \epsilon \cdot A_\epsilon(I) \leqslant \epsilon \cdot \text{OPT}(I) < 1$$

by the definition of ϵ and the assumption on q. But since all solution values are integers, this means that $A_\epsilon(I) = \text{OPT}(I)$, and hence A' finds an optimal solution in pseudo-polynomial time, as desired. ■

A proof technique for ruling out the possibility of a fully polynomial time approximation scheme for Π then follows as a corollary.

Corollary. Let Π be an integer-valued optimization problem satisfying the hypothesis of Theorem 6.8. If Π is NP-hard in the strong sense, then Π cannot be solved by a fully polynomial time approximation scheme unless P=NP.

The two main requirements of the corollary are that Π have integer solution values that are not too large and that Π be NP-hard in the strong sense. These hold for many problems of interest, including, for example, the bin packing problem, the graph coloring problem, the maximum independent set problem, and the minimum vertex cover problem. Thus we have a general method of considerable power for ruling out the possibility of a fully polynomial time approximation scheme.

It is a bit more difficult to rule out approximation schemes that are polynomial but not fully polynomial. The easiest way to do so would seem to be simply to prove that there is some $r > 1$ such that no polynomial time approximation algorithm A for Π can have $R_A < r$, unless P=NP. The graph coloring problem provides an example of a problem where such an r obviously exists. A polynomial time graph coloring algorithm A with $R_A < 4/3$ would have to be able to color any 3-colorable graph with no more than 3 colors and hence would solve the NP-complete GRAPH 3-COLORABILITY problem in polynomial time. Thus, if P≠NP, no such approximation algorithm can exist, and there cannot be a polynomial time approximation scheme for graph coloring. The idea used here is captured by the following easily proved theorem:

Theorem 6.9 Let Π be a minimization problem having all solution values in Z^+, and suppose that for some fixed $K \in Z^+$ the decision problem "Given $I \in D_\Pi$, is $\mathrm{OPT}(I) \leqslant K$?" is NP-hard. Then, if $P \neq NP$, no polynomial time approximation algorithm A for Π can satisfy $R_A < 1+(1/K)$, and Π cannot be solved by a polynomial time approximation scheme.

Of course, the analogous result for maximization problems also holds. In addition to graph coloring, Theorem 6.8 also applies to the bin packing problem (with $K=2$), to the deadline minimization problem based on the PRECEDENCE CONSTRAINED SCHEDULING problem of Section 4.1 (with $K=3$) [Lenstra and Rinnooy Kan, 1978a], and a variety of others. However, many other problems do not contain NP-hard subproblems of the required type, and for them the existence of polynomial time approximation schemes remains open.

Let us now turn to the task of proving lower bounds on $R_{\mathrm{MIN}}(\Pi)$. Notice that Theorem 6.9 does not provide a way for doing this, since it is concerned only with absolute ratios, and a lower bound on the best achievable value of R_A need not hold for R_A^∞. An illustration of this is provided by the bin packing problem. Although Theorem 6.9 tells us we cannot have a polynomial time approximation algorithm A for bin packing that satisfies $R_A < 3/2$ (unless $P=NP$), we have already seen that $R_{\mathrm{FFD}}^\infty = 11/9 < 3/2$. Indeed, even assuming $P \neq NP$, all we can say about the status of bin packing at present is that it satisfies $1 \leqslant R_{\mathrm{MIN}}(\Pi) \leqslant 11/9$.

However, for some problems the bounds obtained via Theorem 6.9 can be strengthened to apply even to asymptotic ratios. We illustrate this for the graph coloring problem.

Theorem 6.10 If $P \neq NP$, then no polynomial time approximation algorithm A for the graph coloring problem can have $R_A^\infty < 4/3$.

Proof: We will show that any such approximation algorithm A could be used to obtain a polynomial time algorithm for the GRAPH 3-COLORABILITY problem, contradicting the assumption that $P \neq NP$. The construction uses the notion of the "composition" of two graphs. If $G_1=(V_1,E_1)$ and $G_2=(V_2,E_2)$, the *composition* $G=G_1[G_2]$ of these two graphs is the graph that has vertex set $V=V_1 \times V_2$ and edge set E defined by

$$E = \{\{(u_1,u_2),(v_1,v_2)\}\text{: either } \{u_1,v_1\} \in E_1 \text{ or } u_1=v_1 \text{ and } \{u_2,v_2\} \in E_2\}$$

An example of the composition of two graphs is shown in Figure 6.6. One convenient way of viewing the composition $G_1[G_2]$ is as being constructed by replacing each vertex of G_1 by a copy of G_2 and then replacing each edge

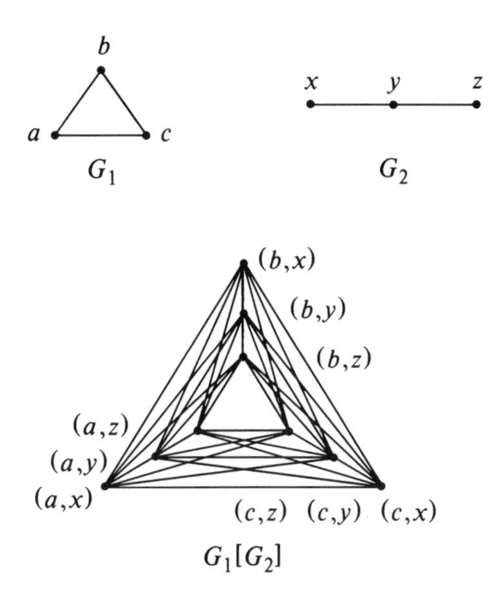

Figure 6.6 The composition $G_1[G_2]$ of two graphs G_1 and G_2.

of G_1 by a complete bipartite subgraph that joins every vertex in the copy corresponding to one endpoint to every vertex in the copy corresponding to the other endpoint.

Our GRAPH 3-COLORABILITY algorithm works as follows: Since $R_A^\infty < 4/3$, there is some $K \in Z^+$ such that $A(G) < (4/3)\text{OPT}(G)$ for all graphs G with $\text{OPT}(G) \geqslant K$. Let $G = (V,E)$ be an arbitrary instance of the GRAPH 3-COLORABILITY problem. Let G' denote the complete graph on K vertices and let $G^* = G'[G]$. Observe that G^* is just K isomorphic copies of G, with every two vertices in different copies being joined by an edge, so that

$$\text{OPT}(G^*) = K \cdot \text{OPT}(G) \geqslant K$$

Moreover, the size of G^* and the time needed to construct it are polynomially bounded in terms of the size of G, since K is independent of G. Thus the time required to apply A to G^* will also be bounded by a polynomial in terms of G. However, if G is 3-colorable, then we have

$$A(G^*) < (4/3) \cdot \text{OPT}(G^*) \leqslant (4/3) \cdot 3K = 4K$$

On the other hand, if G is not 3-colorable, then

$$A(G^*) \geqslant \text{OPT}(G^*) \geqslant 4K$$

Therefore G is 3-colorable if and only if $A(G^*) < 4K$, and the two step procedure of constructing G^* and then applying A to G^* gives a polynomial time algorithm for GRAPH 3-COLORABILITY. This provides the desired contradiction and completes the proof. ■

A similar proof that $R_{\text{MIN}}(\Pi) \geqslant 4/3$ can be given for the precedence constrained scheduling problem cited above. In fact, although no polynomial time approximation algorithms with $R_A^\infty < \infty$ are known for graph coloring, the situation is much better for this scheduling problem. A polynomial time approximation algorithm A with $R_A^\infty = 2$ is given in [Graham, 1966], so that in this case $R_{\text{MIN}}(\Pi)$ satisfies $4/3 \leqslant R_{\text{MIN}}(\Pi) \leqslant 2$ (assuming $\text{P} \neq \text{NP}$).

For graph coloring, the lower bound of 4/3 on $R_{\text{MIN}}(\Pi)$ can be improved nontrivially by a process of "expanding" the ratio. The following result is proved in [Garey and Johnson, 1976a]:

Theorem 6.11 If $\text{P} \neq \text{NP}$, then no polynomial time approximation algorithm A for the graph coloring problem can satisfy $R_A^\infty < 2$.

Proof: Suppose A is such an approximation algorithm. As in the proof of Theorem 6.10, we will show how A can be used to solve the NP-complete GRAPH 3-COLORABILITY problem in polynomial time, and again we will be using the operation of graph composition. However, we also will need one additional concept, that of a graph "multicoloring." A *(k,m)-multicoloring* of a graph $G = (V,E)$ is a function f that assigns to each $v \in V$ a set $f(v) \subseteq \{1,2, \ldots, k\}$ such that $|f(v)| = m$ for all vertices v and such that, whenever $\{u,v\} \in E$, then $f(u)$ and $f(v)$ contain no elements in common. Figure 6.7 shows an example of a (5,2)-multicoloring. The *m-chromatic number* $\chi_m(G)$ is the least integer $k \geqslant 1$ such that there exists a (k,m)-multicoloring of G. Notice that $(k,1)$-multicolorings correspond to ordinary graph colorings, and $\chi_1(G)$ is the same as what we are calling $\text{OPT}(G)$.

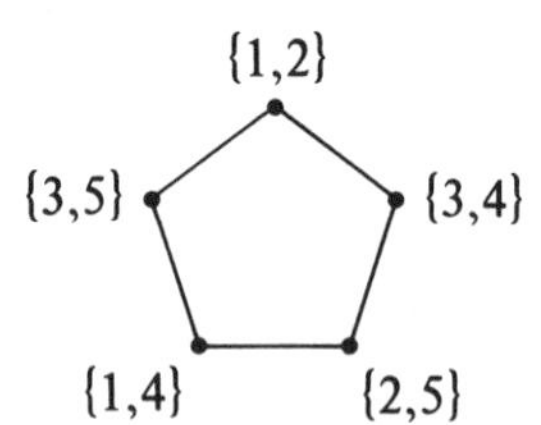

Figure 6.7 A (5,2)-multicoloring of a pentagon C_5, with each vertex v labeled by the corresponding set $f(v)$ of "colors." It can be shown that $\chi_2(C_5) = 5$.

The role of the complete graphs in the proof of Theorem 6.10 is in this proof played by the graphs H_n, $n \geqslant 6$, defined as follows: The graph $H_n = (V_n, E_n)$ has $n(n-1)(n-2)/6$ vertices, each corresponding to a distinct 3-element subset of $\{1,2, \ldots, n\}$, and two such vertices are joined by an edge in E_n if and only if their corresponding subsets have no elements in common. The key property possessed by these graphs, as proved in [Garey and Johnson, 1976a], is that for all $n \geqslant 6$

$$\chi_3(H_n) = n \quad \text{and} \quad \chi_4(H_n) = 2n - 4$$

This property will allow these graphs to serve as our "ratio expanders."

Since $R_A^\infty < 2$, there exists an $\epsilon > 0$ and a $K \in Z^+$ such that $A(G) \leqslant (2-\epsilon)\text{OPT}(G)$ for all graphs G with $\text{OPT}(G) \geqslant K$. Let $N \geqslant \max\{6, K\}$ be an integer chosen so that $2N-4 > (2-\epsilon)N$. Our algorithm for testing 3-colorability proceeds as follows:

Let $G = (V,E)$ be an arbitrary graph to be tested. We first construct the composition graph $G^* = H_N[G]$, which can be done in polynomial time, since N is independent of G. Then we apply A to G^*, which again takes only polynomial time in terms of G. We claim that G is 3-colorable if and only if $A(G^*) < 2N-4$.

Suppose G is 3-colorable. Then there is a way of coloring G^* with only N colors — namely, fix an $(N,3)$-multicoloring of H_N (which exists since $\chi_3(H_N) = N$) and color each copy of G in G^* with the three colors assigned to the corresponding vertex of H_N by this multicoloring. No two vertices in different copies of G that are adjacent in G^* will be assigned the same color, since the corresponding vertices of H_N are adjacent and hence are assigned disjoint sets of colors by the multicoloring. Thus this gives us a coloring of G^* using only N colors, and it follows from the definitions of N and ϵ that

$$A(G^*) \leqslant (2-\epsilon)\text{OPT}(G^*) \leqslant (2-\epsilon)N < 2N-4$$

whenever G is 3-colorable.

On the other hand, suppose G is *not* 3-colorable. Then any coloring of G^* must use at least 4 distinct colors on each copy of G, and by the definition of G^* two copies of G that correspond to adjacent vertices in H_N must use disjoint sets of colors. Thus any such coloring of G^* that uses k colors will induce a $(k,4)$-multicoloring of H_N, formed by assigning to each $v \in V_N$ any four of the colors used on the corresponding copy of G. However, since $\chi_4(H_N) = 2N-4$, it follows that k must be at least $2N-4$, i.e., G^* cannot be colored with fewer than $2N-4$ colors. Thus we have that

$$A(G^*) \geqslant \text{OPT}(G^*) \geqslant 2N-4$$

whenever G is not 3-colorable.

Therefore, G is 3-colorable if and only if $A(G^*) < 2N-4$, and we have a polynomial time algorithm for testing graph 3-colorability. This contradicts the assumption that $P \neq NP$ and completes the proof. ■

As a consequence of Theorem 6.11 we have for the graph coloring problem that $2 \leqslant R_{\text{MIN}}(\Pi) \leqslant \infty$, assuming $P \neq NP$. No stronger bounds are known, even though it is suspected that in this case $R_{\text{MIN}}(\Pi) = \infty$.

Notice that all these lower bound proofs for $R_{MIN}(\Pi)$ depend crucially on the fact that Π has an NP-hard subproblem of the form "Given $I \in D_{\Pi}$, is $\text{OPT}(I) \leqslant K$?" (or "Given $I \in D_{\Pi}$, is $\text{OPT}(I) \geqslant K$?") for some fixed value of K. If Π does not contain such a subproblem, the entire proof technique can be seen to break down. Nevertheless, even in this case it is sometimes possible to salvage something. Consider once more the maximum independent set problem, for which no polynomial time approximation algorithm A with $R_A^{\infty} < \infty$ is known, but which has the property that for each fixed value of K the subproblem "Given G, does G contain an independent set of size at least K?" is solvable in polynomial time. Here we can prove the following result:

Theorem 6.12 Either the maximum independent set problem can be solved with a polynomial time approximation scheme, or else there is no polynomial time approximation algorithm A for it that satisfies $R_A^{\infty} < \infty$.
Proof: Suppose that A is a polynomial time approximation algorithm for this problem satisfying $R_A^{\infty} < \infty$. Since we can assume that A always finds an independent set of size at least 1, we know that we must also have $R_A < \infty$. Let r be the value of R_A. We construct our approximation scheme as follows: For any $\epsilon > 0$, let N_ϵ be the least integer N such that $r^{1/N} < 1+\epsilon$. (For simplicity, we shall use N to denote N_ϵ, since ϵ will be fixed from now on.) Given an arbitrary graph $G=(V,E)$, we must show how to find an independent set of size $M \geqslant \text{OPT}(G)/(1+\epsilon)$, in time polynomial in the size of G.

First we repeatedly apply the composition operation to G to construct a new graph $G_N=(V_N,E_N)$, defined inductively by $G_1=G$ and, for $1<i\leqslant N$, $G_i=G_{i-1}[G]$. Since N depends only on ϵ, G_N can be constructed in polynomial time for a fixed value of ϵ. Moreover, it is straightforward to prove that $\text{OPT}(G_i)=(\text{OPT}(G))^i$, for $1\leqslant i\leqslant N$, and that given any independent set S_N for G_N we can construct an independent set S for G satisfying

$$|S| \geqslant \lceil\, |S_N|^{1/N} \rceil$$

in polynomial time. Thus observe what happens when we apply A to G_N. First, since $R_A=r$, we are guaranteed to obtain an independent set S_N for G_N satisfying

$$|S_N| \geqslant \text{OPT}(G_N)/r = (\text{OPT}(G))^N/r$$

From this we can construct an independent set S for G such that

$$|S| \geqslant \lceil\, ((\text{OPT}(G))^N/r)^{1/N} \rceil \geqslant \text{OPT}(G)/r^{1/N} \geqslant \text{OPT}(G)/(1+\epsilon)$$

(by the definition of N), and for a fixed value of ϵ this is all accomplished in polynomial time with respect to G. Hence we have derived from A a polynomial time approximation scheme for the maximum independent set problem. ■

As a consequence of Theorem 6.12, we have for the maximum independent set problem that either $R_{MIN}(\Pi)=1$ or $R_{MIN}(\Pi)=\infty$, and no value in between is possible. Thus if we could guarantee *some* constant ratio in polynomial time, then we could guarantee *any* fixed ratio within polynomial time. On the other hand, if we could prove for some fixed $\epsilon>0$, no matter how small, that no polynomial time approximation algorithm A for this problem can guarantee $R_A<1+\epsilon$, it would follow that all such polynomial time approximation algorithms must have $R_A=R_A^\infty=\infty$. We suspect that $R_{MIN}(\Pi)=\infty$ but are unaware of any promising approaches toward proving it.

This leads us to our final type of negative result. For some problems it *is* possible to show that $R_{MIN}(\Pi)=\infty$ (unless P=NP). Such proofs look much like those we have already seen for proving lower bounds, only in these cases it is possible to "expand the ratio" much more dramatically. As an example, let us return to the traveling salesman problem, but this time without requiring that the intercity distances obey the triangle inequality. The following theorem appears in [Sahni and Gonzalez, 1976]:

Theorem 6.13 If P$\neq$NP, then no polynomial time approximation algorithm A for the traveling salesman problem can have $R_A^\infty<\infty$.

Proof: Suppose there were such an approximation algorithm. Then there would also be a polynomial time algorithm A satisfying $R_A\leqslant K$, for some positive integer K. We will show how A can be used to solve the NP-complete HAMILTONIAN CIRCUIT problem, using a variant on the HC $\propto$ TS construction from Chapter 2. Let $G=(V,E)$ be an arbitrary graph. We construct a corresponding traveling salesman instance I by letting V be the set of cities and defining the distance $d(u,v)$ between two such cities by

$$d(u,v)=\begin{cases}1 & \text{if } \{u,v\}\in E\\ K\cdot|V| & \text{otherwise}\end{cases}$$

Clearly, we can construct I and apply A to I in time polynomial in the size of G, since K is independent of G. However, if G has a Hamiltonian circuit, then $\text{OPT}(I)=|V|$, whereas if G does not have a Hamiltonian circuit then $\text{OPT}(I)>K\cdot|V|$. Therefore, by the guarantee for A, we will have $A(I)\leqslant K\cdot|V|$ if and only if G has a Hamiltonian circuit. Thus the existence of such an approximation algorithm A implies that the HAMILTONIAN CIRCUIT problem is in P, and this contradicts the assumption that P$\neq$NP. ■

This theorem points up the importance of the triangle inequality to the positive results given in the previous section for the traveling salesman problem. With it we have $R_{MIN}(\Pi)\leqslant 3/2$, whereas without it we have $R_{MIN}(\Pi)=\infty$, unless P=NP. Some results analogous to Theorem 6.13 for other problems can be found in [Sahni and Gonzalez, 1976] and [Rosenthal, 1977].

To summarize, we have seen that, corresponding to each of the types of approximating behavior mentioned in Section 6.1, it is possible to prove negative results showing that such behavior cannot be achieved for a particular problem Π unless P = NP. This suggests that optimization problems can be classified according to the value of $R_{MIN}(\Pi)$ (or even more finely if $R_{MIN}(\Pi) = 1$), on the assumption that P differs from NP. Figure 6.8 indicates the current status of some of the problems we have been discussing with respect to such a classification.

Finally, we remind the reader that, although we have chosen a collection of particularly appealing types of guarantees to study, other possibilities exist, and some work has been done on them. In addition, this is a growing field of study, and new results, both positive and negative, continue to be proved. For further information, [Garey and Johnson, 1976b] provides an annotated bibliography of the field as of mid-1976.

6.3 Performance Guarantees and Behavior "In Practice"

It is not hard to see why it is desirable to have performance guarantees of the sort we have been discussing for our heuristic algorithms. Even if the particular guarantee is not as strong as we would like (a solution that differs from optimal by 50 percent is often not good enough), it still may be worthwhile to begin with a simple algorithm that has such a guarantee and then upgrade it by adding more sophisticated special case heuristics and local optimization techniques. Furthermore, some of the guarantees that have been proved are actually quite good and can be achieved without inordinate computational effort. In particular, the fully polynomial time approximation schemes may be quite attractive for reasonably small values of ϵ. Moreover, guarantees are in their nature worst-case bounds, and approximation algorithms often behave significantly better in practice than their guarantees would suggest.

The key phrase here is "in practice." What we are most interested in is knowing how closely our algorithm will approximate optimal solutions "in practice." As an alternative to the "worst-case" performance guarantee approach, one might therefore attempt to do performance analysis from an "average-case" point of view. Indeed, such analysis has a long history and, until recently, has been performed primarily through empirical studies.

Often this involves simply devising a set of supposedly "typical" instances, running the algorithm and its competitors on them, and comparing the results. Of course, the usefulness of such experiments depends on how "typical" the sample instances actually are, and it is not always easy to generate an appropriate collection of instances.

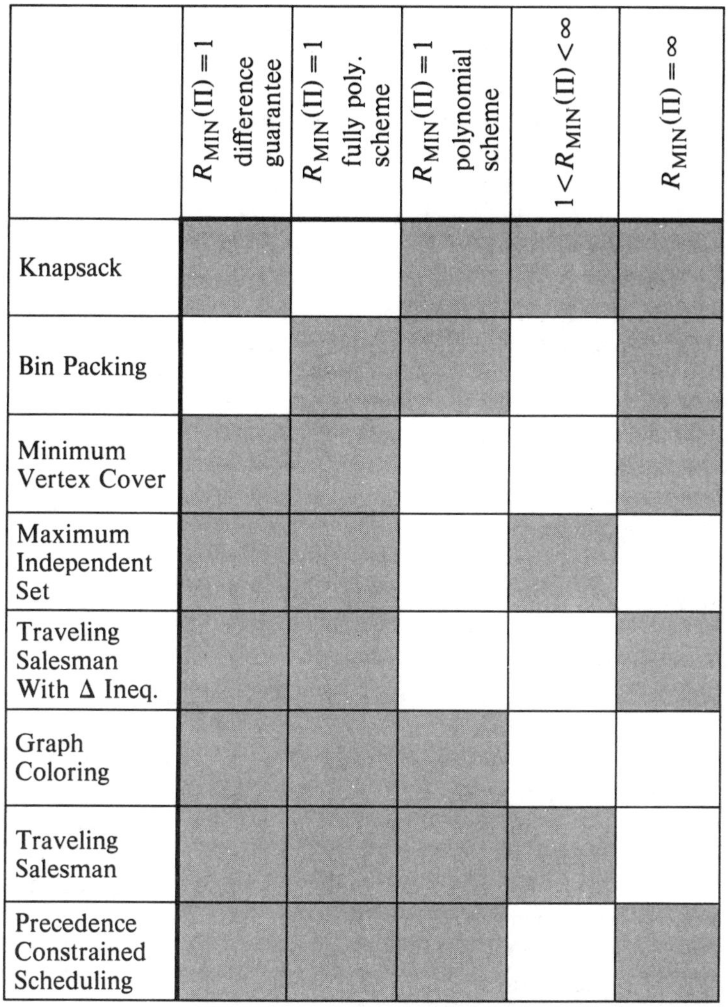

	$R_{MIN}(\Pi) = 1$ difference guarantee	$R_{MIN}(\Pi) = 1$ fully poly. scheme	$R_{MIN}(\Pi) = 1$ polynomial scheme	$1 < R_{MIN}(\Pi) < \infty$	$R_{MIN}(\Pi) = \infty$
Knapsack	shaded	open	shaded	shaded	shaded
Bin Packing	open	shaded	shaded	open	shaded
Minimum Vertex Cover	shaded	shaded	open	open	shaded
Maximum Independent Set	shaded	shaded	open	shaded	open
Traveling Salesman With Δ Ineq.	shaded	shaded	open	open	shaded
Graph Coloring	shaded	shaded	shaded	open	open
Traveling Salesman	shaded	shaded	shaded	shaded	open
Precedence Constrained Scheduling	shaded	shaded	shaded	open	shaded

Figure 6.8 A classification of several optimization problems according to the remaining possibilities for $R_{MIN}(\Pi)$ and different ways of achieving $R_{MIN}(\Pi) = 1$. Open rectangles represent possibilities that are still open, and shaded rectangles represent possibilities that either can be proved not to occur if $P \neq NP$ or cannot occur because a better algorithm is already known.

If the problem instances one expects to encounter in practice can be viewed as obeying some specific probability distribution, then one can use more sophisticated techniques. In this case it may be possible to provide a method for generating "random" instances, with respect to the given distribution, and to run the algorithms on samples obtained in this way. Some experiments that have been done along this line tend to confirm our observation that average behavior is generally much better than worst-case behavior. In [Johnson, 1973] the First Fit and First Fit Decreasing algorithms for bin packing were analyzed by Monte Carlo methods under a number of assumptions about the distribution of item sizes, with the sum of the item sizes used as an estimate of OPT(I). It was discovered that $R_{FF}(I)$ averaged about 1.07 rather than the worst-case bound of 1.70, and $R_{FFD}(I)$ averaged about 1.02 rather than 1.22. (It is interesting to note that the relative rankings of the two algorithms in the "average case" remained much the same as in the worst case.) In [Sahni, 1977], it was found that a particular fully polynomial time approximation scheme for a problem of sequencing with deadlines stayed within a ratio of 1.005, even when ϵ was chosen to guarantee only 1.10, and in fact the algorithm nearly always found an optimal solution. However, the number of such comparisons done so far is quite limited, and it is probably unwise to make any hard and fast generalizations at this point.

An even more sophisticated approach is illustrated by recent results that prove *theorems* about the expected behavior of approximation algorithms under particular probability distributions. In [Grimmet and McDiarmid, 1975], for example, it is shown that, under the assumption that all n vertex graphs are in a sense "equally likely," a simple graph coloring algorithm A yields an average value of $R_A(G)$ equal to 2, even though in the worst case $R_A(G)$ can be arbitrarily large. Similar results have been obtained for the bin packing problem [Shapiro, 1977], [Yao, 1976], and for the maximum independent set problem, the Hamiltonian circuit problem, and geometric versions of the traveling salesman problem and the Steiner tree problem (see [Karp 1975b], [Karp, 1976], [Karp, 1977], [Angluin and Valiant, 1977]). Although such analyses can be quite complicated for even the simplest of algorithms and distributions, this appears to be a growing field of research, and one can certainly expect to see additional results of this sort in the near future.

Of course, such analyses of average case performance have their own drawbacks as predictors of behavior "in practice." In order to compute an average, one must first assume some probability distribution on instances, and it is often not clear what distribution to choose. (An interesting way of avoiding this, using algorithms that do their own randomizing, is discussed in [Rabin, 1976].) The distributions under which we can analyze algorithms, with currently available techniques, can be quite different from the actual distributions that occur in practice, where instances tend to be highly structured (introducing biases that can be quite difficult to capture mathematically), and distributions can change in unpredictable ways as time goes on. Moreover, average case results do not tell us anything about how algorithms will perform for particular instances, whereas worst case guarantees at least provide a bound on this performance. Thus if we are to get as much information as possible about how our approximation algorithms will behave in practice, it is probably best to analyze them in as many ways as possible, from *both* the worst-case and the average-case points of view.

7

Beyond NP-Completeness

In this chapter we survey a number of theoretical topics related to (and inspired by) the theory of NP-completeness. No advanced background on the part of the reader will be assumed, so that even the novice should be able to gain some understanding of the main issues and how they relate to the basic theory. For readers intrigued by results quoted here, more detailed treatments, including proofs, can be found in the articles and books we cite as references. In keeping with the pattern established earlier in this book, we will define new concepts in terms of languages and Turing machines but discuss them informally in terms of problems and algorithms.

Section 7.1 deals with what is known and conjectured about the structure of NP, elaborating on the view presented in Chapter 2 and discussing the existence in NP of intractable problems that are not NP-complete. Sections 7.2 through 7.4 discuss "completeness" for classes of problems that appear to be harder than the NP-complete problems, although these, too, have yet to be *proved* intractable. In particular, Section 7.2 covers the "polynomial hierarchy," and Section 7.3 introduces the notion of "number completeness" for enumeration problems. Section 7.4 begins a discussion of the relationship between time and memory requirements by explaining the concept of polynomial space completeness. Section 7.5 continues the discussion by presenting the concept of "log-space reducibility" and showing how it can be used to investigate the memory requirements of problems in P. Finally, in Section 7.6, we turn once again to the fundamental question, "does $P = NP$?" Although we believe the two classes are *not* equal, there seem to be substantial obstacles to proving such a result with any of the currently known techniques, and we discuss why this is so.

7.1 The Structure of NP

In Chapter 2 we presented the simple diagram of NP shown in Figure 7.1. This class contains the subclass P and the subclass (here abbreviated NPC) made up of the NP-complete languages. Assuming that $P \neq NP$, these two subclasses do not intersect.

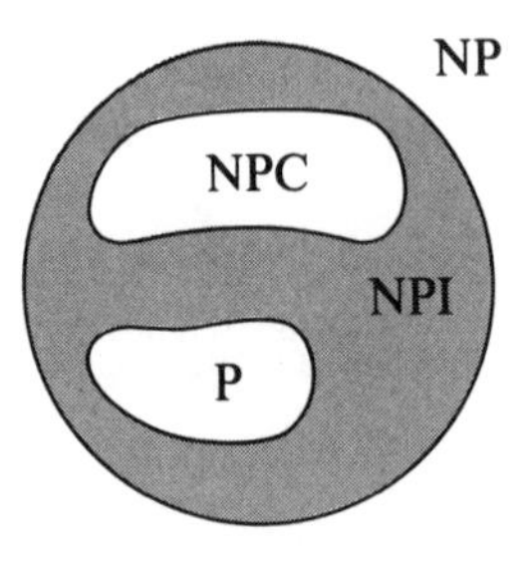

Figure 7.1 The world of NP, reprised (assuming $P \neq NP$).

It is also the case, however, that if $P \neq NP$, then NP cannot be the same as $P \cup NPC$. Let us denote the class $NP - (P \cup NPC)$ by NPI, since it consists of languages having "intermediate" difficulty (between P and NPC). The fact that NPI is not empty if $P \neq NP$ follows as a consequence of a more general result, proved in [Ladner, 1975a]. For the purpose of stating this result, recall that a *recursive* language is any language that can be recognized by a (not necessarily polynomial time) DTM program that halts for all inputs.

Theorem 7.1 Let B be a recursive language such that $B \notin P$. Then there exists a polynomial time recognizable language $D \in P$ such that the language $A = D \cap B$ does not belong to P, $A \propto B$, and yet it is not the case that $B \propto A$.

We apply this theorem as follows: Let B be any NP-complete language. If $P \neq NP$, then $B \notin P$, so the hypothesis of the theorem holds. Moreover, the resulting language A belongs to NP because $D \in P$ and $B \in NP$. The theorem then implies that $B \not\propto A$, so A cannot be NP-complete, and because $A \notin P$, it follows that $A \in NPI$. In more concrete terms, suppose B were the NP-complete HAMILTONIAN CIRCUIT problem. Then the theorem, in effect, tells us that there exists a polynomial time recognizable class of graphs such that HAMILTONIAN CIRCUIT, when restricted to that class of graphs, is neither NP-complete nor in P (assuming P≠NP).

In addition to allowing us to conclude that NPI cannot be empty if P≠NP, Theorem 7.1 also tells us about the structure of this intermediate class. For instance, assuming P≠NP, the class NPI must itself be made up of an infinite collection of distinct equivalence classes of languages, for if B is any problem in NPI, Theorem 7.1 would allow us to construct a new problem that is "easier" than B but still not in P. Ladner [1975a] also shows that, if P≠NP, then NPI must contain pairs of languages C and D such that *neither* $C \propto D$ *nor* $D \propto C$. Thus in the partial order according to "hardness" imposed on NP by the relation $\propto$, there must be incomparable elements if P≠NP.

Given this theoretical framework, it is reasonable to ask if there are any "natural" problems that are candidates for membership in NPI. Under the assumption that P≠NP, Theorem 7.1 *can* be used to exhibit members of this class, but these languages are highly unnatural, due to the complicated "diagonalization" techniques used to construct them. Of course, any "open" problem in NP, that is, one that has not yet been proved either to belong to P or to be NP-complete, can be viewed as a candidate for NPI. However, certain open problems are generally viewed as better candidates, having withstood the test of time and having certain attributes that seem to distinguish them from the types of problems already known to be NP-complete. In particular, three open problems mentioned in the original [Karp, 1972] paper remain open today and are often cited, with varying degrees of conviction, as potential members of NPI.

GRAPH ISOMORPHISM
INSTANCE: Graphs $G=(V,E)$, $G'=(V,E')$.
QUESTION: Are G and G' "isomorphic," that is, is there a one-to-one function $f: V \to V$ such that $\{u,v\} \in E$ if and only if $\{f(u),f(v)\} \in E'$?

COMPOSITE NUMBERS
INSTANCE: Positive integer K.
QUESTION: Are there integers $m,n > 1$ such that $K = mn$?

LINEAR PROGRAMMING
INSTANCE: Integer vectors $V_i = (v_i[1], v_i[2], \ldots, v_i[n])$, $1 \leqslant i \leqslant m$, $D = (d_1, d_2, \ldots, d_m)$, $C = (c_1, c_2, \ldots, c_n)$, and an integer B.
QUESTION: Is there a rational vector $X = (x_1, x_2, \ldots, x_n)$ such that $V_i \cdot X \leqslant d_i$ for $1 \leqslant i \leqslant m$ and such that $C \cdot X \geqslant B$?

Researchers who have attempted to prove that GRAPH ISOMORPHISM is NP-complete have noted that its nature is much more constrained than that of a typical NP-complete problem, such as SUBGRAPH ISOMORPHISM. NP-completeness proofs seem to require a bit of leeway; if the desired structure X (subset, permutation, schedule, etc.) exists, it should still exist even if certain aspects of the instance are locally altered. For example, a function f will be an isomorphism between a graph H and a subgraph of a graph G even if we add edges to G or delete edges not in the image of f. However, if f is an isomorphism between H and G itself, then any change in G must be reflected by a corresponding change in H, or else f will no longer be an isomorphism. In other words, proofs of NP-completeness seem to require a certain amount of redundancy in the target problem, a redundancy that GRAPH ISOMORPHISM lacks. Unfortunately, this lack of redundancy does not seem to be much of a help in designing a polynomial time algorithm for GRAPH ISOMORPHISM either, so perhaps it belongs to NPI.

The arguments as to why COMPOSITE NUMBERS and LINEAR PROGRAMMING might not be NP-complete spring from a different source. Recall that in Chapter 2 we observed that membership for a problem Π in NP does not seem to imply membership in NP for the complementary problem Π^c (the problem with the answers reversed, that is, with $Y_{\Pi^c} = D_\Pi - Y_\Pi$). Let us define

$$\text{co-NP} = \{\Pi^c : \Pi \in \text{NP}\}$$

or in language terms

$$\text{co-NP} = \{\Sigma^* - L : L \text{ is a language over the alphabet } \Sigma \text{ and } L \in \text{NP}\}$$

On the grounds that so many problems in co-NP do not seem to be in NP, one might well conjecture that NP$\neq$co-NP. Note that this conjecture, though justified by arguments similar to those we have presented for P$\neq$NP, is in fact a "stronger" conjecture than P$\neq$NP. The class P *is* closed under complementation (that is, P$=$co-P), so NP$\neq$co-NP would imply P$\neq$NP, although it might be the case that P$\neq$NP even though NP$=$co-NP. Nevertheless, the following easily proved theorem shows that there is a strong link between the NP-complete problems and the conjecture that NP$\neq$co-NP.

Theorem 7.2 If there exists an NP-complete problem Π such that $\Pi^c \in$ NP, then NP$=$co-NP.

The proof is a straightforward argument involving the concatenation of Turing machine programs, as in the proof of Lemma 2.1. Assuming that P$\neq$NP and NP$\neq$co-NP, Theorem 7.2 gives us the new picture of NP and its environs shown in Figure 7.2.

As a consequence of Theorem 7.2, a problem Π for which both Π and Π^c belong to NP cannot be NP-complete unless NP$=$co-NP, contrary to our conjecture. Accordingly, one would not expect such a problem Π to be

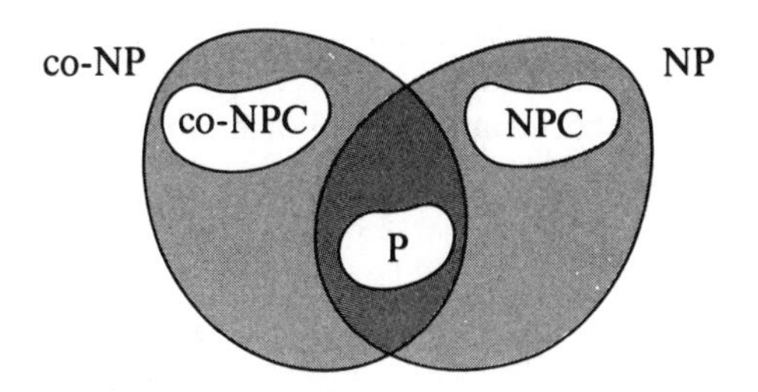

Figure 7.2 The world of NP and vicinity (assuming P$\neq$NP and NP$\neq$ co-NP). It may or may not be the case that P = NP$\cap$ co-NP.

NP-complete. However, both COMPOSITE NUMBERS and LINEAR PROGRAMMING have this property.

COMPOSITE NUMBERS clearly belongs to NP. Its complementary problem is known as PRIMES: Given a positive integer K, is K a prime number? Membership in NP for PRIMES is shown in [Pratt, 1975], where it is proved that for every prime K there exist proofs of primality that are sufficiently short that they can be guessed and checked in time bounded by a polynomial in $\log K$. Similarly, the fundamental duality theorem of linear programming can be used to show that the complementary problem of LINEAR PROGRAMMING is itself a variant of linear programming that belongs to NP.

Thus neither COMPOSITE NUMBERS nor LINEAR PROGRAMMING can be NP-complete unless NP = co-NP. This is taken to be relatively strong evidence that these problems are not NP-complete. If they are not solvable in polynomial time either, then they belong to NPI. Here the evidence is perhaps a bit weaker. As we have already mentioned, the simplex algorithm for linear programming, although it has exponential time complexity, seems to run quickly so often that it is conceivable that an alternative algorithm that runs in polynomial time might be designed. In the case of COMPOSITE NUMBERS we are, in a sense, even closer to a polynomial time algorithm. An algorithm for determining whether a given positive integer is prime or composite is described in [Miller, 1976], and it is proved that this algorithm has polynomial time complexity if the "extended Riemann Hypothesis" of number theory is true. Thus although no polynomial time algorithm is known for either of these problems, we are not nearly as confident that none can be found as we are with the NP-complete problems. (Indeed, discovering that both $\Pi \in$ NP and $\Pi^c \in$ NP might be regarded as suggesting that Π *can* be solved with a polynomial time algorithm, even though we do not yet know whether P = NP $\cap$ co-NP.)

The existence of such (more or less likely) candidates for membership in NPI gives rise to the possibility of constructing new classes of "hard" problems within NP. In particular, the relation $\propto$ of polynomial transformability leads to equivalence classes for each of the problems GRAPH ISOMORPHISM, LINEAR PROGRAMMING (LP), and COMPOSITE NUMBERS, with the obvious consequences. For example, if $\Pi \in NP$ is such that $\Pi \propto LP$ and $LP \propto \Pi$, then Π will be solvable in polynomial time (or NP-complete, or in NPI) if and only if LP is solvable in polynomial time (NP-complete, in NPI). The many alternative formulations of linear programming that appear in linear programming texts can usually be seen to be equivalent under $\propto$ to LP. Reiss and Dobkin [1976] summarize and extend this class, and Itai [1977] shows that certain rational flow problems are similarly equivalent to LP. For the case of GRAPH ISOMORPHISM, polynomially equivalent problems have been discovered by Booth [1978], Babai [1976], Miller [1977], Kozen [1977a], Kozen [1978], and others, although most members of this class found so far are either restricted versions of GRAPH ISOMORPHISM or isomorphism problems for other types of structures and hence are still the same "type" of problem. For COMPOSITE NUMBERS, most number theory texts describe numerous properties that hold if and only if a given number is prime, and an equivalence class can be built from these. Some of these equivalences will be summarized in the portion of our problem list devoted to open problems.

Given a problem in NP that we believe is hard, but which we have not been able to prove NP-complete, there are of course stronger ways of supporting its intractability than merely by showing its equivalence to other open problems. One method, mentioned in Chapter 5, is to use Turing reducibility. If we can show that a known NP-complete problem Π' is polynomial time Turing reducible to our problem Π, then we know that Π shares with Π' the property of being solvable in polynomial time if and only if $P = NP$, even though we will not have shown that Π is NP-complete, since $\Pi' \propto_T \Pi$ is not known to imply $\Pi' \propto \Pi$. (In general, polynomial time Turing reducibility does *not* imply polynomial transformability [Ladner, Lynch, and Selman, 1975], although no examples of this have been found within NP.)

A potentially more useful idea, introduced by [Adelman and Manders, 1977], is that of "γ-reducibility." This provides a way of proving that a problem is intractable under the plausible assumption that $NP \neq$ co-NP, rather than under our standard assumption that $P \neq NP$. Hence it is a weaker notion of reducibility than either Turing reducibility or polynomial transformability, but it still can provide substantial support to the conjecture that a problem is hard.

A γ-reduction, in contrast to our other notions of reducibility, is nondeterministic in nature. Let us first introduce the notion of the *relation* R_M computed by an NDTM program M:

$$R_M = \left\{ <x,y>: \begin{array}{l} \textit{there is a string } z \textit{ such that on input} \\ x \textit{ and guess } z \ M \textit{ has output } y \end{array} \right\}$$

(where the definition of "output" is as in the computation of functions by DTMs).

We say that a language L_1 over alphabet Σ_1 is *γ-reducible* to a language L_2 over Σ_2 (written $L_1 \propto_\gamma L_2$) if there is a polynomial time NDTM program M such that for all $x \in \Sigma_1^*$ there is some $y \in \Sigma_2^*$ for which $<x,y> \in R_M$ and such that, for all $<x,y> \in R_M$, $x \in L_1$ if and only if $y \in L_2$. In other words, there is at least one halting computation for M on every input x and, given an input x, all halting computations on x yield outputs that are in L_2 if and only if $x \in L_1$.

It is a simple matter to observe that, as with the polynomial transformability relation $\propto$, the γ-reducibility relation $\propto_\gamma$ is transitive, and that $L_1 \propto L_2$ implies $L_1 \propto_\gamma L_2$ (although the converse is not known to hold). A language $L \in$ NP is called *γ-complete* if, for every $L' \in$ NP, $L' \propto_\gamma L$. Thus all NP-complete languages are γ-complete, but there might be a language that is γ-complete and not NP-complete. Examples of γ-complete problems not known to be NP-complete can be found in [Plaisted, 1977b] as well as in [Adleman and Manders, 1977]. We describe only one example, taken from the latter paper:

LINEAR DIVISIBILITY
INSTANCE: Positive integers a,c.
QUESTION: Is there a positive integer x such that $ax+1$ divides c?

(It is interesting to note that this problem would be trivially solvable in polynomial time if $ax+1$ were replaced by ax, for then all we would be asking is "does a divide c?" We also note that the problem can be solved in pseudo-polynomial time, and hence its supposed intractability depends heavily on the convention that numbers be represented by strings having length logarithmic in their magnitudes.)

The γ-reduction used to prove that LINEAR DIVISIBILITY is γ-complete is from the problem 3SAT and involves nondeterministically guessing a large number of primes, proofs of primality, and factorizations. Indeed, all the γ-completeness results proved to date are for number theoretic problems and use nondeterminism to guess such number theoretic structures. Thus it remains to be seen how wide the applicability of γ-completeness will be. The implications of γ-completeness for intractability are stated in the following theorem, taken from [Adelman and Manders, 1977]:

Theorem 7.3 If L is γ-complete and $L \in \text{NP} \cap \text{co-NP}$, then $\text{NP} = \text{co-NP}$.

Thus in particular, since $\text{P} \subseteq \text{NP} \cap \text{co-NP}$, if LINEAR DIVISIBILITY is solvable in polynomial time, then NP = co-NP. (Note also that Theorem 7.3 makes it unlikely that either COMPOSITE NUMBERS or LINEAR PROGRAMMING will be γ-complete, since we have seen that both belong to NP $\cap$ co-NP.) Figure 7.3 adds the γ-complete problems to our picture of the world of NP.

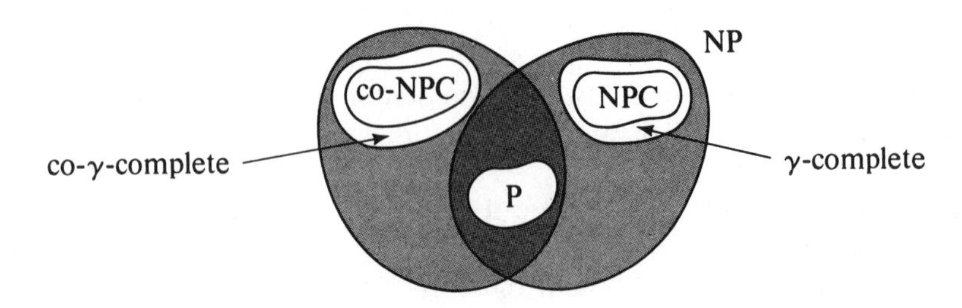

Figure 7.3 The world of NP, once more revised (assuming that both $P \neq NP$ and $NP \neq$ co-NP).

Since we have devoted so much discussion to the structure of NP and in particular the intermediate region NPI, it is only appropriate that we conclude this section with a brief discussion of the structure of NPC, the class of NP-complete problems. Although all these problems are "equivalent" with respect to polynomial transformability, some seem to be much more closely related than others. Recall, for example, the very direct transformations among VERTEX COVER, CLIQUE, and INDEPENDENT SET in Chapter 3. One way of capturing this close relationship is through what is called a "polynomial time isomorphism." We shall say that two languages $L_1 \subseteq \Sigma_1^*$ and $L_2 \subseteq \Sigma_2^*$ are *polynomial time isomorphic* if there exists a one-to-one onto function $f: \Sigma_1^* \rightarrow \Sigma_2^*$ such that f is a polynomial transformation from L_1 to L_2 and such that f^{-1} is a polynomial transformation from L_2 to L_1.

It is easy to verify that VERTEX COVER, CLIQUE, and INDEPENDENT SET are all polynomial time isomorphic, and one might suspect that other equivalence classes with respect to isomorphism could be constructed. Such a project is undertaken in [Berman and Hartmanis, 1977] and [Hartmanis and Berman, 1978], with the surprising conclusion that *all* the NP-complete problems appear to be polynomial time isomorphic! The papers present a number of rather general techniques for converting ordinary polynomial transformations into polynomial isomorphisms, and to date no NP-complete problem has been found to be immune to their application, although the "isomorphisms" constructed are by no means as simple and direct as those among VERTEX COVER, CLIQUE, and INDEPENDENT SET. On the basis of this evidence, Berman and Hartmanis conjecture that all NP-complete problems are polynomial time isomorphic. However, like so many of the other conjectures we have mentioned, this conjecture implies that $P \neq NP$. If P were equal to NP, then all languages in P would be

NP-complete and, by the conjecture, isomorphic. However, P contains both finite and infinite languages, so all languages in P cannot be isomorphic.

7.2 The Polynomial Hierarchy

The last section was concerned with problems that are NP-complete or possibly "easier." In this and the next two sections we discuss problems that are NP-hard and that may not be as "easy" as the NP-complete problems. A standard example of a problem that is NP-hard but does not appear to be NP-easy is the following from [Meyer and Stockmeyer, 1972] and [Stockmeyer, 1976a].

MINIMUM EQUIVALENT EXPRESSION
INSTANCE: A well-formed Boolean expression E involving literals on a set V of variables, the constants T (true) and F (false), and the logical connectives $\wedge$ (and), $\vee$ (or), $\neg$ (not), and $\rightarrow$ (implies), and a nonnegative integer K.
QUESTION: Is there a well-formed Boolean expression E' that contains K or fewer occurrences of literals such that E' is equivalent to E, that is, such that for all truth assignments to V the truth values of E' and E agree?

This problem is NP-hard since SATISFIABILITY is Turing reducible to it (a satisfiable expression E in conjunctive normal form is either an easily recognizable tautology or is not equivalent to any zero literal expression).

However, MINIMUM EQUIVALENT EXPRESSION does not appear to be NP-easy. No one has been able to show that an oracle for SATISFIABILITY (or for any other problem in NP) would enable us to solve it in polynomial time. Such an oracle seems to be useful only for testing whether or not E' is equivalent to E, but not for generating an appropriate choice for E'.

MINIMUM EQUIVALENT EXPRESSION can, however, be solved in polynomial time if we have a *nondeterministic* oracle Turing machine (an NOTM), rather than just the deterministic OTM of Chapter 5. An NOTM is an NDTM augmented with an oracle tape, just as an OTM is a DTM augmented with an oracle tape. Thus an NOTM can both make an initial guess and consult an oracle. Informally, an NOTM with an oracle for problem Π corresponds to a nondeterministic algorithm with a subroutine for Π. Conventions about running times are analogous to those for OTMs and NDTMs.

MINIMUM EQUIVALENT EXPRESSION can be solved by a polynomial time NOTM, with oracle for a problem in NP, as follows: The problem in NP we choose is SATISFIABILITY OF BOOLEAN EXPRESSIONS (given a well-formed Boolean expression E, is there a truth assignment that satisfies it?), a more general version of our standard SATISFIABILITY problem, though still in NP. To obtain the answer for an instance of MINIMUM EQUIVALENT EXPRESSION consisting of E and K, we merely guess an expression E' containing K or fewer occurrences of literals and use our oracle to determine whether $\neg((E' \rightarrow E) \wedge (E \rightarrow E'))$ is satisfiable. If it is not, then E' is the desired reduced expression.

The NOTM program we have just outlined might be called a "(polynomial time) nondeterministic Turing reduction" from MINIMUM EQUIVALENT EXPRESSION to SATISFIABILITY OF BOOLEAN EXPRESSIONS. It suggests the introduction of a new and broader class of problems, those that are polynomial time nondeterministically Turing reducible to problems in NP. For this purpose, let us introduce some notation. Let Y be a class of languages. The classes P^Y and NP^Y of languages are defined as follows:

$$\mathrm{P}^Y = \{L\text{: } \textit{there is a language } L' \in Y \textit{ such that } L \propto_T L'\}$$

$$\mathrm{NP}^Y = \left\{L\text{: } \begin{array}{l}\textit{there is a language } L' \in Y \textit{ such that there is a polynomial} \\ \textit{time nondeterministic Turing reduction from } L \textit{ to } L'\end{array}\right\}$$

Observe that P^{NP} is simply the class of all NP-easy languages. The class $\mathrm{NP}^{\mathrm{NP}}$ contains MINIMUM EQUIVALENT EXPRESSION and may or may not differ from P^{NP}. The containment relationships between P, NP, co-NP, and these two new classes are illustrated in Figure 7.4.

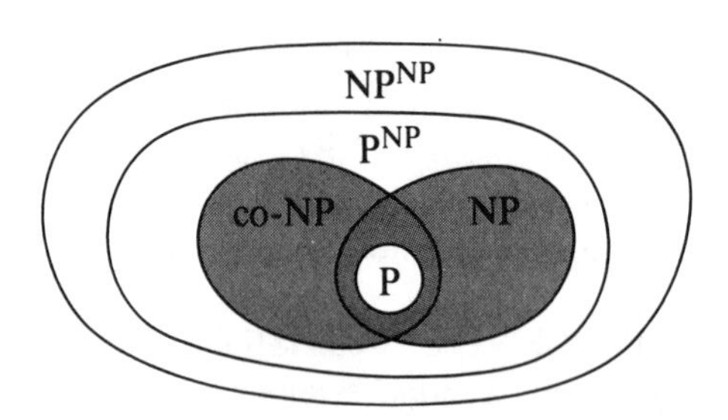

Figure 7.4 Containment relationships between classes of languages, including the new classes P^{NP} and $\mathrm{NP}^{\mathrm{NP}}$ (assuming $\mathrm{P} \neq \mathrm{NP}$).

On the basis of this picture of the world, Meyer and Stockmeyer [1972] observed that this process of defining new classes in terms of old ones could be extended indefinitely, yielding classes of greater and greater apparent difficulty. We thus obtain what is called the *polynomial hierarchy*. The classes in this hierarchy are denoted by Σ_k^p, Π_k^p, and Δ_k^p (where the superscript p is used solely to distinguish these from the analogous sets in the Kleene arithmetical hierarchy [Rogers, 1967]) and are defined as follows:

$$\Sigma_0^p = \Pi_0^p = \Delta_0^p = \mathrm{P}$$

and for all $k \geqslant 0$

$$\Delta_{k+1}^p = \mathrm{P}^{\Sigma_k^p}$$

$$\Sigma_{k+1}^p = \mathrm{NP}^{\Sigma_k^p}$$

$$\Pi_{k+1}^p = \text{co-}\Sigma_{k+1}^p$$

(that is, Π_{k+1}^p consists of all the complementary problems for problems in Σ_{k+1}^p). In fact, Π_1^p= co-NP and Σ_1^p= NP, while Δ_1^p= P. Similarly, $\Delta_2^p = \mathrm{P}^{\mathrm{NP}}$ and $\Sigma_2^p = \mathrm{NP}^{\mathrm{NP}}$. Figure 7.5 indicates the containment relationships between various classes in the hierarchy.

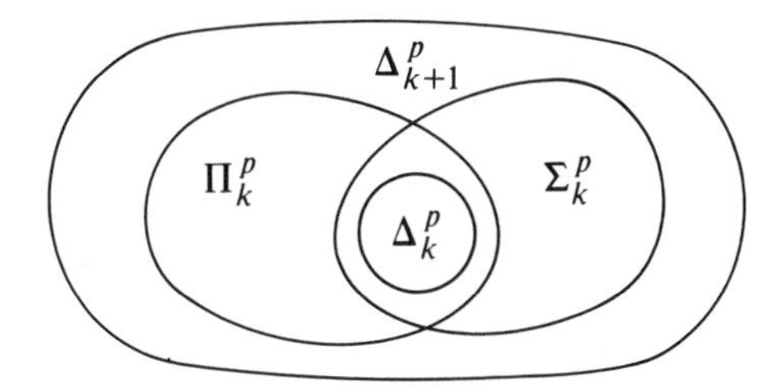

Figure 7.5 Containment relationships within the polynomial hierarchy.

The polynomial hierarchy gives us a more detailed way of classifying NP-hard decision problems. First we might ask if the problem is in the hierarchy at all (in Section 7.4 we will see some problems that appear to be outside the hierarchy, even though they still are not known to be intractable). In order to determine whether a problem is in the hierarchy, it is useful to have a more direct way of showing that a problem is in a particular class than by repeated application of the inductive definitions. This can be done in terms of relations in a manner suggested by Karp [1972] for the definition of NP. If Γ is an alphabet, a relation R of dimension k over Γ^* is a set of k-tuples $<z_1, z_2, \ldots, z_k>$ such that $z_i \in \Gamma^*$ for $1 \leqslant i \leqslant k$. We say that R is recognizable in polynomial time if there is a polynomial time DTM that recognizes the language consisting of exactly those k-tuples in R. The following theorem is from [Wrathall, 1976]:

Theorem 7.4 Let $L \subseteq \Gamma^*$ be a language, with $|\Gamma| \geqslant 2$. For any $k \geqslant 1$, $L \in \Sigma_k^p$ if and only if there exist polynomials $p_1, p_2, \ldots, p_k$ and a polynomial time recognizable relation R of dimension $k+1$ over Γ^* such that for all $x \in \Gamma^*$

$$
\begin{aligned}
x \in L \Leftrightarrow (\exists y_1 \in \Gamma^* \text{ with } |y_1| \leq p_1(|x|)) \\
(\forall y_2 \in \Gamma^* \text{ with } |y_2| \leq p_2(|x|)) \\
\vdots \\
(Q\, y_k \in \Gamma^* \text{ with } |y_k| \leq p_k(|x|)) \\
[< x, y_1, y_2, \ldots, y_k > \in R]
\end{aligned}
$$

where the quantifier Q on y_k is $\exists$ if k is odd and $\forall$ if k is even, and in general the quantifiers alternate.

Note that MINIMUM EQUIVALENT EXPRESSION fits this format for $k=2$, where the relation R is defined by $<(x,K), y_1, y_2> \in R$ if and only if x and y_1 are well-formed Boolean expressions, with y_1 containing K or fewer occurrences of literals, and y_2 is a truth assignment satisfying $(x \rightarrow y_1) \wedge (y_1 \rightarrow x)$. An analogous characterization of Π_k^p can be made by interchanging all the $\exists$ and $\forall$ quantifiers in Theorem 7.4 (see [Wrathall, 1976]).

The above theorem gives us a way of determining an upper bound on the least k such that $L \in \Sigma_k^p$. Determining lower bounds (even conditional ones) is more difficult. A start in that direction has been made by Leggett [1977], who has shown that certain decision problems related to optimization problems cannot be in either NP or co-NP unless NP = co-NP. These decision problems are different from our standard ones in that they make assertions about optimality. A simple example is the following problem:

MAXIMUM CLIQUE SIZE
INSTANCE: Graph G, positive integer K.
QUESTION: Does the largest complete subgraph in G contain exactly K vertices?

Instead of asking "Does there exist a clique of size K or larger?" as in CLIQUE, we are now asking if K is the size of the maximum clique. The optimality decision problem corresponding to MINIMUM EQUIVALENT EXPRESSION is:

MINIMUM EQUIVALENT EXPRESSION SIZE
INSTANCE: Well-formed Boolean expression E, nonnegative integer K.
QUESTION: Is it the case that no well-formed Boolean expression with $K-1$ or fewer occurrences of literals is equivalent to E, but there is such an expression with exactly K occurrences of literals that is equivalent to E?

The key to Leggett's results is the notion of "nondeterministic polynomial time conjunctive truth-table reducibility" described in [Ladner, Lynch, and Selman, 1975] and denoted by $\propto_c^{NP}$. Intuitively, the idea is that

$L \propto_c^{NP} L'$ if a polynomial time NOTM program, restricted so that it halts in the "no-state" whenever its oracle says "no," can test for membership in L using an oracle for L'. The main theorem of [Leggett, 1977], specialized to NP, is stated as follows:

Theorem 7.5 If L_0, L_1, and L_2 are languages such that L_1 and L_2 are NP-complete, $L_1^c \propto_c^{NP} L_0$, and $L_2^c \propto_c^{NP} L_0^c$ (where L^c denotes the complementary language for L), then

$$L_0 \in \text{NP} \cup \text{co-NP} \implies \text{NP} = \text{co-NP}$$

The theorem applies to both MAXIMUM CLIQUE SIZE and MINIMUM EQUIVALENT EXPRESSION SIZE, as well as to many similar decision problems. In the proof for MAXIMUM CLIQUE SIZE, both L_1 and L_2 are taken to be the CLIQUE problem, and one shows that the problem of determining whether a graph does not have a complete subgraph of size K can be reduced in the appropriate way to both MAXIMUM CLIQUE SIZE and its complement, a relatively straightforward exercise in this case.

The combination of Theorems 7.4 and 7.5 gives us a rather precise way of locating problems at the low end of the polynomial hierarchy, where many natural problems reside. For instance, from them we can conclude that, if P $\neq$ co-NP, then MAXIMUM CLIQUE is in $\Delta_2^p - (\Sigma_1^p \cup \Pi_1^p)$, that is, it is NP-easy but not in NP or co-NP.

Efforts to classify problems higher in the hierarchy rely on a generalization of the notion of NP-completeness to these higher levels. If we can identify a "hardest" problem in Σ_k^p, then that problem will certainly be in $\Sigma_k^p - \Sigma_{k-1}^p$ unless the two classes are equal. Our notion of a "hardest" problem should be quite familiar: A language L is complete for Σ_k^p (with respect to polynomial transformability) if $L \in \Sigma_k^p$ and, for all $L' \in \Sigma_k^p$, $L' \propto L$. Analogous definitions of completeness can be made for Π_k^p and Δ_k^p. It is easy to show that, if L is complete for Σ_k^p and $L \in \Sigma_{k-1}^p$, then $\Sigma_k^p = \Sigma_{k-1}^p$ (and the analogous statements hold for Π_k^p and Δ_k^p). Thus proving a problem complete for a class in the hierarchy is about as effective a way for showing where that problem lies as might be expected.

In analogy with the case of NP-completeness, it would be much easier to prove completeness for a class in the hierarchy if we knew some "first" problem to be complete for that class. This is the role played by SATISFIABILITY for the class NP. Such problems were first identified in [Meyer and Stockmeyer, 1972], where a problem B_k complete for Σ_k^p is defined for each $k \geqslant 1$. An instance of B_k is a well-formed Boolean expression E over a set of variables $X = \{x[i,j]: 1 \leqslant i \leqslant k, 1 \leqslant j \leqslant m_i\}$ for some integers $m_1, m_2, \ldots, m_k \geqslant 0$. The question is whether the following expression is true:

$$(\exists x[1,1]) \cdots (\exists x[1,m_1])$$
$$(\forall x[2,1]) \cdots (\forall x[2,m_2])$$
$$\vdots$$
$$(Qx[k,1]) \cdots (Qx[k,m_k])E$$

where the quantifier Q is $\exists$ if k is odd and $\forall$ if k is even, and in general the quantifiers alternate with respect to the first parameter of x.

We observe that $B_k \in \Sigma_k^p$ by Theorem 7.4. The following result is proved in [Meyer and Stockmeyer, 1972] and [Wrathall, 1976]:

Theorem 7.6 For all $k \geqslant 1$, B_k is complete for Σ_k^p, and the complementary problem B_k^c is complete for Π_k^p with respect to polynomial transformability.

Using the transitivity of $\propto$, we can therefore prove the completeness of a problem Π for class Σ_k^p merely by showing that $\Pi \in \Sigma_k^p$ and $B_k \propto \Pi$. However, few results of this type have been proven so far. Kozen [1977b] demonstrates the existence of an analogous hierarchy of complete problems involving finitely presented algebras rather than Boolean expressions, but results about individual problems of independent interest are few and far between. One that appears in [Stockmeyer, 1976a] concerns the problem of INTEGER EXPRESSION INEQUIVALENCE. If $n \in Z^+$, then the binary representation of n is an integer expression representing the set $\{n\}$. If e and f are integer expressions representing the sets E and F, then $(e \cup f)$ is an integer expression representing $E \cup F$, and $(e+f)$ is an integer expression representing the set $\{m+n: m \in E \text{ and } n \in F\}$. An instance of INTEGER EXPRESSION INEQUIVALENCE is a pair $<e,f>$ of integer expressions, and the question asked is "Do e and f represent different subsets of Z^+?" Stockmeyer shows this problem to be complete for Σ_2^p. Beyond this, little is known, even about Σ_2^p. It is still an open question whether MINIMUM EQUIVALENT EXPRESSION is complete for Σ_2^p.

This lack of results need not be loudly lamented, however, since the distinctions implied by the structure of the polynomial hierarchy appear to have more theoretical than computational significance, as shown by the following theorem from [Stockmeyer, 1976a]:

Theorem 7.7 If for some $k \geqslant 1$ we have $\Sigma_k^p = \Pi_k^p$, then $\Sigma_j^p = \Pi_j^p = \Sigma_k^p$ for all $j \geqslant k$.

In particular, if P = NP, then NP = $\Sigma_1^p = \Pi_1^p$, and so Σ_j^p = P for all $j \geqslant 0$. If we let PH = $\bigcup_{j=1}^{\infty} \Sigma_j^p$ be the set of all languages in the polynomial hierarchy, we thus have P = NP if and only if PH = P, that is, if and only if all languages in the hierarchy can be recognized in polynomial time and the entire hierarchy collapses into P.

Hence, showing that an NP-hard decision problem Π is in PH, although it is not equivalent to showing that Π is NP-easy, has the same consequences: Π can be solved in polynomial time if and only if P = NP. Finer distinctions as to where Π might lie in the hierarchy will have little apparent computational significance, especially since it can be shown that every problem in PH can be solved by exhaustive search in deterministic time $O(2^{q(n)})$ for some polynomial q in the input length n (the same result we had for NP in Theorem 2.1).

Nevertheless, assuming P $\neq$ NP, the polynomial hierarchy remains of theoretical interest, and Theorem 7.7 adds to our list of basic open problems the following: Is PH a true infinite hierarchy, that is, is $\Sigma_{k+1}^p - \Sigma_k^p$ nonempty for all $k \geqslant 1$? If the hierarchy collapses at some point, what is the least k such that $\Sigma_k^p = \Sigma_{k+1}^p$? Observe that we could have P $\neq$ NP, NP $\neq$ co-NP, and still have the hierarchy all collapse into Σ_2^p, although such a result would perhaps be a bit surprising.

7.3 The Complexity of Enumeration Problems

Enumeration problems provide natural candidates for the type of problem that might be intractable even if P = NP. To define what we mean by an enumeration problem, let us first recall the definition of a search problem from Chapter 5. In a search problem Π, each instance $I \in D_\Pi$ has an associated solution set $S_\Pi(I)$, and, given I, we are required to find *one* element of $S_\Pi(I)$ (the corresponding decision problem asks whether or not $S_\Pi(I)$ is empty). The *enumeration problem* based on the search problem Π is "Given I, what is the cardinality of $S_\Pi(I)$, that is, how *many* solutions are there?"

Such enumeration problems are naturally associated with many of the decision problems we have been discussing. For example, associated with HAMILTONIAN CIRCUIT is the enumeration problem "Given a graph G, how many distinct Hamiltonian circuits are there for G?" Associated with the SATISFIABILITY problem is the enumeration problem "Given a set C of clauses over a set V of variables, how many truth assignments for V simultaneously satisfy all the clauses in C?"

Observe that enumeration problems do not require that we display all the members of $S_\Pi(I)$, but merely that we determine how many there are. Thus even though the number of Hamiltonian circuits in G can be exponentially large in terms of the number of vertices, and an exponential amount of time would be required to list them all, the answer to the enumeration problem can be written down with a polynomially bounded number of binary digits. Hence the "size" of the answer does not by itself prevent the enumeration problem from being solved in polynomial time.

Indeed, some nontrivial enumeration problems can be solved in polynomial time. In [Harary and Palmer, 1973], for example, it is shown that the enumeration problem "Given a graph G, how many distinct spanning trees are there for G?" can be solved in polynomial time using Kirchoff's "matrix tree theorem" and evaluating a certain determinant. Similarly, the problem "Given a graph G, how many Eulerian paths are there for G?" can also be solved with a polynomial time algorithm.

Nevertheless, many enumeration problems appear to be quite difficult. The enumeration problems associated with NP-complete problems are clearly NP-hard, since if we know the cardinality of $S_\Pi(I)$ we can easily tell whether or not $S_\Pi(I)$ is empty. Moreover, some enumeration problems seem to be even harder than the corresponding existence problems. Even if P=NP, and we could tell in polynomial time whether an arbitrary graph contains a Hamiltonian circuit, it is not apparent that this would enable us to count how many Hamiltonian circuits are contained in G in polynomial time.

On the basis of such observations, Valiant [1977a] proposes that we consider a new class of polynomial time equivalent problems, the "#P-complete problems" (read "number-P-complete"), which includes many of these enumeration problems and is designed to reflect the additional difficulty of enumeration. (A similar proposal, with different terminology, is made in [Simon, 1977].) An enumeration problem Π belongs to #P if there is a nondeterministic algorithm such that for each $I \in D_\Pi$ the number of distinct "guesses" that lead to acceptance of I is exactly $|S_\Pi(I)|$ and such that the length of the longest accepting computation is bounded by a polynomial in Length[I]. (Valiant [1977a] defines this class in terms of what he calls "counting TMs," which are similar in concept to the "threshold machines" of [Simon, 1975] and [Simon, 1977] and to the probabilistic Turing machines of [Gill, 1977].)

It should be clear that the enumeration problems we have mentioned so far are in #P. In fact, if Π is any search problem for which there is a polynomial p such that for all $I \in D_\Pi$ and all $\sigma \in S_\Pi(I)$ the "length" of σ is less than or equal to p(Length[I]) and for which one can determine in polynomial time for given I and σ whether $\sigma \in S_\Pi(I)$, then the enumeration problem for Π can easily be seen to belong to #P. We thus see that #P contains at least a large fraction of the enumeration problems one might wish to consider and certainly many apparently difficult ones. The concept of "completeness" for #P is once again used to capture the notion of a "hardest" problem in the class. An enumeration problem Π will be called *#P-complete* if $\Pi \in$ #P and, for all $\Pi' \in$ #P, $\Pi' \propto_T \Pi$.

Note that we use polynomial time Turing reducibility rather than polynomial transformability in this definition. This is because the problems being related are not simply decision problems but have answers that are numbers. However, consider the following variation on polynomial transformability. Given two search problems Π and Π', a (polynomial time) *parsimonious transformation* from Π to Π' is a function $f: D_\Pi \rightarrow D_{\Pi'}$ that can be computed in polynomial time and that satisfies, for all $I \in D_\Pi$, $|S_\Pi(I)| = |S_{\Pi'}(f(I))|$. Notice that a parsimonious transformation from Π to Π' is also a polynomial transformation from one associated decision problem to the other. More important, however, is the observation that a parsimonious transformation from Π to Π' automatically gives rise to a Turing reduction between the associated enumeration problems. Thus parsimonious transformations can be a valuable tool for proving #P-completeness.

The first use of this tool was by Simon [1975], who observed that the generic transformation in the proof of Cook's theorem could be made parsimonious, thus making it possible to derive the following result:

Theorem 7.8 The problem of counting the number of satisfying truth assignments for an instance of SATISFIABILITY is #P-complete.

Simon [1977] also observes that many of the transformations appearing in the literature for proving NP-completeness are parsimonious, and that when they are not, parsimonious alternatives can often be found, as in [Valiant, 1976b] and [Galil, 1974]. We thus can conclude, for example, that the enumeration problems associated with the six basic NP-complete problems of Chapter 3 are #P-complete (although we do not claim that the transformations given there are parsimonious).

What is perhaps surprising is that some enumeration problems are #P-complete even when the associated search problems *can* be solved in polynomial time. Consider the following problem: Given a bipartite graph G, how many distinct perfect matchings does it contain? (Recall that a bipartite graph $G=(V,E)$ is one in which the vertex set V is partitioned into two sets V_1 and V_2, and no edge has both endpoints in the same set. A perfect matching is a set of edges $E' \subseteq E$ such that every vertex in V is included in exactly one edge in E'.) This enumeration problem is particularly interesting because it is equivalent to the problem of computing the "permanent" of a 0-1 matrix, where the permanent is the variant on the determinant in which all summands are given positive signs. The underlying search problem is also well known, since it is just the "marriage problem" referred to in Section 3.1.2, and, as we mentioned there, it can be solved in polynomial time. Nevertheless, in [Valiant, 1977a] the following result is proved.

Theorem 7.9 The problem of counting the number of distinct perfect matchings in a given bipartite graph is #P-complete.

Thus for some search problems that can be solved in polynomial time, the corresponding enumeration problems cannot be solved in polynomial time unless P=NP (and perhaps not even then), whereas for others, such as the spanning tree and Euler path problems mentioned at the beginning of this section, the enumeration problems *can* be solved in polynomial time. It would be interesting to see this classification extended. Partial results in this direction are contained in [Johnson and Kashdan, 1976], which shows that several enumeration problems based on polynomial time solvable search problems (such as the Kth LARGEST SUBSET problem discussed in Chapter 5) are NP-hard. Although this paper does not address questions of #P-completeness, it is not difficult to show that the problems studied there are in fact #P-complete. See also [Valiant, 1977b] where additional #P-complete problems are identified.

7.4 Polynomial Space Completeness

Throughout this book, our emphasis has been on just one of the "resources" required by a computation, the *time* it takes the computation to be performed. In practice, another resource is often just as important, the amount of computer memory or storage required by the computation, which we call the "space" requirement. In a Turing machine computation, the time used is the number of steps taken before a halt state is entered. The *space* used is defined to be the number of distinct tape squares visited by the read-write head. Since the number of tape squares visited cannot be more than the number of steps in the computation, it follows that any problem solvable in polynomial time is also solvable in polynomial space. Moreover, polynomial space also shares with polynomial time the property of being a model-independent concept. For all realistic computer models, the class of problems solvable in polynomial space is the same. (In fact, for space the model independence is even more strict, with the class of problems solvable using space $O(n^k)$, for any integer $k>0$, remaining the same for all the standard models [Hopcroft and Ullman, 1969].) In this section we examine more closely the relationship between polynomial time and polynomial space.

Although all problems solvable in polynomial time can be solved in polynomial space, it is still an unresolved question whether there exist problems solvable in polynomial space that *cannot* be solved in polynomial time. The conjecture that there might be such problems is quite plausible, since all problems in NP, all problems in the polynomial hierarchy, and all problems in #P can be solved in polynomial space. Thus if $P \neq NP$ or $P \neq \#P$, the inequivalence of polynomial time and polynomial space would immediately follow. In addition, there are many problems that can be solved in polynomial space that appear to be even "harder" than the problems in the above-mentioned classes. To single out the "hardest" such problems, we introduce the notion of "completeness" for polynomial space.

To do this formally, we once again restrict attention to languages (decision problems) and define PSPACE to be the class of all languages recognizable by polynomial space bounded DTM programs that halt on all inputs.

Our definitions then follow the standard format: a language L is PSPACE-*complete* (with respect to polynomial transformability) if $L \in$ PSPACE and, for all $L' \in$ PSPACE, $L' \propto L$. From this definition it follows that if L is PSPACE-complete, then $L \in$ P if and only if P = PSPACE, and an analogous statement can be made with P replaced by NP. Thus the fact that a problem is PSPACE-complete is an even stronger indication that it is intractable than if it were NP-complete; we could have P = NP even if P $\neq$ PSPACE. It is also considered as evidence that the problem is not in NP or even in the polynomial hierarchy. As shown in [Wrathall, 1976], if a PSPACE-complete problem is in Σ_k^p for some k, then the hierarchy collapses at that point and $\Sigma_j^p = \Sigma_k^p =$ PSPACE for all $j \geqslant k$.

A fundamental PSPACE-complete problem was identified in [Stockmeyer and Meyer, 1973] and bears an interesting relationship to the complete problems $B_k \in \Sigma_k^p$ discussed in Section 7.2. In particular, viewing the problems B_k as languages under the same encoding scheme, this PSPACE-complete language is just $B_\omega = \bigcup_{k=1}^{\infty} B_k$. However, for clarity, let us view B_ω as a decision problem and give it a more meaningful name and definition.

QUANTIFIED BOOLEAN FORMULAS (QBF)
INSTANCE: A well-formed quantified Boolean formula

$$F = (Q_1 x_1)(Q_2 x_2) \cdots (Q_n x_n) E$$

where E is a Boolean expression involving the variables $x_1, x_2, \ldots, x_n$ and each Q_i is either "$\exists$" or "$\forall$."
QUESTION: Is F true?

Observe that the essential difference between QBF and the problems B_k is that, whereas each B_k only allowed instances with a particular bounded number of quantifier alternations (a "$\forall$" followed by "$\exists$," or a "$\exists$" followed by "$\forall$"), in QBF the number of alternations is unconstrained. That QBF is in PSPACE follows from the fact that we can check whether F is true by cycling through all the possible truth assignments for the variables $x_1, x_2, \ldots, x_n$ and evaluating E for each. Recording the current assignment, testing E, and keeping track of where we are in the process can all be done in polynomial space, even though exponential time will be required to examine all 2^n truth assignments. That each language $L' \in$ PSPACE can be transformed to QBF follows from an analogue of Cook's theorem, in which one simulates a polynomial space bounded computation instead of a polynomial time bounded computation. The fact that the number of steps in the computation might be as large as $O(2^{p(n)})$, where p is a polynomial in the input length n (by an analogue of Theorem 2.1 it can be no larger), is handled by a clever trick spelled out in [Stockmeyer and Meyer, 1973] and [Stockmeyer, 1976a]. We then can conclude the following.

Theorem 7.10 QBF is PSPACE-complete.

Starting with QBF, the class of PSPACE-complete problems has already undergone considerable exploration. Even though many of its members are readily seen to be NP-hard, the added "evidence" for intractability provided by PSPACE-completeness is worthy of attention. Not only is a PSPACE-complete problem not likely to be in P, it is also not likely to be in NP. Hence a property whose existence question is PSPACE-complete probably cannot even be *verified* in polynomial time using a polynomial length "guess." Moreover, there is a certain theoretical satisfaction in precisely locating the complexity of a problem via a "completeness" result that cannot be obtained with a "hardness" result.

As the class of known PSPACE-complete problems has expanded, it has developed distinct characteristics. The problems it contains are usually quite different from the type of problems known to be NP-complete. In the remainder of the section, we shall mention a number of these so that the reader can gain some insight into what such problems look like.

First, we remark that, just as SATISFIABILITY could be transformed into the restricted version 3SAT, QUANTIFIED BOOLEAN FORMULAS can be transformed into the restricted version QUANTIFIED 3SAT, where the Boolean expression E is restricted to being an instance of 3SAT (a conjunction of 3-literal disjunctive clauses). Thus this restricted problem is PSPACE-complete [Stockmeyer and Meyer, 1973]. The proof uses a trick described in [Bauer, Brand, Fischer, Meyer, and Paterson, 1973] to convert an arbitrary Boolean expression, possibly using the connectives "$\rightarrow$" and "$\neg$", into one of the desired form. The role of a "basic" PSPACE-complete problem, corresponding to our seven basic NP-complete problems, has been played by QUANTIFIED 3SAT.

One especially rich source of PSPACE-complete problems has been the area of combinatorial games. The games of interest here are two-person games that can be specified concisely by presenting a combinatorial description of a generic "position" in the set of all possible positions, along with criteria for identifying who can "move" in a given position, how a "move" changes one position into another, which positions are final (and who wins), and which position is the initial position. If we denote the players as WHITE and BLACK, let P_0 be the initial position, and make the convention that, whenever a final position is reached, all later moves leave that position unchanged, then the statement that WHITE as first player has a forced win in n moves (n even) can be formalized as follows:

There is a move for WHITE from P_0 to a position P_1 such that,
for all moves of BLACK from P_1 to a position P_2,
there is a move for WHITE from P_2 to a position P_3 such that,
for all moves of BLACK from P_3 to a position P_4,

$\vdots$

> there is a move for WHITE from P_{n-2} to a position P_{n-1} such that,
> for all moves of BLACK from P_{n-1} to a position P_n,
> position P_n is a win for WHITE.

Observe that this statement displays the same "alternation of quantifiers" as made the problem QBF PSPACE-complete. It is thus not surprising that decision problems of the form "Given an initial position of a particular game, does WHITE have a forced win?" should be PSPACE-complete.

The first such result proved appears in [Even and Tarjan, 1976], and the problem they discuss is defined as follows:

GENERALIZED HEX
INSTANCE: An undirected graph $G=(V,E)$, specified vertices $s,t \in V$.
QUESTION: Does WHITE have a forced win in the following game played on G? *Positions* are partitions of V into three sets $< V_1, V_2, V_3 >$ such that $\{s,t\} \subseteq V_3$. The initial position is $< \phi,\phi, V >$, and all positions of the form $< V_1, V_2, \{s,t\} >$ are final. In a position $< V_1, V_2, V_3 >$ where $V_3 \neq \{s,t\}$, it is WHITE's move if $|V_1|+|V_2|$ is even, BLACK's move otherwise. WHITE makes a move by choosing a vertex $u \in V_3-\{s,t\}$ and changing the position to $< V_1 \cup \{u\}, V_2, V_3-\{u\} >$. BLACK moves analogously, removing a vertex from V_3 and adding it to V_2. A final position $< V_1, V_2, \{s,t\} >$ is a win for WHITE if and only if the subgraph of G induced by $V_1 \cup \{s,t\}$ contains a path from s to t.

The basic idea of the game is that WHITE and BLACK alternately choose vertices from $V-\{s,t\}$, WHITE having a goal of constructing a path from s to t, while BLACK's goal is to prevent this. The actual game of HEX is essentially the restriction of GENERALIZED HEX to a particular graph having a regular, boardlike structure. The problem is in PSPACE because a game can last only a polynomial number of moves ($|V|-2$ to be exact). It is proved PSPACE-complete by a transformation from QUANTIFIED 3SAT.

A number of other games have since been proved PSPACE-complete, including generalizations of the "geography" game in which players alternate naming countries, each new country having to start with the last letter of the previous country [Schaefer, 1978a], a game based on a generalization of Parker Brothers "Instant Insanity®" puzzle [Robertson and Munro, 1978], and generalizations of Checkers and Go to $n \times n$ boards (with these last two only known to be PSPACE-hard unless a rule ensuring polynomial length games is adopted) [Fraenkel, Garey, Johnson, Schaefer, and Yesha, 1978], [Lichtenstein and Sipser, 1978]. Additional examples are included in the lists.

Outside of the world of games, PSPACE-complete problems also appear in areas associated with automata, programming, and languages, where they often are restricted versions of problems already known to be intractable or undecidable. One such problem appears in [Meyer and Stockmeyer, 1972], [Aho, Hopcroft, and Ullman, 1974]. It concerns "regular expressions" and is interesting because it does not involve alternation of quantifiers in such an obvious way as the problems we have mentioned so far. Let us first define a regular expression and the language it represents. Let Σ be a finite alphabet. *Regular expressions* over Σ are defined inductively as follows:

(1) ϕ is a regular expression representing the empty set of strings;

(2) ϵ is a regular expression representing the set consisting of the empty string ϵ;

(3) for each $a \in \Sigma$, a is a regular expression representing the set $\{a\}$;

(4) If α and β are regular expressions representing sets A and B, then

   (i) $\alpha + \beta$ is a regular expression representing $A \cup B$,

   (ii) $\alpha\beta$ is a regular expression representing the language $AB = \{xy : x \in A, y \in B\}$,

   (iii) α^* is a regular expression representing $A^* = \bigcup_{i=0}^{\infty} A^i$, where $A^0 = \{\epsilon\}$ and $A^{i+1} = A^i A$ for $i \geqslant 0$.

The PSPACE-complete problem concerning regular expressions is the following:

REGULAR EXPRESSION NON-UNIVERSALITY
INSTANCE: A regular expression α over a finite alphabet Σ.
QUESTION: Does the set represented by α differ from Σ^*?

This problem remains PSPACE-complete even if Σ is restricted to the fixed alphabet $\{0,1\}$. However, if it is generalized by expanding the definition of regular expression to allow the abbreviation $(\alpha)^2$ to stand for $\alpha \cdot \alpha$ (and thus, for example, $(((\alpha)^2)^2)^2 = (\alpha)^8$), the resulting problem becomes *provably* intractable and can be shown to require *exponential space* [Meyer and Stockmeyer, 1972]. The PSPACE-completeness proof given in [Aho, Hopcroft, and Ullman, 1974] is especially interesting in that it constructs a generic transformation for an arbitrary language in PSPACE in which regular expressions are designed to represent *non-accepting* computations rather than accepting computations. Thus it is, in some sense, dual to the QBF proof; if any string fails to be in the set represented by the constructed regular expression, that string must correspond to an accepting computation.

As a final example, we consider a problem whose PSPACE-completeness yields an interesting corollary about the complexity of PSPACE itself. Given any DTM program, it is possible to modify it in a standard way so that no computation uses more than $n+1$ tape squares, where n is the length of the input string, and so that computations that previously obeyed that space bound stay essentially unchanged. The resulting program may not recognize the same language in general, but it will definitely obey that space bound, and programs that already obeyed that space bound *will* recognize the same language as before. We shall call such a DTM program *linear bounded*. By a well known padding argument (for example, see [Karp, 1972]), one can prove that the following problem is PSPACE-complete.

LINEAR SPACE ACCEPTANCE
INSTANCE: A linear bounded DTM program M and a finite string x over its input alphabet.
QUESTION: Does M accept x?

This result may seem quite surprising at first. It says that, in some sense, the problems that can be solved in polynomial space are no harder than those solvable in linear space. This would seem to contradict the fact that (as we shall see in Section 7.6) there are languages recognizable in polynomial space that cannot be recognized by any linear space-bounded DTM program. However, there is actually no contradiction, for the correct corollary to be drawn from the PSPACE-completeness of LINEAR SPACE ACCEPTANCE is not that PSPACE equals linear space, but only that linear space is contained in P *if and only if* PSPACE $\subseteq$ P. In fact, one has the more general result:

Theorem 7.11 If for any $k \geqslant 1$ the class of languages recognizable by DTM programs using no more than $O(n^k)$ space is contained in P (or NP), then PSPACE = P (*NP*).

(A similar result can be obtained for time: If for any $k \geqslant 1$, we have nondeterministic $O(n^k)$-time contained in P, then P = NP.)

Having defined the concept of PSPACE-completeness, one is naturally inclined to go further and consider the class of NPSPACE, consisting of those languages that can be recognized by *nondeterministic* Turing machine programs obeying polynomial space bounds. Before we can talk about space bounds for nondeterministic computations, however, we must deal with the issue of the space used by the "guess" in our model of an NDTM. For many computations, it is not actually necessary to remember all the symbols of the guess once they have been seen. Rather than penalize such computations by charging them for the space required to write down the entire guess at the beginning of the computation (as would be done in our NDTM model), the standard NDTM models used for measuring space can be viewed as being equipped with an additional device from which the program can request the "next" symbol of the guess at any time and that does not use any space itself. Only if the program chooses to remember the symbol for some later operation need it actually record the symbol on its tape and thus use "space" for it. This altered model of an NDTM, despite its differing measure of space, can be shown easily to be equivalent to the NDTM model of Chapter 2 with respect to polynomial time complexity, so by using it to define NPSPACE we will not upset any of our previous results about nondeterministic polynomial time. Defining NPSPACE to be the class of languages recognized by programs for this modified NDTM model that use only polynomially bounded space in their accepting computations, we then can ask about the relationship between PSPACE and NPSPACE. The perhaps surprising answer is that PSPACE = NPSPACE, which is implied by the following result of Savitch [1970]:

Theorem 7.12 If L can be recognized by an NDTM program in space bounded by $T(n)$, where $T(n) \geqslant \log n$ for all $n \geqslant 1$, then L can be recognized by a DTM program in space bounded by $T^2(n)$.

As a consequence of Theorem 7.12, PSPACE-completeness is the strongest type of completeness result we currently have, short of those that imply intractability. Of course, there still remains the possibility of proving a problem to be PSPACE-hard (by showing that some known PSPACE-complete problem is Turing reducible to it, without asserting that it is itself in PSPACE), but beyond this we move into the area of provably intractable problems, which will be discussed in Section 7.6.

We close this section by mentioning an open problem related to Theorem 7.12, which was perhaps the most famous open problem in complexity theory before the P vs. NP question arose. It is known as "the LBA problem." Let us call a DTM program that obeys a space bound of $n+1$ a *deterministic linear bounded automaton* (DLBA) and an NDTM program obeying the same bound a *nondeterministic linear bounded automaton* (NLBA). The set of languages recognizable by NLBAs is precisely the set of "context sensitive languages" (a class having a special grammatical description, for example, see [Hopcroft and Ullman, 1969]). By Theorem 7.12 all such languages can be recognized by DTM programs obeying a space bound of $(n+1)^2$. The question is, can every such language be recognized by a DLBA? (Or, even more strongly, can the exponent of 2 be removed from Theorem 7.12?) This question involves the same issues of determinism vs. nondeterminism as does the P vs. NP question, albeit at a more detailed level, and it has remained open since it was first posed in the mid-1960's. For a thorough discussion of this question, its consequences, and some related conjectures (including an analogue of the NP vs. co-NP question), see [Hartmanis and Hunt, 1974].

7.5 Logarithmic Space

In the preceding section we considered the relationships between the time complexity classes P and NP and the space complexity class PSPACE. In this section we continue our discussion of the ties between time and space complexity, this time looking within P and NP to the class DLOGSPACE of languages recognizable by DTM programs using space bounded by the logarithm of the input length.

The notion of logarithmic space might at first seem to be vacuous, since an input string of length n takes up n tape squares by itself, so any DTM program that looks at the entire input would use at least linear space. However, it is often useful in practice to make a distinction between the space required by the input string and the additional space in which the computation is carried out. For example, if we are merging two sorted lists that are stored on tapes, and the output is to be written onto another tape, we need only enough internal memory to store one element from each list at a time. The fact that we do not need an amount of internal memory comparable to the lengths of the input strings is quite significant here.

For the purpose of considering the possibility of using less than linear space, we alter our basic DTM model to that pictured in Figure 7.6. Here we have added an input tape with a two-way read-only head and an output tape with a write-only head. The input string will be written on the input tape beginning with square 1, leaving square 0 blank so that the left end can be recognized. The computation itself is performed on the usual two-way work tape using its read-write head. The *space* used by a computation in this model is simply the number of squares visited by the read-write head. It should be observed that the classes P, NP, and PSPACE are left unchanged by this model, with the only difference being that we now have the possibility of using less than linear space.

Research into the use of less than linear space has concentrated on the class DLOGSPACE of all languages recognizable by DTM programs that obey a space bound of $\lceil \log_2 n + 1 \rceil$, where n is the length of the input string. By a standard "speed-up" theorem (for example, see [Hopcroft and Ullman, 1969]), this can be seen to be the same as the class of all languages recognizable in space $c \lceil \log_2 n + 1 \rceil$ for any $c > 0$, so it is not quite so limited a class as it might seem. Furthermore, it is not hard to see that DLOGSPACE $\subseteq$ P. Our first open question is, of course, does DLOGSPACE $=$ P?

Although there are a variety of nontrivial problems that *can* be solved in logarithmic space (for example, see [Lipton and Zalcstein, 1977], [Lynch, 1977]), it appears to be the case that DLOGSPACE $\neq$ P. Many problems in

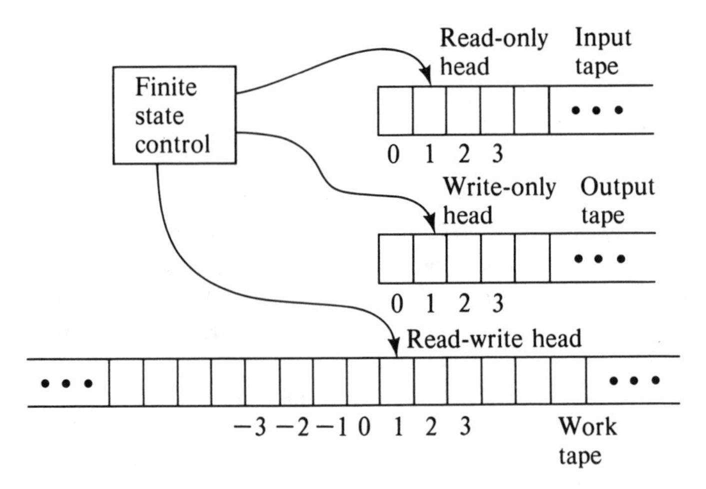

Figure 7.6 Schematic representation of a DTM model in which it is possible to use less than linear space.

P seem to require considerably more than logarithmic space. For example, consider the PARTITION problem. In Chapter 4 we showed that this problem can be solved in polynomial time if all item sizes are required to be no larger than, say, the square of the number of items. However, our algorithm in this case would require space proportional to the cube of the number of items, which is more than quadratic in the input length, and there does not seem to be any way of reducing this amount significantly. In [Cook, 1974] and [Cook and Sethi, 1976], examples of problems in P are given for which space at least proportional to $n^{1/2}$ or $n^{1/4}$ must be used by *any* algorithm chosen from a fairly general class. Moreover, it is known that we cannot have *both* P = DLOGSPACE and P = PSPACE (this is a consequence of Theorem 7.15, in the next section). Thus it seems likely that some problems in P require more than logarithmic space, and we can once again use completeness results to identify the most likely candidates.

For this new kind of completeness we will need a new kind of reducibility, since polynomial transformations cannot make distinctions within P. Let L_1 and L_2 be languages over alphabets Σ_1 and Σ_2 respectively. We say that a function $f: \Sigma_1^* \rightarrow \Sigma_2^*$ is a *log-space transformation* from L_1 to L_2 if

(i) f can be computed by a DTM program using space bounded by $\lceil \log_2 n + 1 \rceil$ for input strings of length n, and

(ii) $x \in L_1$ if and only if $f(x) \in L_2$.

If there is a log-space transformation from L_1 to L_2, we shall denote this by $L_1 \propto_{\text{LOG}} L_2$. Notice that since all log-space computable functions are also computable in polynomial time, we have $L_1 \propto_{\text{LOG}} L_2$ implies $L_1 \propto L_2$. In analogy with $\propto$, the following properties of $\propto_{\text{LOG}}$ can be proved (the proofs are not so trivial, however; see [Stockmeyer and Meyer, 1973]).

Theorem 7.13 Suppose $L_1 \propto_{\text{LOG}} L_2$ and $L_2 \propto_{\text{LOG}} L_3$. Then

(1) $L_1 \propto_{\text{LOG}} L_3$

(2) $L_2 \in$ DLOGSPACE $\Rightarrow L_1 \in$ DLOGSPACE.

We say that a language $L \in$ P is *log-space complete for* P if, for all $L' \in$ P, $L' \propto_{\text{LOG}} L$. Thus we have the desired consequence that if L is log-space complete for P, then $L \in$ DLOGSPACE if and only if DLOGSPACE = P. Moreover, the following more general result has been proved in [Jones, 1975], where DLOG^k-SPACE ($k>1$) is defined similarly to DLOGSPACE, with the bound of $\lceil \log_2 n + 1 \rceil$ being replaced by $(\lceil \log_2 n + 1 \rceil)^k$:

Theorem 7.14 If L is log-space complete for P and if $L \in$ DLOG^k-SPACE, then P $\subseteq$ DLOG^k-SPACE.

The first log-space complete problem to be explicitly identified is the following, from [Cook, 1974]:

PATH SYSTEM ACCESSIBILITY
INSTANCE: A finite set X of "nodes," a relation $R \subseteq X \times X \times X$, and two sets $S, T \subseteq X$ of "source" and "terminal" nodes.
QUESTION: Is there an "accessible" terminal node, where a node $x \in X$ is accessible if $x \in S$ or if there exist accessible nodes y, z such that $<x, y, z> \in R$?

This problem is in P since the set of all accessible nodes can be constructed trivially in polynomial time. The proof of log-space completeness for P, as expected, involves a generic log-space transformation of an arbitrary problem in P to this problem. A similar proof of log-space completeness for P is given in [Jones and Laaser, 1976] for the following problem:

UNIT RESOLUTION
INSTANCE: A set C of clauses on a set $X = \{x_1, x_2, \ldots, x_n\}$ of variables (that is, the same as an instance of SATISFIABILITY).
QUESTION: Can the empty clause (indicating a contradiction) be derived from C by "unit resolution," that is, does there exist a sequence $c_1, c_2, \ldots, c_m$ of clauses, with c_m being the empty clause, such that each c_i is either a clause from C or there exist two previously derived clauses c_k and c_l with $k, l < i$ of the forms $c_k = \{x_j\}$, $c_l = \{\overline{x}_j\} \cup c_i$ (or $c_k = \{\overline{x}_j\}$, $c_l = \{x_j\} \cup c_i$) for some $x_j \in X$?

Other problems that are log-space complete for P can be found in [Jones and Laaser, 1976], [Ladner, 1975b], [Galil, 1976], [van Leeuwen, 1976b], [Goldschlager, 1977], and [Kozen, 1977a].

The uses of log-space reducibility have not been limited solely to proving log-space completeness for P. For instance, LINEAR PROGRAMMING [Dobkin, Lipton, and Reiss, 1976] has been shown to be "log-space hard for P." That is, even though we don't know that a polynomial time algorithm for this problem would imply P = NP, we do know that a log-space algorithm would imply DLOGSPACE = P. Furthermore, it has been observed [Jones, 1973], [Stockmeyer and Meyer, 1973] that most (if not all) of the transformations used in proving NP-completeness results are also log-space transformations (the main need for memory in such constructions is for counting up to $p(n)$ for some polynomial p in the input length n, and this can be done in logarithmic space provided the tape alphabet has enough symbols). Thus the class of languages that are "log-space complete for NP" is at least a large subclass of the NP-complete problems, and has the additional property that if any one of them belongs to DLOG^k-SPACE for some $k \geq 1$, then all problems in NP belong to DLOG^k-SPACE.

Similarly, it has been observed that the transformations that have been used for proving PSPACE-completeness are also all log-space transformations (indeed, some references define PSPACE-completeness in terms of log-space transformations rather than polynomial transformations). This observation does not have consequences analogous to those for NP since we already know that PSPACE $\neq$ DLOG^k-SPACE for any k, and indeed that

$$\text{PSPACE} \neq \text{POLYLOGSPACE} = \bigcup_{k=1}^{\infty} \text{DLOG}^k\text{-SPACE}$$

However, if a language L is log-space complete for PSPACE, we can conclude that there exists a constant $r > 0$ such that any DTM program recognizing L must require space at least proportional to n^r infinitely often (for example, see [Stockmeyer, 1976a]), a conclusion that cannot be drawn if all we know is that L is PSPACE-complete in the sense defined in Section 7.4.

Finally, log-space transformability can be used to address another question of determinism versus nondeterminism, this time at the logarithmic level. Let NLOGSPACE be the set of all languages recognizable using space bounded by $\lceil \log_2 n + 1 \rceil$ by NDTMs (suitably defined so that the space for the input and the guess are not counted). The question is "does DLOGSPACE = NLOGSPACE?" The conjecture is that it does *not*, and candidates for languages in NLOGSPACE but not DLOGSPACE are provided by those languages that are log-space complete for NLOGSPACE. Examples and further details can be found in [Savitch, 1974], [Jones, 1975], [Sudborough, 1975], and [Jones, Lien, and Laaser, 1976].

We conclude this section with a brief look at the containment relationships between the classes we have been discussing. We have already observed that DLOGSPACE $\subseteq$ P. A less obvious result from [Cook, 1971b] is that NLOGSPACE $\subseteq$ P. However, it does not appear to be the case that POLYLOGSPACE $\subseteq$ P, or even that DLOG^2-SPACE $\subseteq$ NP. Although neither of these non-containment results has yet been proved, it is shown in [Book, 1976] that P$\neq$POLYLOGSPACE and NP$\neq$POLYLOGSPACE. The results presented there are especially intriguing because they do not tell us whether one of the sets is contained in the other or whether the two sets are incomparable with respect to containment, even though they do tell us that the two sets cannot be identical. Additional results concerning the relationships between various time and space complexity classes are discussed in [Book, 1972], [Book, 1974], and [Meyer and Shamos, 1977].

7.6 Proofs of Intractability and P vs. NP

In conjecturing that P$\neq$NP, we have cited as "evidence" the fact that so much effort has been spent in unsuccessful attempts to find polynomial time algorithms for what are now known to be NP-complete problems. It is perhaps more accurate to say that this is at least evidence that, if P$=$NP, that fact will be hard to prove. In this section we cite evidence on the other side of the question, showing why P$\neq$NP has also been hard to prove.

Let us begin by examining some of the partial results that *have* been proved. Rather than attack the question in full generality by trying to show that all algorithms for a particular NP-complete problem require exponential time, several researchers have restricted their attention to certain limited classes of algorithms, with the aim of showing that no algorithm in the class can both solve the problem and run in polynomial time. For instance, Galil [1977] (extending work of Tseitin [1970]) has shown that certain classes of algorithms for SATISFIABILITY based on resolution techniques must have exponential time complexity. For the INDEPENDENT SET problem, Chvátal [1977] considers a rather broad class of algorithms, which includes the algorithm of [Tarjan and Trojanowski, 1977] mentioned in Chapter 6, and shows that any such algorithm must take exponential time (indeed that for each there must exist a $c>1$ such that the algorithm requires time c^n for "almost all" n vertex graphs). McDiarmid [1976] presents similar results about algorithms for GRAPH K-COLORABILITY.

Such results have practical significance because they rule out certain approaches to designing polynomial time algorithms that one might otherwise be tempted to try. However, even though further research may widen the class of untenable approaches, it does not seem likely that we can eliminate *all* possible approaches by proceeding in such a piecemeal fashion. If we are to prove that P$\neq$NP, we must prove a lower bound on time complexity that holds for all solution algorithms. Unfortunately, the strongest lower bounds of this form proved to date leave a great deal to be desired.

One method that has been tried is that of bounding the number of logical gates required by combinational circuits that compute finite functions. Such a function is obtained when one restricts an NP-complete problem to instances of some fixed size n, and if the number of gates required can be shown to grow exponentially with n, then it can be concluded that the problem is intractable. The strongest lower bounds that have been proved within this model, using a great deal of ingenuity, are only as large as $2.5n$ [Paul, 1977]. Indeed, it is currently quite an accomplishment to prove bounds that are even slightly more than linear for a general model of computation, as in [Pippenger and Valiant, 1976]. Proving exponential lower bounds by such methods seems a long way off.

Nevertheless, as we mentioned in Chapter 1, intractability results *have* been proved for some problems, if not those in NP. Might not the techniques used in these proofs be applicable for showing that $P \neq NP$? This, too, seems to be unlikely, as we shall see after first examining these techniques.

The proofs of intractability that have been obtained up to now all rest essentially on two basic results from complexity theory. Let us say that a function $F(n)$ is *space constructible* if there exists a DTM program that when given a string of n 1's as input, halts after its read-write head has visited exactly $F(n)$ tape squares (we assume the DTM model in which there is a separate read-only input tape). A function $F(n)$ is said to be *time constructible* if there is a DTM program that, when given a string of n 1's, halts after taking exactly $F(n)$ steps. Most common functions, such as $n^k, 2^{cn}, k^{cn}, n!$, etc., for positive integer constants c and k, are both space and time constructible. The functions $\lceil \log_2 n \rceil^k$, for each positive integer k, are space constructible but, for obvious reasons, not time constructible. The two basic results are as follows:

Theorem 7.15 If $F_1(n)$ and $F_2(n)$ are space constructible functions, with $F_2(n) \geqslant \log_2 n$ for all $n \geqslant 1$, and if

$$\liminf_{n \to \infty} \frac{F_1(n)}{F_2(n)} = 0$$

then there exists a language L that can be recognized by a DTM program with space complexity bounded by $F_2(n)$, but not by any DTM program with space complexity bounded by $F_1(n)$.

Theorem 7.16 If $F_1(n)$ and $F_2(n)$ are time constructible functions and if

$$\liminf_{n \to \infty} \frac{F_1(n)}{F_2(n)} = 0$$

then there exists a language L that can be recognized by a DTM program with time complexity bounded by $F_2(n) \cdot \log_2 F_2(n)$, but not by any DTM program with time complexity bounded by $F_1(n)$.

These results were first proved in [Hartmanis, Lewis, and Stearns, 1965] and [Hartmanis and Stearns, 1965] (see also [Hopcroft and Ullman, 1969]). The main conclusions to be drawn from them are that certain classes of languages must contain some languages whose recognition problems are intractable. For example, the class EXPTIME (consisting of all languages with time complexity bounded above by $2^{p(n)}$ for some polynomial p of the input length n) must contain, for any $k \geqslant 1$, some languages whose time complexity is at least 2^{n^k}. However, it should be stressed that such languages are constructed by "diagonalization" arguments, and cannot be said to correspond to "natural" problems, nor are they particularly amenable to direct manipulation. For this reason, we use them to prove intractability of natural problems by an indirect route, based on our notions of "completeness" or "hardness" for a class. Recall that we say a language L is complete for a class C of languages (with respect to polynomial transformability) if $L \in C$ and, for all $L' \in C$, $L' \propto L$ (we say L is C-hard if the second property holds, but L is not necessarily itself in C). We then have the following general analogue of Lemma 2.1:

Lemma 7.1 If L is complete for C (or C-hard), and C contains an intractable problem, then L is intractable.

Proofs of intractability for natural problems rely on this basic lemma with various specific classes substituted for C. In [Meyer and Stockmeyer, 1972], it is shown that the problem of inequivalence for regular expressions with "squaring" is complete for EXPSPACE, which contains EXPTIME and hence contains intractable problems. In [Fischer and Rabin, 1974], it is shown that the problem of recognizing true statements in the theory of Presburger arithmetic is C-hard for C = NEXPTIME (the nondeterministic counterpart of EXPTIME). In [Chandra and Stockmeyer, 1976] and [Stockmeyer and Chandra, 1978], various combinatorial games are proved to be complete for EXPTIME. In [Jazayeri, Ogden, and Rounds, 1975], the "circularity problem for attribute grammars" is likewise proved to be complete for EXPTIME.

Indeed, many of these proofs yield stronger lower bounds on complexity than "mere" intractability, and some also prove lower bounds on space complexity. Moreover, if the class C contains difficult enough problems, one can prove such astronomical lower bounds as that given in [Meyer, 1975] for the "weak monadic second order theory of successor," where it is shown that any DTM program for recognizing true statements in this theory must have time complexity greater than

$$2^{2^{2^{\cdot^{\cdot^{\cdot^{n}}}}}}$$

for any stack of 2's in the exponent. We shall not go into the specific proof techniques since, as we shall see, this approach is not likely to be useful for proving that NP contains intractable problems.

To prove that P≠NP using Lemma 7.1, one would have to show that some language $L \in$ NP is complete for a class C containing intractable problems. However, observe that such a result would have certain consequences beyond proving that P≠NP. In particular, we have the following easily proved lemma:

Lemma 7.2 If $L \in$ NP, and L is complete for a class C (with respect to polynomial transformability), then C is a subset of NP.

Thus if $L \in$ NP, there are insurmountable roadblocks to proving that L is complete for EXPSPACE or NEXPTIME. By Theorem 7.15, we cannot have EXPSPACE ⊆ NP, since EXPSPACE strictly contains PSPACE which itself contains NP. A nondeterministic version of Theorem 7.16 follows from results in [Cook, 1973] (see also [Seiferas, Fischer, and Meyer, 1978]), and this implies that we cannot have NEXPTIME ⊆ NP. Furthermore, it is unlikely (though not disproved) that we will have EXPTIME contained in NP, as would be the case if a problem in NP were complete for this class, so proofs of P≠NP using completeness for EXPTIME are not to be expected either. We note that Lemma 7.2 also holds if NP is replaced by PSPACE, and similar observations can be made about trying to prove P≠PSPACE in this way.

It thus appears to be the case that Lemma 7.1 is too general for proving intractability at the "low" level of complexity corresponding to NP. Therefore, rather than trying to prove P≠NP by showing (albeit indirectly) that some intractable problem is transformable to some problem in NP, we might try to construct an intractable problem that belongs to NP directly, using diagonalization methods like those used to prove Theorem 7.15 and Theorem 7.16. Unfortunately, no one knows how to do this. The standard techniques for constructing such problems do not seem to be applicable to proving that P≠NP, and the reasons for this can be demonstrated using the notion of "relativization."

In [Baker, Gill, and Solovay, 1975], it is observed that ordinary diagonalization techniques (as well as "simulation" techniques such as used to prove Theorem 2.1) continue to work when "relativized" to any language L_0, that is, when all Turing machine programs are assumed to be programs for an oracle machine using L_0 as the oracle set. For example, the relativized version of Theorem 7.16 would be the following:

Theorem 7.16 (relativized) If $F_1(n)$ and $F_2(n)$ are time constructible functions such that

$$\liminf_{n\to\infty} \frac{F_1(n)}{F_2(n)} = 0$$

then there exists a language L that can be recognized by an OTM program with an oracle for L_0 having time complexity bounded by $F_2(n)\cdot\log_2 F_2(n)$, but not by any OTM program with an oracle for L_0 having time complexity bounded by $F_1(n)$.

This relativized theorem is true for any language L_0, as is the corresponding relativized version of Theorem 7.15. Thus we would expect in general that if we could prove $\mathrm{P}\neq\mathrm{NP}$ using a diagonalization argument, it would follow that for any language L, we would have $\mathrm{P}^L \neq \mathrm{NP}^L$ (where P^L and NP^L are the versions of P and NP defined in terms of OTMs having an oracle for L). On the other hand, if we could prove $\mathrm{P}=\mathrm{NP}$ by simulation techniques, we would also expect that this would show that $\mathrm{P}^L=\mathrm{NP}^L$ for any language L, again because these techniques seem to relativize easily. Since these are the main techniques known for proving results of this sort, it is rather disconcerting to be confronted with the following theorem from [Baker, Gill, and Solovay, 1975].

Theorem 7.18 There exist recursive languages A and B such that

(1) $\mathrm{P}^A = \mathrm{NP}^A$, and

(2) $\mathrm{P}^B \neq \mathrm{NP}^B$

This provides impressive evidence that the techniques that are currently available will not suffice for proving that $\mathrm{P}\neq\mathrm{NP}$ or that $\mathrm{P}=\mathrm{NP}$. The following results from the same paper also suggest that these techniques will not be useful for resolving many of our other open problems either.

Theorem 7.19 There exist recursive languages D, E, F, and G such that:

(1) $\mathrm{NP}^D \neq \text{co-NP}^D$

(2) $\mathrm{NP}^E = \text{co-NP}^E$, but $\mathrm{P}^E \neq \mathrm{NP}^E$

(3) $\mathrm{P}^F \neq \mathrm{NP}^F$, but $\mathrm{P}^F = \mathrm{NP}^F \cap \text{co-NP}^F$

(4) $\mathrm{P}^G \neq \mathrm{NP}^G \cap \text{co-NP}^G$ and $\mathrm{NP}^G \neq \text{co-NP}^G$

Furthermore, these results (along with results in [Baker and Selman, 1976]) can be used to make similar statements about the likelihood of answering questions about the polynomial hierarchy using currently available techniques, and much the same can be said about the P vs. PSPACE question. (In the case of space, however, there is some question as to what is an appropriate model for relativization results — see [Ladner and Lynch, 1976] and [Lynch, 1978].)

We conclude that substantially new proof techniques will probably be required to resolve the P vs. NP question, as well as the other related questions we have mentioned. That is, *if* we can resolve them. Hartmanis and Hopcroft [1976] use relativization techniques like those mentioned above to indicate at least the possibility that questions like this one might be independent of set theory, and hence not resolvable in our standard system of logic. Although we and other researchers in this area are not quite so pessimistic about the ultimate resolvability of the P vs. NP question, no one expects the answer to come very soon.

Appendix: A List of NP-Complete Problems

It has now become more or less standard practice to address questions of NP-completeness (or NP-hardness) whenever a problem is studied from a computational point of view. As a consequence, many NP-completeness results have appeared in the technical literature, and perhaps an even larger number have been obtained but not published. In this appendix we collect together a large number of these results, presenting an extensive list of known NP-complete and NP-hard problems.

The list contains more than 300 main entries, and our comments on these extend the total number of results in the list to several times this many. A typical entry consists of four parts: (1) a problem name, (2) a problem definition, (3) a reference for the result, and (4) a section of additional comments.

The problem name is intended as a convenient shorthand for referring to the problem. For reasons of consistency, we have on occasion assigned names that differ slightly from those in previous usage. An asterisk in parentheses "(*)" following a name indicates that the problem is not known to belong to NP, so the claimed result should be taken as one of NP-hardness rather than NP-completeness.

Problem definitions are given in the format used throughout this book, by specifying a generic *instance* and a *question* asked about that instance. (We restrict our attention to decision problems for main entries, although other types of problems are sometimes mentioned in the comments.) In most cases we have been able to give a set-theoretic definition that describes the problem completely, although the need for conciseness has often led to some obscuring of the intuition behind the problem; the reference can be consulted in such cases. Certain terms that are widely known have not been defined, and lengthy definitions common to an entire group of problems have been given only once, the first time the need for them arises. In a few cases the complete problem definition is so complex, or requires so much background information, that we have merely provided a sketchy definition using the appropriate "buzzwords," explicitly referring the reader to the references for details.

The *reference* section includes both a bibliographical citation for the main result and an indication of an appropriate "known NP-complete (or NP-hard) problem" that can be used for proving the result. Problems mentioned in this respect have been chosen on the basis of our own view of the most natural proof, and often come from proofs cited in the comments or from unpublished proofs of our own, rather than from the main citation. Several results have been deemed so straightforward that no one has been assigned the credit (or blame) for them.

The *comments* section, present for most entries, elaborates on the main result. This normally takes the form of citing various results concerning the polynomial time solvability and NP-hardness of subproblems and variants of the main problem. Thus the list can serve as a useful source of information on what can and cannot be solved in polynomial time, assuming $P \neq NP$. However, we have not attempted to cite the "best" polynomial time algorithms for problems, but merely to justify our claim that they can be solved in polynomial time. Whenever appropriate, we also mention provably intractable and undecidable generalizations and variants of the main problem. For main entries not known to be in NP, we provide any additional complexity information that is known, such as whether the problem is PSPACE-complete or whether the corresponding enumeration problem is #P-complete.

The list itself is organized according to subject matter. The entries are divided into the following twelve categories:

A1 Graph Theory
A2 Network Design
A3 Sets and Partitions
A4 Storage and Retrieval
A5 Sequencing and Scheduling
A6 Mathematical Programming
A7 Algebra and Number Theory
A8 Games and Puzzles
A9 Logic
A10 Automata and Languages
A11 Program Optimization
A12 Miscellaneous

The problems within each category are numbered individually, each number being preceded by a two-letter abbreviation derived from the section heading (such as GT7 for the seventh problem under Graph Theory). A thirteenth section is devoted to a small collection of open problems, selected primarily on the basis of our own judgment of their importance and difficulty. Additional open problems can be found scattered throughout the comments sections in the main list, and still more should be apparent from obvious gaps in the list.

Many of the problems in the list could have been placed in any one of a number of categories, so the reader is advised to make a general perusal of the list in order to get some idea of where particular types of problems can be found. One should also keep in mind that many results are mentioned only in the comments sections for other problems, so these should be examined carefully. We have attempted to provide helpful "navigational" information in the index.

Finally, it is appropriate that we make some mention of the standards and methods used in compiling this list. As with all such compilations, the collection is somewhat idiosyncratic, but we have attempted to be moderately thorough. In addition to plundering the published literature, we have sought, by word of mouth and by advertisements in appropriate periodicals and at relevant meetings, to make it widely known that we were compiling such a list and were interested in receiving any relevant unpublished results. However, no result has been included on the basis of an unsupported claim. Those that have not been published are backed up in our files by a manuscript or a proof sketch, checked either by ourselves or by other "trusted experts." On occasion we have verified the result, rather than the proof, by coming up with our own alternative proof (this is often easier than checking someone else's proof).

A certain amount of selectivity has been exercised in deciding what to exclude (although we may still be accused of including some rather obscure problems). For the most part, we have ignored problems with hopelessly convoluted descriptions unless they seem likely to be of interest to the relevant experts, although we usually do include pointers to the references where such results are presented. We also have felt free to exclude problems whose NP-completeness follows more or less immediately from that of other problems on the list, unless the problem is of substantial independent interest or has subproblems that we would like to comment on. Thus the list should not be regarded as an encyclopedia containing all known NP-completeness results. Rather, it is part annotated bibliography, serving as an access point to the literature on NP-completeness, and part data base, providing a large collection of known NP-complete problems that can be used for proving other problems NP-complete.

A1 GRAPH THEORY

A1.1 COVERING AND PARTITIONING

[GT1] VERTEX COVER

INSTANCE: Graph $G=(V,E)$, positive integer $K \leqslant |V|$.
QUESTION: Is there a vertex cover of size K or less for G, i.e., a subset $V' \subseteq V$ with $|V'| \leqslant K$ such that for each edge $\{u,v\} \in E$ at least one of u and v belongs to V'?

Reference: [Karp, 1972]. Transformation from 3SAT (see Chapter 3).
Comment: Equivalent complexity to INDEPENDENT SET with respect to restrictions on G. Variation in which the subgraph induced by V' is required to be connected is also NP-complete, even for planar graphs with no vertex degree exceeding 4 [Garey and Johnson, 1977a]. Easily solved in polynomial time if V' is required to be both a vertex cover and an independent set for G. The related EDGE COVER problem, in which one wants the smallest set $E' \subseteq E$ such that every $v \in V$ belongs to at least one $e \in E'$, can be solved in polynomial time by graph matching (e.g., see [Lawler, 1976a]).

[GT2] DOMINATING SET

INSTANCE: Graph $G=(V,E)$, positive integer $K \leqslant |V|$.
QUESTION: Is there a dominating set of size K or less for G, i.e., a subset $V' \subseteq V$ with $|V'| \leqslant K$ such that for all $u \in V-V'$ there is a $v \in V'$ for which $\{u,v\} \in E$?

Reference: Transformation from VERTEX COVER.
Comment: Remains NP-complete for planar graphs with maximum vertex degree 3 and planar graphs that are regular of degree 4 [Garey and Johnson, ——]. Variation in which the subgraph induced by V' is required to be connected is also NP-complete, even for planar graphs that are regular of degree 4 [Garey and Johnson, ——]. Also NP-complete if V' is required to be both a dominating set and an independent set. Solvable in polynomial time for trees [Cockayne, Goodman, and Hedetniemi, 1975]. The related EDGE DOMINATING SET problem, where we ask for a set $E' \subseteq E$ of K or fewer edges such that every edge in E shares at least one endpoint with some edge in E', is NP-complete, even for planar or bipartite graphs of maximum degree 3, but can be solved in polynomial time for trees [Yannakakis and Gavril, 1978], [Mitchell and Hedetniemi, 1977].

[GT3] DOMATIC NUMBER

INSTANCE: Graph $G=(V,E)$, positive integer $K \leqslant |V|$.
QUESTION: Is the domatic number of G at least K, i.e., can V be partitioned into $k \geqslant K$ disjoint sets $V_1, V_2, \ldots, V_k$ such that each V_i is a dominating set for G?

Reference: [Garey, Johnson, and Tarjan, 1976b]. Transformation from 3SAT. The problem is discussed in [Cockayne and Hedetniemi, 1975].
Comment: Remains NP-complete for any fixed $K \geqslant 3$. (The domatic number is always at least 2 unless G contains an isolated vertex.)

[GT4] GRAPH K-COLORABILITY (CHROMATIC NUMBER)

INSTANCE: Graph $G=(V,E)$, positive integer $K \leqslant |V|$.

QUESTION: Is G K-colorable, i.e., does there exist a function $f: V \to \{1,2, \ldots, K\}$ such that $f(u) \neq f(v)$ whenever $\{u,v\} \in E$?

Reference: [Karp, 1972]. Transformation from 3SAT.

Comment: Solvable in polynomial time for $K=2$, but remains NP-complete for all fixed $K \geqslant 3$ and, for $K=3$, for planar graphs having no vertex degree exceeding 4 [Garey, Johnson, and Stockmeyer, 1976]. Also remains NP-complete for $K=3$ if G is an intersection graph for straight line segments in the plane [Ehrlich, Even, and Tarjan, 1976]. For arbitrary K, the problem is NP-complete for circle graphs and circular arc graphs (even given their representation as families of arcs), although for circular arc graphs the problem is solvable in polynomial time for any fixed K (given their representation) [Garey, Johnson, Miller, and Papadimitriou, 1978]. The general problem can be solved in polynomial time for comparability graphs [Even, Pnueli, and Lempel, 1972], for chordal graphs [Gavril, 1972], for (3,1) graphs [Walsh and Burkhard, 1977], and for graphs having no vertex degree exceeding 3 [Brooks, 1941].

[GT5] ACHROMATIC NUMBER

INSTANCE: Graph $G=(V,E)$, positive integer $K \leqslant |V|$.

QUESTION: Does G have achromatic number K or greater, i.e., is there a partition of V into disjoint sets $V_1, V_2, \ldots, V_k$, $k \geqslant K$, such that each V_i is an independent set for G (no two vertices in V_i are joined by an edge in E) and such that, for each pair of distinct sets V_i, V_j, $V_i \cup V_j$ is *not* an independent set for G?

Reference: [Yannakakis and Gavril, 1978]. Transformation from MINIMUM MAXIMAL MATCHING.

Comment: Remains NP-complete even if G is the complement of a bipartite graph and hence has no independent set of more than two vertices.

[GT6] MONOCHROMATIC TRIANGLE

INSTANCE: Graph $G=(V,E)$.

QUESTION: Is there a partition of E into two disjoint sets E_1, E_2 such that neither $G_1=(V,E_1)$ nor $G_2=(V,E_2)$ contains a triangle?

Reference: [Burr, 1976]. Transformation from 3SAT.

Comment: Variants in which "triangle" is replaced by any larger fixed complete graph are also NP-complete [Burr, 1976]. Variants in which "triangle" is replaced by "k-star" (a single degree k vertex adjacent to k degree one vertices) is solvable in polynomial time [Burr, Erdös, and Lovasz, 1976].

[GT7] FEEDBACK VERTEX SET

INSTANCE: Directed graph $G=(V,A)$, positive integer $K \leqslant |V|$.

QUESTION: Is there a subset $V' \subseteq V$ with $|V'| \leqslant K$ such that V' contains at least one vertex from every directed cycle in G?

Reference: [Karp, 1972]. Transformation from VERTEX COVER.

Comment: Remains NP-complete for digraphs having no in- or out-degree exceeding 2, for planar digraphs with no in- or out-degree exceeding 3 [Garey and

Johnson, ——], and for edge digraphs [Gavril, 1977a], but can be solved in polynomial time for reducible graphs [Shamir, 1977]. The corresponding problem for undirected graphs is also NP-complete.

[GT8] FEEDBACK ARC SET

INSTANCE: Directed graph $G=(V,A)$, positive integer $K \leqslant |A|$.
QUESTION: Is there a subset $A' \subseteq A$ with $|A'| \leqslant K$ such that A' contains at least one arc from every directed cycle in G?

Reference: [Karp, 1972]. Transformation from VERTEX COVER.
Comment: Remains NP-complete for digraphs in which no vertex has total indegree and out-degree more than 3, and for edge digraphs [Gavril, 1977a]. Solvable in polynomial time for planar digraphs [Luchesi, 1976]. The corresponding problem for undirected graphs is trivially solvable in polynomial time.

[GT9] PARTIAL FEEDBACK EDGE SET

INSTANCE: Graph $G=(V,E)$, positive integers $K \leqslant |E|$ and $L \leqslant |V|$.
QUESTION: Is there a subset $E' \subseteq E$ with $|E'| \leqslant K$ such that E' contains at least one edge from every circuit of length L or less in G?

Reference: [Yannakakis, 1978b]. Transformation from VERTEX COVER.
Comment: Remains NP-complete for any fixed $L \geqslant 3$ and for bipartite graphs (with fixed $L \geqslant 4$). However, if $L=|V|$, i.e., if we ask that E' contain an edge from *every* cycle in G, then the problem is trivially solvable in polynomial time.

[GT10] MINIMUM MAXIMAL MATCHING

INSTANCE: Graph $G=(V,E)$, positive integer $K \leqslant |E|$.
QUESTION: Is there a subset $E' \subseteq E$ with $|E'| \leqslant K$ such that E' is a maximal matching, i.e., no two edges in E' share a common endpoint and every edge in $E-E'$ shares a common endpoint with some edge in E'?

Reference: [Yannakakis and Gavril, 1978]. Transformation from VERTEX COVER for cubic graphs.
Comment: Remains NP-complete for planar graphs and for bipartite graphs, in both cases even if no vertex degree exceeds 3. The problem of finding a maximum "maximal matching" is just the usual graph matching problem and is solvable in polynomial time (e.g., see [Lawler, 1976a]).

[GT11] PARTITION INTO TRIANGLES

INSTANCE: Graph $G=(V,E)$, with $|V|=3q$ for some integer q.
QUESTION: Can the vertices of G be partitioned into q disjoint sets $V_1, V_2, \ldots, V_q$, each containing exactly 3 vertices, such that for each $V_i=\{u_i,v_i,w_i\}$, $1 \leqslant i \leqslant q$, all three of the edges $\{u_i,v_i\}$, $\{u_i,w_i\}$, and $\{v_i,w_i\}$ belong to E?

Reference: [Schaefer, 1974]. Transformation from 3DM (see Chapter 3).
Comment: See next problem for a generalization.

[GT12] PARTITION INTO ISOMORPHIC SUBGRAPHS

INSTANCE: Graphs $G=(V,E)$ and $H=(V',E')$ with $|V|=q|V'|$ for some $q \in Z^+$.

QUESTION: Can the vertices of G be partitioned into q disjoint sets $V_1, V_2, \ldots, V_q$ such that, for $1 \leqslant i \leqslant q$, the subgraph of G induced by V_i is isomorphic to H?

Reference: [Kirkpatrick and Hell, 1978]. Transformation from 3DM.

Comment: Remains NP-complete for any fixed H that contains at least 3 vertices. The analogous problem in which the subgraph induced by V_i need only have the same number of vertices as H and contain a subgraph isomorphic to H is also NP-complete, for any fixed H that contains a connected component of three or more vertices. Both problems can be solved in polynomial time (by matching) for any H not meeting the stated restrictions.

[GT13] PARTITION INTO HAMILTONIAN SUBGRAPHS

INSTANCE: Directed graph $G=(V,A)$.

QUESTION: Can the vertices of G be partitioned into disjoint sets $V_1, V_2, \ldots, V_k$, for some k, such that each V_i contains at least three vertices and induces a subgraph of G that contains a Hamiltonian circuit?

Reference: [Valiant, 1977a]. Transformation from 3SAT. (See also [Herrmann, 1973]).

Comment: Solvable in polynomial time by matching techniques if each V_i need only contain at least 2 vertices [Edmonds and Johnson, 1970]. The analogous problem for undirected graphs can be similarly solved, even with the requirement that $|V_i| \geqslant 3$. However, it becomes NP-complete if we require that $|V_i| \geqslant 6$ [Papadimitriou, 1978d] or if the instance includes an upper bound K on k.

[GT14] PARTITION INTO FORESTS

INSTANCE: Graph $G=(V,E)$, positive integer $K \leqslant |V|$.

QUESTION: Can the vertices of G be partitioned into $k \leqslant K$ disjoint sets $V_1, V_2, \ldots, V_k$ such that, for $1 \leqslant i \leqslant k$, the subgraph induced by V_i contains no circuits?

Reference: [Garey and Johnson, ——]. Transformation from GRAPH 3-COLORABILITY.

[GT15] PARTITION INTO CLIQUES

INSTANCE: Graph $G=(V,E)$, positive integer $K \leqslant |V|$.

QUESTION: Can the vertices of G be partitioned into $k \leqslant K$ disjoint sets $V_1, V_2, \ldots, V_k$ such that, for $1 \leqslant i \leqslant k$, the subgraph induced by V_i is a complete graph?

Reference: [Karp, 1972] (there called CLIQUE COVER). Transformation from GRAPH K-COLORABILITY.

Comment: Remains NP-complete for edge graphs [Arjomandi, 1977], for graphs containing no complete subgraphs on 4 vertices (see construction for PARTITION INTO TRIANGLES in Chapter 3), and for all fixed $K \geqslant 3$. Solvable in polynomial time for $K \leqslant 2$, for graphs containing no complete subgraphs on 3 vertices (by matching), for circular arc graphs (given their representations as families of arcs) [Gavril, 1974a], for chordal graphs [Gavril, 1972], and for comparability graphs [Golumbic, 1977].

[GT16] PARTITION INTO PERFECT MATCHINGS

INSTANCE: Graph $G=(V,E)$, positive integer $K \leqslant |V|$.

QUESTION: Can the vertices of G be partitioned into $k \leqslant K$ disjoints sets $V_1, V_2, \ldots, V_k$ such that, for $1 \leqslant i \leqslant k$, the subgraph induced by V_i is a perfect matching (consists entirely of vertices with degree one)?

Reference: [Schaefer, 1978b]. Transformation from NOT-ALL-EQUAL 3SAT.
Comment: Remains NP-complete for $K=2$ and for planar cubic graphs.

[GT17] COVERING BY CLIQUES

INSTANCE: Graph $G=(V,E)$, positive integer $K \leqslant |E|$.

QUESTION: Are there $k \leqslant K$ subsets $V_1, V_2, \ldots, V_k$ of V such that each V_i induces a complete subgraph of G and such that for each edge $\{u,v\} \in E$ there is some V_i that contains both u and v?

Reference: [Kou, Stockmeyer, and Wong, 1978], [Orlin, 1976]. Transformation from PARTITION INTO CLIQUES.

[GT18] COVERING BY COMPLETE BIPARTITE SUBGRAPHS

INSTANCE: Bipartite graph $G=(V,E)$, positive integer $K \leqslant |E|$.

QUESTION: Are there $k \leqslant K$ subsets $V_1, V_2, \ldots, V_k$ of V such that each V_i induces a complete bipartite subgraph of G and such that for each edge $\{u,v\} \in E$ there is some V_i that contains both u and v?

Reference: [Orlin, 1976]. Transformation from PARTITION INTO CLIQUES.

A1.2 SUBGRAPHS AND SUPERGRAPHS

[GT19] CLIQUE

INSTANCE: Graph $G=(V,E)$, positive integer $K \leqslant |V|$.

QUESTION: Does G contain a clique of size K or more, i.e., a subset $V' \subseteq V$ with $|V'| \geqslant K$ such that every two vertices in V' are joined by an edge in E?

Reference: [Karp, 1972]. Transformation from VERTEX COVER (see Chapter 3).
Comment: Solvable in polynomial time for graphs obeying any fixed degree bound d, for planar graphs, for edge graphs, for chordal graphs [Gavril, 1972], for comparability graphs [Even, Pnueli, and Lempel, 1972], for circle graphs [Gavril, 1973], and for circular arc graphs (given their representation as families of arcs) [Gavril, 1974a]. The variant in which, for a given r, $0<r<1$, we are asked whether G contains a clique of size $r|V|$ or more is NP-complete for any fixed value of r.

[GT20] INDEPENDENT SET

INSTANCE: Graph $G=(V,E)$, positive integer $K \leqslant |V|$.

QUESTION: Does G contain an independent set of size K or more, i.e., a subset

$V' \subseteq V$ such that $|V'| \geqslant K$ and such that no two vertices in V' are joined by an edge in E?

Reference: Transformation from VERTEX COVER (see Chapter 3).
Comment: Remains NP-complete for cubic planar graphs [Garey, Johnson, and Stockmeyer, 1976], [Garey and Johnson, 1977a], [Maier and Storer, 1977], for edge graphs of directed graphs [Gavril, 1977a], for total graphs of bipartite graphs [Yannakakis and Gavril, 1978], and for graphs containing no triangles [Poljak, 1974]. Solvable in polynomial time for bipartite graphs (by matching, e.g., see [Harary, 1969]), for edge graphs (by matching), for graphs with no vertex degree exceeding 2, for chordal graphs [Gavril, 1972], for circle graphs [Gavril, 1973], for circular arc graphs (given their representation as families of arcs) [Gavril, 1974a], for comparability graphs [Golumbic, 1977}, and for claw-free graphs [Minty, 1977].

[GT21] **INDUCED SUBGRAPH WITH PROPERTY Π (*)**
INSTANCE: Graph $G=(V,E)$, positive integer $K \leqslant |V|$.
QUESTION: Is there a subset $V' \subseteq V$ with $|V'| \geqslant K$ such that the subgraph of G induced by V' has property Π (see comments for possible choices for Π)?

Reference: [Yannakakis, 1978a], [Yannakakis, 1978b], [Lewis, 1978]. Transformation from 3SAT.
Comment: NP-hard for any property Π that holds for arbitrarily large graphs, does not hold for all graphs, and is "hereditary," i.e., holds for all induced subgraphs of G whenever it holds for G. If in addition one can determine in polynomial time whether Π holds for a graph, then the problem is NP-complete. Examples of such properties Π include "G is a clique," "G is an independent set," "G is planar," "G is bipartite," "G is outerplanar," "G is an edge graph," "G is chordal," "G is a comparability graph," and "G is a forest." The same general results hold if G is restricted to planar graphs and Π satisfies the above constraints for planar graphs, or if G is restricted to acyclic directed graphs and Π satisfies the above constraints for such graphs. A weaker result holds when G is restricted to bipartite graphs [Yannakakis, 1978b].

[GT22] **INDUCED CONNECTED SUBGRAPH WITH PROPERTY Π (*)**
INSTANCE: Graph $G=(V,E)$, positive integer $K \leqslant |V|$.
QUESTION: Is there a subset $V' \subseteq V$ with $|V'| \geqslant K$ such that the subgraph of G induced by V' is connected and has property Π (see comments for possible choices for Π)?

Reference: [Yannakakis, 1978b]. Transformation from 3SAT.
Comment: NP-hard for any hereditary property that holds for arbitrarily large connected graphs but not for all connected graphs. If, in addition, one can determine in polynomial time whether Π holds for a graph, then the problem is NP-complete. Examples include all the properties mentioned for the preceding problem except "G is an independent set". The related question "Is the maximum induced subgraph of G having property Π also connected?" is not in NP or co-NP unless NP = co-NP [Yannakakis, 1978b].

[GT23] INDUCED PATH

INSTANCE: Graph $G=(V,E)$, positive integer $K \leqslant |V|$.

QUESTION: Is there a subset $V' \subseteq V$ with $|V'| \geqslant K$ such that the subgraph induced by V' is a simple path on $|V'|$ vertices?

Reference: [Yannakakis, 1978c]. Transformation from 3SAT.

Comment: Note that this is not a hereditary property, so the result is not implied by either of the previous two results. Remains NP-complete if G is bipartite. The same result holds for the variant in which "simple path" is replaced by "simple cycle." The problems of finding the longest simple path or longest simple cycle (not necessarily induced) are also NP-complete.

[GT24] BALANCED COMPLETE BIPARTITE SUBGRAPH

INSTANCE: Bipartite graph $G=(V,E)$, positive integer $K \leqslant |V|$.

QUESTION: Are there two disjoint subsets $V_1, V_2 \subseteq V$ such that $|V_1|=|V_2|=K$ and such that $u \in V_1$, $v \in V_2$ implies that $\{u,v\} \in E$?

Reference: [Garey and Johnson, ——]. Transformation from CLIQUE.

Comment: The related problem in which the requirement "$|V_1|=|V_2|=K$" is replaced by "$|V_1|+|V_2|=K$" is solvable in polynomial time for bipartite graphs (because of the connection between matchings and independent sets in such graphs, e.g., see [Harary, 1969]), but is NP-complete for general graphs [Yannakakis, 1978b].

[GT25] BIPARTITE SUBGRAPH

INSTANCE: Graph $G=(V,E)$, positive integer $K \leqslant |E|$.

QUESTION: Is there a subset $E' \subseteq E$ with $|E'| \geqslant K$ such that $G'=(V,E')$ is bipartite?

Reference: [Garey, Johnson, and Stockmeyer, 1976]. Transformation from MAXIMUM 2-SATISFIABILITY.

Comment: Remains NP-complete for graphs with no vertex degree exceeding 3 and no triangles and/or if we require that the subgraph be connected [Yannakakis, 1978b]. Solvable in polynomial time if G is planar [Hadlock, 1975], [Orlova and Dorfman, 1972], or if $K=|E|$.

[GT26] DEGREE-BOUNDED CONNECTED SUBGRAPH

INSTANCE: Graph $G=(V,E)$, non-negative integer $d \leqslant |V|$, positive integer $K \leqslant |E|$.

QUESTION: Is there a subset $E' \subseteq E$ with $|E'| \geqslant K$ such that the subgraph $G'=(V,E')$ is connected and has no vertex with degree exceeding d?

Reference: [Yannakakis, 1978b]. Transformation from HAMILTONIAN PATH.

Comment: Remains NP-complete for any fixed $d \geqslant 2$. Solvable in polynomial time if G' is not required to be connected (by matching techniques, see [Edmonds and Johnson, 1970]). The corresponding induced subgraph problem, where we ask for a subset $V' \subseteq V$ with $|V'| \geqslant K$ such that the subgraph of G induced by V' has no vertex with degree exceeding d, is NP-complete for any fixed $d \geqslant 0$ [Lewis, 1976] and for any fixed $d \geqslant 2$ if we require that G' be connected [Yannakakis, 1978b].

[GT27] PLANAR SUBGRAPH

INSTANCE: Graph $G=(V,E)$, positive integer $K \leqslant |E|$.
QUESTION: Is there a subset $E' \subseteq E$ with $|E'| \geqslant K$ such that $G'=(V,E')$ is planar?

Reference: [Liu and Geldmacher, 1978]. Transformation from HAMILTONIAN PATH restricted to bipartite graphs.
Comment: Corresponding problem in which G' is the subgraph induced by a set V' of at least K vertices is also NP-complete [Krishnamoorthy and Deo, 1977a], [Yannakakis, 1978b]. The former can be solved in polynomial time when $K=|E|$, and the latter when $K=|V|$, since planarity testing can be done in polynomial time (e.g., see [Hopcroft and Tarjan, 1974]). The related problem in which we ask if G contains a connected "outerplanar" subgraph with K or more edges is also NP-complete [Yannakakis, 1978b].

[GT28] EDGE-SUBGRAPH

INSTANCE: Graph $G=(V,E)$, positive integer $K \leqslant |E|$.
QUESTION: Is there a subset $E' \subseteq E$ with $|E'| \geqslant K$ such that the subgraph $G'=(V,E')$ is an edge graph, i.e., there exists a graph $H=(U,F)$ such that G' is isomorphic to the graph having vertex set F and edge set consisting of all pairs $\{e,f\}$ such that the edges e and f share a common endpoint in H?

Reference: [Yannakakis, 1978b]. Transformation from 3SAT.
Comment: Remains NP-complete even if G has no vertex with degree exceeding 4. If we require that the subgraph be connected, the degree bound for NP-completeness can be reduced to 3. Edge graphs can be recognized in polynomial time, e.g., see [Harary, 1969] (under the term "line graphs").

[GT29] TRANSITIVE SUBGRAPH

INSTANCE: Directed graph $G=(V,A)$, positive integer $K \leqslant |A|$.
QUESTION: Is there a subset $A' \subseteq A$ with $|A'| \geqslant K$ such that $G'=(V,A')$ is transitive, i.e., for all pairs $u,v \in V$, if there exists a $w \in V$ for which $(u,w),(w,v) \in A'$, then $(u,v) \in A'$?

Reference: [Yannakakis, 1978b] Transformation from BIPARTITE SUBGRAPH with no triangles.
Comment: The variant in which G is undirected and we ask for a subgraph that is a "comparability graph," i.e., can be made into a transitive digraph by directing each of its edges in one of the two possible directions, is also NP-complete, even if G has no vertex with degree exceeding 3. For both problems, the variant in which we require the subgraph to be connected is also NP-complete.

[GT30] UNICONNECTED SUBGRAPH

INSTANCE: Directed graph $G=(V,A)$, positive integer $K \leqslant |A|$.
QUESTION: Is there a subset $A' \subseteq A$ with $|A'| \geqslant K$ such that $G'=(V,A')$ has at most one directed path between any pair of vertices?

Reference: [Maheshwari, 1976]. Transformation from VERTEX COVER.
Comment: Remains NP-complete for acyclic directed graphs.

[GT31] MINIMUM K-CONNECTED SUBGRAPH

INSTANCE: Graph $G=(V,E)$, positive integers $K \leqslant |V|$ and $B \leqslant |E|$.
QUESTION: Is there a subset $E' \subseteq E$ with $|E'| \leqslant B$ such that $G'=(V,E')$ is K-connected, i.e., cannot be disconnected by removing fewer than K vertices?

Reference: [Chung and Graham, 1977]. Transformation from HAMILTONIAN CIRCUIT.
Comment: Corresponding edge-connectivity problem is also NP-complete. Both problems remain NP-complete for any fixed $K \geqslant 2$ and can be solved trivially in polynomial time for $K=1$.

[GT32] CUBIC SUBGRAPH

INSTANCE: Graph $G=(V,E)$.
QUESTION: Is there a nonempty subset $E' \subseteq E$ such that in the graph $G'=(V,E')$ every vertex has either degree 3 or degree 0?

Reference: [Chvátal, 1976]. Transformation from GRAPH 3-COLORABILITY.

[GT33] MINIMUM EQUIVALENT DIGRAPH

INSTANCE: Directed graph $G=(V,A)$, positive integer $K \leqslant |A|$.
QUESTION: Is there a subset $A' \subseteq A$ with $|A'| \leqslant K$ such that, for every ordered pair of vertices $u,v \in V$, the graph $G'=(V,A')$ contains a directed path from u to v if and only if G does?

Reference: [Sahni, 1974]. Transformation from DIRECTED HAMILTONIAN CIRCUIT for strongly connected graphs (see Chapter 3).
Comment: Corresponding problem in which $A' \subseteq V \times V$ instead of $A' \subseteq A$ (called TRANSITIVE REDUCTION) can be solved in polynomial time, e.g., see [Aho, Garey, and Ullman, 1972].

[GT34] HAMILTONIAN COMPLETION

INSTANCE: Graph $G=(V,E)$, non-negative integer $K \leqslant |V|$.
QUESTION: Is there a superset E' containing E such that $|E'-E| \leqslant K$ and the graph $G'=(V,E')$ has a Hamiltonian circuit?

Reference: Transformation from HAMILTONIAN CIRCUIT.
Comment: Remains NP-complete for any fixed $K \geqslant 0$. Corresponding "completion" versions of HAMILTONIAN PATH, DIRECTED HAMILTONIAN PATH, and DIRECTED HAMILTONIAN CIRCUIT are also NP-complete. HAMILTONIAN COMPLETION and HAMILTONIAN PATH COMPLETION can be solved in polynomial time if G is a tree [Boesch, Chen, and McHugh, 1974].

[GT35] INTERVAL GRAPH COMPLETION

INSTANCE: Graph $G=(V,E)$, non-negative integer K.
QUESTION: Is there a superset E' containing E such that $|E'-E| \leqslant K$ and the graph $G'=(V,E')$ is an interval graph?

Reference: [Garey, Gavril, and Johnson, 1977]. Transformation from OPTIMAL LINEAR ARRANGEMENT.
Comment: Remains NP-complete when G is restricted to be an edge graph. Solv-

able in polynomial time for $K=0$ [Fulkerson and Gross, 1965],[Booth and Lueker, 1976].

[GT36] **PATH GRAPH COMPLETION**

INSTANCE: Graph $G=(V,E)$, non-negative integer K.

QUESTION: Is there a superset E' containing E such that $|E'-E| \leqslant K$ and the graph $G'=(V,E')$ is the intersection graph of a family of paths on an undirected tree?

Reference: [Gavril, 1977b]. Transformation from INTERVAL GRAPH COMPLETION.

Comment: Corresponding problem in which G' must be the intersection graph of a family of directed paths on an oriented tree (i.e., rooted, with all arcs directed away from the root) is also NP-complete.

A1.3 VERTEX ORDERING

[GT37] **HAMILTONIAN CIRCUIT**

INSTANCE: Graph $G=(V,E)$.

QUESTION: Does G contain a Hamiltonian circuit?

Reference: [Karp, 1972]. Transformation from VERTEX COVER (see Chapter 3).

Comment: Remains NP-complete (1) if G is planar, cubic, 3-connected, and has no face with fewer than 5 edges [Garey, Johnson, and Tarjan, 1976a], (2) if G is bipartite [Krishnamoorthy, 1975], (3) if G is the square of a graph [Chvátal, 1976], and (4) if a Hamiltonian path for G is given as part of the instance [Papadimitriou and Stieglitz, 1976]. Solvable in polynomial time if G has no vertex with degree exceeding 2 or if G is an edge graph (e.g., see [Liu, 1968]). The cube of a nontrivial connected graph always has a Hamiltonian circuit [Karaganis, 1968].

[GT38] **DIRECTED HAMILTONIAN CIRCUIT**

INSTANCE: Directed graph $G=(V,A)$.

QUESTION: Does G contain a directed Hamiltonian circuit?

Reference: [Karp, 1972]. Transformation from VERTEX COVER (see Chapter 3).

Comment: Remains NP-complete if G is planar and has no vertex involved in more than three arcs [Plesnik, 1978]. Solvable in polynomial time if no in-degree (no out-degree) exceeds 1, if G is a tournament [Morrow and Goodman, 1976], or if G is an edge digraph (e.g., see [Liu, 1968]).

[GT39] **HAMILTONIAN PATH**

INSTANCE: Graph $G=(V,E)$.

QUESTION: Does G contain a Hamiltonian path?

Reference: Transformation from VERTEX COVER (see Chapter 3).

Comment: Remains NP-complete under restrictions (1) and (2) for HAMILTONIAN CIRCUIT and is polynomially solvable under the same restrictions as HC. Corresponding DIRECTED HAMILTONIAN PATH problem is also NP-complete, and the comments for DIRECTED HC apply to it as well. The variants in which either the starting point or the ending point or both are specified in the instance are also NP-complete. DIRECTED HAMILTONIAN PATH can be solved in polynomial time for acyclic digraphs, e.g., see [Lawler, 1976a].

[GT40] **BANDWIDTH**

INSTANCE: Graph $G=(V,E)$, positive integer $K \leqslant |V|$.

QUESTION: Is there a linear ordering of V with bandwidth K or less, i.e., a one-to-one function $f: V \rightarrow \{1,2,\ldots,|V|\}$ such that, for all $\{u,v\} \in E$, $|f(u)-f(v)| \leqslant K$?

Reference: [Papadimitriou, 1976a]. Transformation from 3-PARTITION.

Comment: Remains NP-complete for trees with no vertex degree exceeding 3 [Garey, Graham, Johnson, and Knuth, 1978]. This problem corresponds to that of minimizing the "bandwidth" of a symmetric matrix by simultaneous row and column permutations.

[GT41] **DIRECTED BANDWIDTH**

INSTANCE: Directed graph $G=(V,A)$, positive integer $K \leqslant |V|$.

QUESTION: Is there a one-to-one function $f: V \rightarrow \{1,2,\ldots,|V|\}$ such that, for all $(u,v) \in A$, $f(u) < f(v)$ and $(f(v)-f(u)) \leqslant K$?

Reference: [Garey, Graham, Johnson, and Knuth, 1978]. Transformation from 3-PARTITION.

Comment: Remains NP-complete for rooted directed trees with maximum in-degree 1 and maximum out-degree at most 2. This problem corresponds to that of minimizing the "bandwidth" of an upper triangular matrix by simultaneous row and column permutations.

[GT42] **OPTIMAL LINEAR ARRANGEMENT**

INSTANCE: Graph $G=(V,E)$, positive integer K.

QUESTION: Is there a one-to-one function $f: V \rightarrow \{1,2,\ldots,|V|\}$ such that $\sum_{\{u,v\} \in E} |f(u)-f(v)| \leqslant K$?

Reference: [Garey, Johnson, and Stockmeyer, 1976]. Transformation from SIMPLE MAX CUT.

Comment: Remains NP-complete if G is bipartite [Even and Shiloach, 1975]. Solvable in polynomial time if G is a tree [Shiloach, 1976], [Gol'dberg and Klipker, 1976].

[GT43] **DIRECTED OPTIMAL LINEAR ARRANGEMENT**

INSTANCE: Directed graph $G=(V,A)$, positive integer K.

QUESTION: Is there a one-to-one function $f: V \rightarrow \{1,2,\ldots,|V|\}$ such that $f(u) < f(v)$ whenever $(u,v) \in A$ and such that $\sum_{(u,v) \in A} (f(v)-f(u)) \leqslant K$?

Reference: [Even and Shiloach, 1975]. Transformation from OPTIMAL LINEAR ARRANGEMENT.

Comment: Solvable in polynomial time if G is a tree, even if each edge has a given integer weight and the cost function is a weighted sum [Adolphson and Hu, 1973].

[GT44] MINIMUM CUT LINEAR ARRANGEMENT

INSTANCE: Graph $G=(V,E)$, positive integer K.

QUESTION: Is there a one-to-one function $f: V \rightarrow \{1,2,\ldots,|V|\}$ such that for all i, $1<i<|V|$,

$$|\{\{u,v\} \in E: f(u) \leqslant i < f(v)\}| \leqslant K \ ?$$

Reference: [Stockmeyer, 1974b], [Gavril, 1977a]. Transformation from SIMPLE MAX CUT.

[GT45] ROOTED TREE ARRANGEMENT

INSTANCE: Graph $G=(V,E)$, positive integer K.

QUESTION: Is there a rooted tree $T=(U,F)$, with $|U|=|V|$, and a one-to-one function $f: V \rightarrow U$ such that for every edge $\{u,v\} \in E$ there is a simple path from the root that includes both $f(u)$ and $f(v)$ and such that if $d(x,y)$ is the number of edges on the path from x to y in T, then $\sum_{\{u,v\} \in E} d(f(u),f(v)) \leqslant K$?

Reference: [Gavril, 1977a]. Transformation from OPTIMAL LINEAR ARRANGEMENT.

[GT46] DIRECTED ELIMINATION ORDERING

INSTANCE: Directed graph $G=(V,A)$, non-negative integer K.

QUESTION: Is there an elimination ordering for G with fill-in K or less, i.e., a one-to-one function $f: V \rightarrow \{1,2,\ldots,|V|\}$ such that there are at most K pairs of vertices $(u,v) \in (V \times V)-A$ with the property that G contains a directed path from u to v that only passes through vertices w satisfying $f(w) < \min\{f(u),f(v)\}$?

Reference: [Rose and Tarjan, 1978]. Transformation from 3SAT.

Comment: Problem arises in performing Gaussian elimination on sparse matrices. Solvable in polynomial time for $K=0$. The analogous problem for undirected graphs (symmetric matrices) is equivalent to CHORDAL GRAPH COMPLETION and is open as to complexity.

[GT47] ELIMINATION DEGREE SEQUENCE

INSTANCE: Graph $G=(V,E)$, sequence $<d_1,d_2,\ldots,d_{|V|}>$ of non-negative integers not exceeding $|V|-1$.

QUESTION: Is there a one-to-one function $f: V \rightarrow \{1,2,\ldots,|V|\}$ such that, for $1 \leqslant i \leqslant |V|$, if $f(v)=i$ then there are exactly d_i vertices u such that $f(u)>i$ and $\{u,v\} \in E$?

Reference: [Garey, Johnson, and Papadimitriou, 1977]. Transformation from EXACT COVER BY 3-SETS.

Comment: The variant in which it is required that f be such that, for $1 \leqslant i \leqslant |V|$, if $f(v)=i$ then there are exactly d_i vertices u such that $\{u,v\} \in E$, is trivially solvable in polynomial time.

A1.4 ISO- AND OTHER MORPHISMS

[GT48] SUBGRAPH ISOMORPHISM

INSTANCE: Graphs $G=(V_1,E_1)$, $H=(V_2,E_2)$.
QUESTION: Does G contain a subgraph isomorphic to H, i.e., a subset $V \subseteq V_1$ and a subset $E \subseteq E_1$ such that $|V|=|V_2|$, $|E|=|E_2|$, and there exists a one-to-one function $f: V_2 \rightarrow V$ satisfying $\{u,v\} \in E_2$ if and only if $\{f(u),f(v)\} \in E$?

Reference: [Cook, 1971a]. Transformation from CLIQUE.
Comment: Contains CLIQUE, COMPLETE BIPARTITE SUBGRAPH, HAMILTONIAN CIRCUIT, etc., as special cases. Can be solved in polynomial time if G is a forest and H is a tree [Edmonds and Matula, 1975] (see also [Reyner, 1977]), but remains NP-complete if G is a tree and H is a forest (see Chapter 4) or if G is a graph and H is a tree (HAMILTONIAN PATH). Variant for directed graphs is also NP-complete, even if G is acyclic and H is a directed tree [Aho and Sethi, 1977], but can be solved in polynomial time if G is a directed forest and H is a directed tree [Reyner, 1977]. If $|V_1|=|V_2|$ and $|E_1|=|E_2|$ we have the GRAPH ISOMORPHISM problem, which is open for both directed and undirected graphs.

[GT49] LARGEST COMMON SUBGRAPH

INSTANCE: Graphs $G=(V_1,E_1)$, $H=(V_2,E_2)$, positive integer K.
QUESTION: Do there exist subsets $E_1' \subseteq E_1$ and $E_2' \subseteq E_2$ with $|E_1'|=|E_2'| \geqslant K$ such that the two subgraphs $G'=(V_1,E_1')$ and $H'=(V_2,E_2')$ are isomorphic?

Reference: Transformation from CLIQUE.
Comment: Can be solved in polynomial time if both G and H are trees [Edmonds and Matula, 1975].

[GT50] MAXIMUM SUBGRAPH MATCHING

INSTANCE: Directed graphs $G=(V_1,A_1)$, $H=(V_2,A_2)$, positive integer K.
QUESTION: Is there a subset $R \subseteq V_1 \times V_2$ with $|R| \geqslant K$ such that, for all $<u,u'>,<v,v'> \in R$, $(u,v) \in A_1$ if and only if $(u',v') \in A_2$?

Reference: [Garey and Johnson, ——]. Transformation from CLIQUE. Problem is discussed in [Barrow and Burstall, 1976].

[GT51] GRAPH CONTRACTABILITY

INSTANCE: Graphs $G=(V_1,E_1)$, $H=(V_2,E_2)$.
QUESTION: Can a graph isomorphic to H be obtained from G by a sequence of edge contractions, i.e., a sequence in which each step replaces two adjacent vertices u,v by a single vertex w adjacent to exactly those vertices that were previously adjacent to at least one of u and v?

Reference: [Statman, 1976]. Transformation from 3SAT.
Comment: Can be solved in polynomial time if H is a triangle.

[GT52] GRAPH HOMOMORPHISM

INSTANCE: Graphs $G=(V_1,E_1)$, $H=(V_2,E_2)$.
QUESTION: Can a graph isomorphic to H be obtained from G by a sequence of

identifications of non-adjacent vertices, i.e., a sequence in which each step replaces two non-adjacent vertices u,v by a single vertex w adjacent to exactly those vertices that were preciously adjacent to at least one of u and v?

Reference: [Levin, 1973]. Transformation from GRAPH K-COLORABILITY.
Comment: Remains NP-complete for H fixed to be a triangle, but can be solved in polynomial time if H is just a single edge.

[GT53] **DIGRAPH D-MORPHISM**

INSTANCE: Directed graphs $G=(V_1,A_1)$, $H=(V_2,A_2)$.
QUESTION: Is there a D-morphism from G to H, i.e., a function $f\colon V_1 \rightarrow V_2$ such that for all $(u,v) \in A_1$ either $(f(u),f(v)) \in A_2$ or $(f(v),f(u)) \in A_2$ and such that for all $u \in V_1$ and $v' \in V_2$ if $(f(u),v') \in A_2$ then there exists a $v \in f^{-1}(v')$ for which $(u,v) \in A_1$?

Reference: [Fraenkel and Yesha, 1977]. Transformation from GRAPH GRUNDY NUMBERING.

A1.5 MISCELLANEOUS

[GT54] **PATH WITH FORBIDDEN PAIRS**

INSTANCE: Directed graph $G=(V,A)$, specified vertices $s,t \in V$, collection $C=\{(a_1,b_1), \ldots, (a_n,b_n)\}$ of pairs of vertices from V.
QUESTION: Is there a directed path from s to t in G that contains at most one vertex from each pair in C?

Reference: [Gabow, Maheshwari, and Osterweil, 1976]. Transformation from 3SAT.

Comment: Remains NP-complete even if G is acyclic with no in- or out-degree exceeding 2. Variant in which the "forbidden pairs" are arcs instead of vertices is also NP-complete under the same restrictions. Both problems remain NP-complete even if all the given pairs are required to be disjoint.

[GT55] **MULTIPLE CHOICE MATCHING**

INSTANCE: Graph $G=(V,E)$, partition of E into disjoint sets $E_1,E_2, \ldots, E_J$, positive integer K.
QUESTION: Is there a subset $E' \subseteq E$ with $|E'| \geqslant K$ such that no two edges in E' share a common vertex and such that E' contains at most one edge from each E_i, $1 \leqslant i \leqslant J$?

Reference: [Valiant, 1977c], [Itai and Rodeh, 1977a], [Itai, Rodeh, and Tanimota, 1978]. Transformation from 3SAT.
Comment: Remains NP-complete even if G is bipartite, each E_i contains at most 2 edges, and $K=|V|/2$. If each E_i contains only a single edge, this becomes the ordinary graph matching problem and is solvable in polynomial time.

[GT56] **GRAPH GRUNDY NUMBERING**

INSTANCE: Directed graph $G=(V,A)$.
QUESTION: Is there a function $f\colon V \rightarrow Z^+$ such that, for each $v \in V$, $f(v)$ is the

least non-negative integer not contained in the set $\{f(u): u \in V, (v,u) \in A\}$?

Reference: [van Leeuwen, 1976a]. Transformation from 3SAT.

Comment: Remains NP-complete when restricted to planar graphs in which no vertex has in- or out-degree exceeding 5 [Fraenkel and Yesha, 1977].

[GT57] **KERNEL**

INSTANCE: Directed graph $G=(V,A)$.

QUESTION: Does G have a kernel, i.e., a subset $V' \subseteq V$ such that no two vertices in V' are joined by an arc in A and such that for every vertex $v \in V-V'$ there is a vertex $u \in V'$ for which $(u,v) \in A$?

Reference: [Chvátal, 1973]. Transformation from 3SAT.

[GT58] **K-CLOSURE**

INSTANCE: Directed graph $G=(V,A)$, positive integer $K \leqslant |V|$.

QUESTION: Is there a subset $V' \subseteq V$ with $|V'| \leqslant K$ such that for all $(u,v) \in A$ either $u \in V'$ or $v \notin V'$?

Reference: [Queyranne, 1976]. Transformation from CLIQUE.

[GT59] **INTERSECTION GRAPH BASIS**

INSTANCE: Graph $G=(V,E)$, positive integer $K \leqslant |E|$.

QUESTION: Is G the intersection graph for a family of sets whose union has cardinality K or less, i.e., is there a K-element set S and for each $v \in V$ a subset $S[v] \subseteq S$ such that $\{u,v\} \in E$ if and only if $S[u]$ and $S[v]$ are not disjoint?

Reference: [Kou, Stockmeyer, and Wong, 1978]. Transformation from COVERING BY CLIQUES.

[GT60] **PATH DISTINGUISHERS**

INSTANCE: Acyclic directed graph $G=(V,A)$, specified vertices $s,t \in V$, positive integer $K \leqslant |A|$.

QUESTION: Is there a subset $A' \subseteq A$ with $|A'| \leqslant K$ such that, for any pair p_1,p_2 of paths from s to t in G, there is some arc in A' that is in one of p_1 and p_2 but not both?

Reference: [Maheshwari, 1976]. Transformation from VERTEX COVER.

[GT61] **METRIC DIMENSION**

INSTANCE: Graph $G=(V,E)$, positive integer $K \leqslant |V|$.

QUESTION: Is there a metric basis for G of cardinality K or less, i.e., a subset $V' \subseteq V$ with $|V'| \leqslant K$ such that for each pair $u,v \in V$ there is a $w \in V'$ such that the length of the shortest path from u to w is different from the length of the shortest path from v to w?

Reference: [Garey and Johnson, ——]. Transformation from 3DM. The definition of metric dimension appears in [Harary and Melter, 1976].

[GT62] NESETRIL-RÖDL DIMENSION

INSTANCE: Graph $G=(V,E)$, positive integer $K \leqslant |E|$.
QUESTION: Is there a one-to-one function $f: V \rightarrow \{(a_1,a_2, \ldots, a_K): 1 \leqslant a_i \leqslant |V| \text{ for } 1 \leqslant i \leqslant K\}$ such that, for all $u,v \in V$, $\{u,v\} \in E$ if and only if $f(u)$ and $f(v)$ disagree in all K components?

Reference: [Nesetril and Pultr, 1977]. Transformation from GRAPH 3-COLORABILITY. The definition appears in [Nesetril and Rödl, 1977].

[GT63] THRESHOLD NUMBER

INSTANCE: Graph $G=(V,E)$, positive integer $K \leqslant |E|$.
QUESTION: Is there a partition of E into disjoint sets $E_1,E_2, \ldots, E_K$ such that each of the graphs $G_i=(V,E_i)$, $1 \leqslant i \leqslant K$, is a "threshold graph"?

Reference: [Chvátal and Hammer, 1975]. Transformation from INDEPENDENT SET restricted to triangle free graphs.
Comment: Solvable in polynomial time for $K=1$.

[GT64] ORIENTED DIAMETER

INSTANCE: Graph $G=(V,E)$, positive integer $K \leqslant |V|$.
QUESTION: Can the edges of G be directed in such a way that the resulting directed graph is strongly connected and has diameter no more than K?

Reference: [Chvátal and Thomassen, 1978]. Transformation from SET SPLITTING.
Comment: The variation in which "diameter" is replaced by "radius" is also NP-complete. Both problems remain NP-complete for $K=2$.

[GT65] WEIGHTED DIAMETER

INSTANCE: Graph $G=(V,E)$, collection C of $|E|$ not necessarily distinct non-negative integers, positive integer K.
QUESTION: Is there a one-to-one function $f: E \rightarrow C$ such that, if $f(e)$ is taken as the length of edge e, then G has diameter K or less, i.e., every pair of points $u,v \in V$ is joined by a path in G of length K or less.

Reference: [Perl and Zaks, 1978]. Transformation from 3-PARTITION.
Comment: NP-complete in the strong sense, even if G is a tree. The variant in which "diameter" is replaced by "radius" has the same complexity. If C consists entirely of 0's and 1's, then both the diameter and radius versions are solvable in polynomial time for trees, but are NP-complete for general graphs, even if K is fixed at 2 (diameter) or 1 (radius). The variant in which we ask for an assignment yielding diameter K or *greater* is NP-complete in the strong sense for general graphs, is solvable in polynomial time for trees in the diameter case, and is NP-complete for trees in the radius case.

A2 NETWORK DESIGN

A2.1 SPANNING TREES

[ND1] DEGREE CONSTRAINED SPANNING TREE

INSTANCE: Graph $G=(V,E)$, positive integer $K \leqslant |V|$.
QUESTION: Is there a spanning tree for G in which no vertex has degree larger than K?

Reference: Transformation from HAMILTONIAN PATH.
Comment: Remains NP-complete for any fixed $K \geqslant 2$.

[ND2] MAXIMUM LEAF SPANNING TREE

INSTANCE: Graph $G=(V,E)$, positive integer $K \leqslant |V|$.
QUESTION: Is there a spanning tree for G in which K or more vertices have degree 1?

Reference: [Garey and Johnson, ——]. Transformation from DOMINATING SET.
Comment: Remains NP-complete if G is regular of degree 4 or if G is planar with no degree exceeding 4.

[ND3] SHORTEST TOTAL PATH LENGTH SPANNING TREE

INSTANCE: Graph $G=(V,E)$, integer bound $B \in Z^+$.
QUESTION: Is there a spanning tree T for G such that the sum, over all pairs of vertices $u,v \in V$, of the length of the path in T from u to v is no more than K?

Reference: [Johnson, Lenstra, and Rinnooy Kan, 1978]. Transformation from EXACT COVER BY 3-SETS.

[ND4] BOUNDED DIAMETER SPANNING TREE

INSTANCE: Graph $G=(V,E)$, weight $w(e) \in Z^+$ for each $e \in E$, positive integer $D \leqslant |V|$, positive integer B.
QUESTION: Is there a spanning tree T for G such that the sum of the weights of the edges in T does not exceed B and such that T contains no simple path with more than D edges?

Reference: [Garey and Johnson, ——]. Transformation from EXACT COVER BY 3-SETS.
Comment: Remains NP-complete for any fixed $D \geqslant 4$, even if all edge weights are either 1 or 2. Can be solved easily in polynomial time if $D \leqslant 3$, or if all edge weights are equal.

[ND5] CAPACITATED SPANNING TREE

INSTANCE: Graph $G=(V,E)$, specified vertex $v_0 \in V$, capacity $c(e) \in Z_0^+$ and length $l(e) \in Z_0^+$ for each $e \in E$, requirement $r(v) \in Z_0^+$ for each $v \in V-\{v_0\}$, and a bound $B \in Z_0^+$.
QUESTION: Is there a spanning tree T for G such that the sum of the lengths of the edges in T does not exceed B and such that for each edge e in T, if $U(e)$ is

the set of vertices whose path to v_0 in T contains e, then $\sum_{u \in U(e)} r(u) \leqslant c(e)$?

Reference: [Papadimitriou, 1976c]. Transformation from 3SAT.

Comment: NP-complete in the strong sense, even if all requirements are 1 and all capacities are equal to 3. Solvable in polynomial time by weighted matching techniques if all requirements are 1 and all capacities 2. Can also be solved in polynomial time (by minimum cost network flow algorithms, e.g., see [Edmonds and Karp, 1972]) if all capacities are 1 and all requirements are either 0 or 1, but remains NP-complete if all capacities are 2, all requirements 0 or 1, and all edge lengths 0 or 1 [Even and Johnson, 1977].

[ND6] GEOMETRIC CAPACITATED SPANNING TREE

INSTANCE: Set $P \subseteq Z \times Z$ of points in the plane, specified point $p_0 \in P$, requirement $r(p) \in Z_0^+$ for each $p \in P - p_0$, capacity $c \in Z^+$, bound $B \in Z^+$.

QUESTION: Is there a spanning tree $T=(P,E')$ for the complete graph $G=(P,E)$ such that $\sum_{e \in E'} d(e) \leqslant B$, where $d((x_1,y_1),(x_2,y_2))$ is the discretized Euclidean distance $\lceil ((x_1-x_2)^2+(y_1-y_2)^2)^{1/2} \rceil$, and such that for each $e \in E'$, if $U(e)$ is the set of vertices whose paths to p_0 pass through e, then $\sum_{u \in U(e)} r(u) \leqslant c$?

Reference: [Papadimitriou, 1976c]. Transformation from X3C.

Comment: Remains NP-complete even if all requirements are equal.

[ND7] OPTIMUM COMMUNICATION SPANNING TREE

INSTANCE: Complete graph $G=(V,E)$, weight $w(e) \in Z_0^+$ for each $e \in E$, requirement $r(\{u,v\}) \in Z_0^+$ for each pair $\{u,v\}$ of vertices from V, bound $B \in Z_0^+$.

QUESTION: Is there a spanning tree T for G such that, if $W(\{u,v\})$ denotes the sum of the weights of the edges on the path joining u and v in T, then

$$\sum_{u,v \in V} \left[W(\{u,v\}) \cdot r(\{u,v\}) \right] \leqslant B \ ?$$

Reference: [Johnson, Lenstra, and Rinnooy Kan, 1978]. Transformation from X3C.

Comment: Remains NP-complete even if all requirements are equal. Can be solved in polynomial time if all edge weights are equal [Hu, 1974].

[ND8] ISOMORPHIC SPANNING TREE

INSTANCE: Graph $G=(V,E)$, tree $T=(V_T,E_T)$.

QUESTION: Does G contain a spanning tree isomorphic to T?

Reference: Transformation from HAMILTONIAN PATH.

Comment: Remains NP-complete even if (a) T is a path, (b) T is a full binary tree [Papadimitriou and Yannakakis, 1978], or if (c) T is a 3-star (that is, $V_T=\{v_0\} \cup \{u_i,v_i,w_i: 1 \leqslant i \leqslant n\}$, $E_T=\{\{v_0,u_i\},\{u_i,v_i\},\{v_i,w_i\}: 1 \leqslant i \leqslant n\}$) [Garey and Johnson, ——]. Solvable in polynomial time by graph matching if G is a 2-star. For a classification of the complexity of this problem for other types of trees, see [Papadimitriou and Yannakakis, 1978].

[ND9] Kth BEST SPANNING TREE (*)

INSTANCE: Graph $G=(V,E)$, weight $w(e) \in Z_0^+$ for each $e \in E$, positive integers K and B.

QUESTION: Are there K distinct spanning trees for G, each having total weight B or less?

Reference: [Johnson and Kashdan, 1976]. Turing reduction from HAMILTONIAN PATH.

Comment: Not known to be in NP. Can be solved in pseudo-polynomial time (polynomial in $|V|$, K, $\log B$, $\max\{\log w(e): e \in E\}$) [Lawler, 1972], and hence in polynomial time for any fixed value of K. The corresponding enumeration problem is #P-complete. However, the unweighted case of the enumeration problem is solvable in polynomial time (e.g., see [Harary and Palmer, 1973]).

[ND10] BOUNDED COMPONENT SPANNING FOREST

INSTANCE: Graph $G=(V,E)$, weight $w(v) \in Z_0^+$ for each $v \in V$, positive integers $K \leqslant |V|$ and B.

QUESTION: Can the vertices in V be partitioned into $k \leqslant K$ disjoint sets $V_1, V_2, \ldots, V_k$ such that, for $1 \leqslant i \leqslant k$, the subgraph of G induced by V_i is connected and the sum of the weights of the vertices in V_i does not exceed B?

Reference: [Hadlock, 1974]. Transformation from PARTITION INTO PATHS OF LENGTH 2.

Comment: Remains NP-complete even if all weights equal 1 and B is any fixed integer larger than 2 [Garey and Johnson, ——]. Can be solved in polynomial time if G is a tree or if all weights equal 1 and $B=2$ [Hadlock, 1974].

[ND11] MULTIPLE CHOICE BRANCHING

INSTANCE: Directed graph $G=(V,A)$, a weight $w(a) \in Z^+$ for each arc $a \in A$, a partition of A into disjoint sets $A_1, A_2, \ldots, A_m$, and a positive integer K.

QUESTION: Is there a subset $A' \in A$ with $\sum_{a \in A'} w(a) \geqslant K$ such that no two arcs in A' enter the same vertex, A' contains no cycles, and A' contains at most one arc from each of the A_i, $1 \leqslant i \leqslant m$?

Reference: [Garey and Johnson, ——]. Transformation from 3SAT.

Comment: Remains NP-complete even if G is strongly connected and all weights are equal. If all A_i have $|A_i|=1$, the problem becomes simply that of finding a "maximum weight branching," a 2-matroid intersection problem that can be solved in polynomial time (e.g., see [Tarjan, 1977]). (In a strongly connected graph, a maximum weight branching can be viewed as a maximum weight directed spanning tree.) Similarly, if the graph is symmetric, the problem becomes equivalent to the "multiple choice spanning tree" problem, another 2-matroid intersection problem that can be solved in polynomial time [Suurballe, 1975].

[ND12] STEINER TREE IN GRAPHS

INSTANCE: Graph $G=(V,E)$, a weight $w(e) \in Z_0^+$ for each $e \in E$, a subset $R \subseteq V$, and a positive integer bound B.

QUESTION: Is there a subtree of G that includes all the vertices of R and such that the sum of the weights of the edges in the subtree is no more than B?

Reference: [Karp, 1972]. Transformation from EXACT COVER BY 3-SETS.
Comment: Remains NP-complete if all edge weights are equal, even if G is a bipartite graph having no edges joining two vertices in R or two vertices in $V-R$ [Berlekamp, 1976] or G is planar [Garey and Johnson, 1977a].

[ND13] GEOMETRIC STEINER TREE

INSTANCE: Set $P \subseteq Z \times Z$ of points in the plane, positive integer K.
QUESTION: Is there a finite set $Q \subseteq Z \times Z$ such that there is a spanning tree of total weight K or less for the vertex set $P \cup Q$, where the weight of an edge $\{(x_1,y_1),(x_2,y_2)\}$ is the discretized Euclidean length $\lceil ((x_1-x_2)^2+(y_1-y_2)^2)^{1/2} \rceil$?
Reference: [Garey, Graham, and Johnson, 1977]. Transformation from X3C.
Comment: NP-complete in the strong sense. Remains so if the distance measure is replaced by the L_1 "rectilinear" metric, $|x_1-x_2|+|y_1-y_2|$, [Garey and Johnson, 1977a] or the L_∞ metric, $\max \{|x_1-x_2|,|y_1-y_2|\}$, which is equivalent to L_1 under a 45° rotation. Problem remains NP-hard in the strong sense if the (nondiscretized) Euclidean metric $((x_1-x_2)^2+(y_1-y_2)^2)^{1/2}$ is used, but is not known to be in NP [Garey, Graham, and Johnson, 1977]. Some polynomial time algorithms for special cases of the rectilinear case are presented in [Aho, Garey, and Hwang, 1977].

A2.2 CUTS AND CONNECTIVITY

[ND14] GRAPH PARTITIONING

INSTANCE: Graph $G=(V,E)$, weights $w(v) \in Z^+$ for each $v \in V$ and $l(e) \in Z^+$ for each $e \in E$, positive integers K and J.
QUESTION: Is there a partition of V into disjoint sets $V_1, V_2, \cdots, V_m$ such that $\sum_{v \in V_i} w(v) \leqslant K$ for $1 \leqslant i \leqslant m$ and such that if $E' \subseteq E$ is the set of edges that have their two endpoints in two different sets V_i, then $\sum_{e \in E'} l(e) \leqslant J$?
Reference: [Hyafil and Rivest, 1973]. Transformation from PARTITION INTO TRIANGLES.
Comment: Remains NP-complete for fixed $K \geqslant 3$ even if all vertex and edge weights are 1. Can be solved in polynomial time for $K=2$ by matching.

[ND15] ACYCLIC PARTITION

INSTANCE: Directed graph $G=(V,A)$, weight $w(v) \in Z^+$ for each $v \in V$, cost $c(a) \in Z^+$ for each $a \in A$, positive integers B and K.
QUESTION: Is there a partition of V into disjoint sets $V_1, V_2, \ldots, V_m$ such that the directed graph $G'=(V',A')$, where $V'=\{V_1, V_2, \ldots, V_m\}$, and $(V_i,V_j) \in A'$ if and only if $(v_i,v_j) \in A$ for some $v_i \in V_i$ and some $v_j \in V_j$, is acyclic, such that the sum of the weights of the vertices in each V_i does not exceed B, and such that the sum of the costs of all those arcs having their endpoints in different sets does not exceed K?
Reference: [Garey and Johnson, ——]. Transformation from X3C.
Comment: Remains NP-complete even if all $v \in V$ have $w(v)=1$ and all $a \in A$ have $c(a)=1$. Can be solved in polynomial time if G contains a Hamiltonian path (a

property that can be verified in polynomial time for acyclic digraphs) [Kernighan, 1971]. If G is a tree the general problem is NP-complete in the ordinary sense, but can be solved in pseudo-polynomial time [Lukes, 1974]. The tree problem can be solved in polynomial time if all edge weights are equal (see [Hadlock, 1974]) or if all vertex weights are equal [Garey and Johnson, ——].

[ND16] **MAX CUT**

INSTANCE: Graph $G=(V,E)$, weight $w(e) \in Z^+$ for each $e \in E$, positive integer K.

QUESTION: Is there a partition of V into disjoint sets V_1 and V_2 such that the sum of the weights of the edges from E that have one endpoint in V_1 and one endpoint in V_2 is at least K?

Reference: [Karp, 1972]. Transformation from MAXIMUM 2-SATISFIABILITY.

Comment: Remains NP-complete if $w(e)=1$ for all $e \in E$ (the SIMPLE MAX CUT problem) [Garey, Johnson, and Stockmeyer, 1976], and if, in addition, no vertex has degree exceeding 3 [Yannakakis, 1978b]. Can be solved in polynomial time if G is planar [Hadlock, 1975], [Orlova and Dorfman, 1972].

[ND17] **MINIMUM CUT INTO BOUNDED SETS**

INSTANCE: Graph $G=(V,E)$, weight $w(e) \in Z^+$ for each $e \in E$, specified vertices $s,t \in V$, positive integer $B \leqslant |V|$, positive integer K.

QUESTION: Is there a partition of V into disjoint sets V_1 and V_2 such that $s \in V_1$, $t \in V_2$, $|V_1| \leqslant B$, $|V_2| \leqslant B$, and such that the sum of the weights of the edges from E that have one endpoint in V_1 and one endpoint in V_2 is no more than K?

Reference: [Garey, Johnson, and Stockmeyer, 1976]. Transformation from SIMPLE MAX CUT.

Comment: Remains NP-complete for $B=|V|/2$ and $w(e)=1$ for all $e \in E$. Can be solved in polynomial time for $B=|V|$ by standard network flow techniques.

[ND18] **BICONNECTIVITY AUGMENTATION**

INSTANCE: Graph $G=(V,E)$, weight $w(\{u,v\}) \in Z^+$ for each unordered pair $\{u,v\}$ of vertices from V, positive integer B.

QUESTION: Is there a set E' of unordered pairs of vertices from V such that $\sum_{e \in E'} w(e) \leqslant B$ and such that the graph $G'=(V,E \cup E')$ is biconnected, i.e., cannot be disconnected by removing a single vertex?

Reference: [Eswaran and Tarjan, 1976]. Transformation from HAMILTONIAN CIRCUIT.

Comment: The related problem in which G' must be bridge connected, i.e., cannot be disconnected by removing a single edge, is also NP-complete. Both problems remain NP-complete if all weights are either 1 or 2 and E is empty. Both can be solved in polynomial time if all weights are equal.

[ND19] **STRONG CONNECTIVITY AUGMENTATION**

INSTANCE: Directed graph $G=(V,A)$, weight $w(u,v) \in Z^+$ for each ordered pair $(u,v) \in V \times V$, positive integer B.

QUESTION: Is there a set A' of ordered pairs of vertices from V such that

$\sum_{a \in A'} w(a) \leqslant B$ and such that the graph $G' = (V, A \cup A')$ is strongly connected?

Reference: [Eswaran and Tarjan, 1976]. Transformation from HAMILTONIAN CIRCUIT.

Comment: Remains NP-complete if all weights are either 1 or 2 and A is empty. Can be solved in polynomial time if all weights are equal.

[ND20] NETWORK RELIABILITY (*)

INSTANCE: Graph $G=(V,E)$, subset $V' \subseteq V$, a rational "failure probability" $p(e)$, $0 \leqslant p(e) \leqslant 1$, for each $e \in E$, a positive rational number $q \leqslant 1$.

QUESTION: Assuming edge failures are independent of one another, is the probability q or greater that each pair of vertices in V' is joined by at least one path containing no failed edge?

Reference: [Rosenthal, 1974]. Transformation from STEINER TREE IN GRAPHS.

Comment: Not known to be in NP. Remains NP-hard even if $|V'|=2$ [Valiant, 1977b]. The related problem in which we want *two* disjoint paths between each pair of vertices in V' is NP-hard even if $V'=V$ [Ball, 1977b]. If G is directed and we ask for a directed path between each *ordered* pair of vertices in V', the one-path problem is NP-hard for both $|V'|=2$ [Valiant, 1977b] and $V'=V$ [Ball, 1977a]. Many of the underlying subgraph enumeration problems are #P-complete (see [Valiant, 1977b]).

[ND21] NETWORK SURVIVABILITY (*)

INSTANCE: Graph $G=(V,E)$, a rational "failure probability" $p(x)$, $0 \leqslant p(x) \leqslant 1$, for each $x \in V \cup E$, a positive rational number $q \leqslant 1$.

QUESTION: Assuming all edge and vertex failures are independent of one another, is the probability q or greater that for all $\{u,v\} \in E$ at least one of u, v, or $\{u,v\}$ will fail?

Reference: [Rosenthal, 1974]. Transformation from VERTEX COVER.

Comment: Not known to be in NP.

A2.3 ROUTING PROBLEMS

[ND22] TRAVELING SALESMAN

INSTANCE: Set C of m cities, distance $d(c_i, c_j) \in Z^+$ for each pair of cities $c_i, c_j \in C$, positive integer B.

QUESTION: Is there a tour of C having length B or less, i.e., a permutation $< c_{\pi(1)}, c_{\pi(2)}, \ldots, c_{\pi(m)} >$ of C such that

$$\left(\sum_{i=1}^{m-1} d(c_{\pi(i)}, c_{\pi(i+1)}) \right) + d(c_{\pi(m)}, c_{\pi(1)}) \leqslant B \ ?$$

Reference: Transformation from HAMILTONIAN CIRCUIT.

Comment: Remains NP-complete even if $d(c_i, c_j) \in \{1,2\}$ for all $c_i, c_j \in C$. Special cases that can be solved in polynomial time are discussed in [Gilmore and Gomory, 1964], [Garfinkel, 1977], and [Syslo, 1973]. The variant in which we ask for a tour with "mean arrival time" of B or less is also NP-complete [Sahni and Gonzalez, 1976].

[ND23] GEOMETRIC TRAVELING SALESMAN

INSTANCE: Set $P \subseteq Z \times Z$ of points in the plane, positive integer B.
QUESTION: Is there a tour of length B or less for the TRAVELING SALESMAN instance with $C = P$ and $d((x_1,y_1),(x_2,y_2))$ equal to the discretized Euclidean distance $\lceil ((x_1 - x_2)^2 + (y_1 - y_2)^2)^{1/2} \rceil$?

Reference: [Papadimitriou, 1977] [Garey, Graham, and Johnson, 1976]. Transformation from X3C.
Comment: NP-complete in the strong sense. Remains NP-complete in the strong sense if the distance measure is replaced by the L_1 "rectilinear" metric [Garey, Graham, and Johnson, 1976] or the L_∞ metric, which is equivalent to L_1 under a 45° rotation. Problem remains NP-hard in the strong sense if the (nondiscretized) Euclidean metric is used, but is not known to be in NP [Garey, Graham, and Johnson, 1976].

[ND24] BOTTLENECK TRAVELING SALESMAN

INSTANCE: Set C of m cities, distance $d(c_i,c_j) \in Z^+$ for each pair of cities $c_i,c_j \in C$, positive integer B.
QUESTION: Is there a tour of C whose longest edge is no longer than B, i.e., a permutation $< c_{\pi(1)}, c_{\pi(2)}, \ldots, c_{\pi(m)} >$ of C such that $d(c_{\pi(i)},c_{\pi(i+1)}) \leqslant B$ for $1 \leqslant i < m$ and such that $d(c_{\pi(m)},c_{\pi(1)}) \leqslant B$?

Reference: Transformation from HAMILTONIAN CIRCUIT.
Comment: Remains NP-complete even if $d(c_i,c_j) \in \{1,2\}$ for all $c_i,c_j \in C$. An important special case that is solvable in polynomial time can be found in [Gilmore and Gomory, 1964].

[ND25] CHINESE POSTMAN FOR MIXED GRAPHS

INSTANCE: Mixed graph $G = (V,A,E)$, where A is a set of directed edges and E is a set of undirected edges on V, length $l(e) \in Z_0^+$ for each $e \in A \cup E$, bound $B \in Z^+$.
QUESTION: Is there a cycle in G that includes each directed and undirected edge at least once, traversing directed edges only in the specified direction, and that has total length no more than B?

Reference: [Papadimitriou, 1976b]. Transformation from 3SAT.
Comment: Remains NP-complete even if all edge lengths are equal, G is planar, and the maximum vertex degree is 3. Can be solved in polynomial time if either A or E is empty (i.e., if G is either a directed or an undirected graph) [Edmonds and Johnson, 1973].

[ND26] STACKER-CRANE

INSTANCE: Mixed graph $G = (V,A,E)$, length $l(e) \in Z_0^+$ for each $e \in A \cup E$, bound $B \in Z^+$.
QUESTION: Is there a cycle in G that includes each directed edge in A at least

once, traversing such edges only in the specified direction, and that has total length no more than B?

Reference: [Frederickson, Hecht, and Kim, 1978]. Transformation from HAMILTONIAN CIRCUIT.

Comment: Remains NP-complete even if all edge lengths equal 1. The analogous path problem (with or without specified endpoints) is also NP-complete.

[ND27] **RURAL POSTMAN**

INSTANCE: Graph $G=(V,E)$, length $l(e) \in Z_0^+$ for each $e \in E$, subset $E' \subseteq E$, bound $B \in Z^+$.

QUESTION: Is there a circuit in G that includes each edge in E' and that has total length no more than B?

Reference: [Lenstra and Rinnooy Kan, 1976]. Transformation from HAMILTONIAN CIRCUIT.

Comment: Remains NP-complete even if $l(e)=1$ for all $e \in E$, as does the corresponding problem for directed graphs.

[ND28] **LONGEST CIRCUIT**

INSTANCE: Graph $G=(V,E)$, length $l(e) \in Z^+$ for each $e \in E$, positive integer K.

QUESTION: Is there a simple circuit in G of length K or more, i.e., whose edge lengths sum to at least K?

Reference: Transformation from HAMILTONIAN CIRCUIT.

Comment: Remains NP-complete if $l(e)=1$ for all $e \in E$, as does the corresponding problem for directed circuits in directed graphs. The directed problem with all $l(e)=1$ can be solved in polynomial time if G is a "tournament" [Morrow and Goodman, 1976]. The analogous directed and undirected problems, which ask for a simple circuit of length K or less, can be solved in polynomial time (e.g., see [Itai and Rodeh, 1977b]), but are NP-complete if negative lengths are allowed.

[ND29] **LONGEST PATH**

INSTANCE: Graph $G=(V,E)$, length $l(e) \in Z^+$ for each $e \in E$, positive integer K, specified vertices $s,t \in V$.

QUESTION: Is there a simple path in G from s to t of length K or more, i.e., whose edge lengths sum to at least K?

Reference: Transformation from HAMILTONIAN PATH BETWEEN TWO VERTICES.

Comment: Remains NP-complete if $l(e)=1$ for all $e \in E$, as does the corresponding problem for directed paths in directed graphs. The general problem can be solved in polynomial time for acyclic digraphs, e.g., see [Lawler, 1976a]. The analogous directed and undirected "shortest path" problems can be solved for arbitrary graphs in polynomial time (e.g., see [Lawler, 1976a]), but are NP-complete if negative lengths are allowed.

[ND30] SHORTEST WEIGHT-CONSTRAINED PATH

INSTANCE: Graph $G=(V,E)$, length $l(e) \in Z^+$, and weight $w(e) \in Z^+$ for each $e \in E$, specified vertices $s,t \in V$, positive integers K,W.
QUESTION: Is there a simple path in G from s to t with total weight W or less and total length K or less?

Reference: [Megiddo, 1977]. Transformation from PARTITION.
Comment: Also NP-complete for directed graphs. Both problems are solvable in polynomial time if all weights are equal or all lengths are equal.

[ND31] K[th] SHORTEST PATH (*)

INSTANCE: Graph $G=(V,E)$, length $l(e) \in Z^+$ for each $e \in E$, specified vertices $s,t \in V$, positive integers B and K.
QUESTION: Are there K or more distinct simple paths from s to t in G, each having total length B or less?

Reference: [Johnson and Kashdan, 1976]. Turing reduction from HAMILTONIAN PATH.
Comment: Not known to be in NP. Corresponding K[th] shortest circuit problem is also NP-hard. Both remain NP-hard if $l(e)=1$ for all $e \in E$, as do the corresponding problems for directed graphs. However, all versions can be solved in pseudo-polynomial time (polynomial in $|V|$, K, and $\log B$) and hence in polynomial time for any fixed value of K. The corresponding enumeration problems are #P-complete.

A2.4 FLOW PROBLEMS

[ND32] MINIMUM EDGE-COST FLOW

INSTANCE: Directed graph $G=(V,A)$, specified vertices s and t, capacity $c(a) \in Z^+$ and price $p(a) \in Z_0^+$ for each $a \in A$, requirement $R \in Z^+$, bound $B \in Z^+$.
QUESTION: Is there a flow function $f: A \rightarrow Z_0^+$ such that
(1) $f(a) \leqslant c(a)$ for all $a \in A$,
(2) for each $v \in V-\{s,t\}$, $\sum_{(u,v) \in A} f((u,v)) = \sum_{(v,u) \in A} f((v,u))$, i.e., flow is "conserved" at v,
(3) $\sum_{(u,t) \in A} f((u,t)) - \sum_{(t,u) \in A} f((t,u)) \geqslant R$, i.e., the net flow into t is at least R, and
(4) if $A'=\{a \in A: f(a) \neq 0\}$, then $\sum_{a \in A'} p(a) \leqslant B$?

Reference: [Even and Johnson, 1977]. Transformation from X3C.
Comment: Remains NP-complete if $c(a)=2$ and $p(a) \in \{0,1\}$ for all $a \in A$. Solvable in polynomial time if $c(a)=1$ for all $a \in A$ [Even and Johnson, 1977] or if (4) is replaced by $\sum_{a \in A} p(a) \cdot f(a) \leqslant B$ (e.g., see [Lawler, 1976a]). However, becomes NP-complete once more if (4) is replaced by $\sum_{a \in A} (p_1(a) f(a)^2 + p_2(a) f(a)) \leqslant B$ [Herrmann, 1973].

[ND33] INTEGRAL FLOW WITH MULTIPLIERS

INSTANCE: Directed graph $G=(V,A)$, specified vertices s and t, multiplier $h(v) \in Z^+$ for each $v \in V-\{s,t\}$, capacity $c(a) \in Z^+$ for each $a \in A$, requirement $R \in Z^+$.
QUESTION: Is there a flow function $f: A \rightarrow Z_0^+$ such that
(1) $f(a) \leqslant c(a)$ for all $a \in A$,
(2) for each $v \in V-\{s,t\}$, $\sum_{(u,v) \in A} h(v) \cdot f((u,v)) = \sum_{(v,u) \in A} f((v,u))$, and
(3) the net flow into t is at least R?

Reference: [Sahni, 1974]. Transformation from PARTITION.
Comment: Can be solved in polynomial time by standard network flow techniques if $h(v)=1$ for all $v \in V-\{s,t\}$. Corresponding problem with non-integral flows allowed can be solved by linear programming.

[ND34] PATH CONSTRAINED NETWORK FLOW

INSTANCE: Directed graph $G=(V,A)$, specified vertices s and t, a capacity $c(a) \in Z^+$ for each $a \in A$, a collection P of directed paths in G, and a requirement $R \in Z^+$.
QUESTION: Is there a function $g: P \rightarrow Z_0^+$ such that if $f: A \rightarrow Z_0^+$ is the flow function defined by $f(a) = \sum_{p \in P(a)} g(p)$, where $P(a) \subseteq P$ is the set of all paths in P containing the arc a, then f is such that
(1) $f(a) \leqslant c(a)$ for all $a \in A$,
(2) for each $v \in V-\{s,t\}$, flow is conserved at v, and
(3) the net flow into t is at least R?

Reference: [Prömel, 1978]. Transformation from 3SAT.
Comment: Remains NP-complete even if all $c(a)=1$. The corresponding problem with non-integral flows is equivalent to LINEAR PROGRAMMING, but the question of whether the best rational flow fails to exceed the best integral flow is NP-complete.

[ND35] INTEGRAL FLOW WITH HOMOLOGOUS ARCS

INSTANCE: Directed graph $G=(V,A)$, specified vertices s and t, capacity $c(a) \in Z^+$ for each $a \in A$, requirement $R \in Z^+$, set $H \subseteq A \times A$ of "homologous" pairs of arcs.
QUESTION: Is there a flow function $f: A \rightarrow Z_0^+$ such that
(1) $f(a) \leqslant c(a)$ for all $a \in A$,
(2) for each $v \in V-\{s,t\}$, flow is conserved at v,
(3) for all pairs $<a,a'> \in H$, $f(a)=f(a')$, and
(4) the net flow into t is at least R?

Reference: [Sahni, 1974]. Transformation from 3SAT.
Comment: Remains NP-complete if $c(a)=1$ for all $a \in A$ (by modifying the construction in [Even, Itai, and Shamir, 1976]). Corresponding problem with non-integral flows is polynomially equivalent to LINEAR PROGRAMMING [Itai, 1977].

[ND36] INTEGRAL FLOW WITH BUNDLES

INSTANCE: Directed graph $G=(V,A)$, specified vertices s and t, "bundles" $I_1,I_2, \cdots ,I_k \subseteq A$ such that $\bigcup_{1 \leqslant j \leqslant k} I_j = A$, bundle capacities $c_1,c_2, \cdots ,c_k \in Z^+$, requirement $R \in Z^+$.
QUESTION: Is there a flow function $f: A \rightarrow Z_0^+$ such that
(1) for $1 \leqslant j \leqslant k$, $\sum_{a \in I_j} f(a) \leqslant c_j$,
(2) for each $v \in V-\{s,t\}$, flow is conserved at v, and
(3) the net flow into t is at least R?

Reference: [Sahni, 1974]. Transformation from INDEPENDENT SET.
Comment: Remains NP-complete if all capacities are 1 and all bundles have two arcs. Corresponding problem with non-integral flows allowed can be solved by linear programming.

[ND37] UNDIRECTED FLOW WITH LOWER BOUNDS

INSTANCE: Graph $G=(V,E)$, specified vertices s and t, capacity $c(e) \in Z^+$ and lower bound $l(e) \in Z_0^+$ for each $e \in E$, requirement $R \in Z^+$.
QUESTION: Is there a flow function f: $\{(u,v),(v,u): \{u,v\} \in E\} \rightarrow Z_0^+$ such that
(1) for all $\{u,v\} \in E$, either $f((u,v))=0$ or $f((v,u))=0$,
(2) for each $e=\{u,v\} \in E$, $l(e) \leqslant \max\{f((u,v)),f((v,u))\} \leqslant c(e)$,
(3) for each $v \in V-\{s,t\}$, flow is conserved at v, and
(4) the net flow into t is at least R?

Reference: [Itai, 1977]. Transformation from SATISFIABILITY.
Comment: Problem is NP-complete in the strong sense, even if non-integral flows are allowed. Corresponding problem for directed graphs can be solved in polynomial time, even if we ask that the total flow be R or less rather than R or more [Ford and Fulkerson, 1962] (see also [Lawler, 1976a]). The analogous DIRECTED M-COMMODITY FLOW WITH LOWER BOUNDS problem is polynomially equivalent to LINEAR PROGRAMMING for all $M \geqslant 2$ if non-integral flows are allowed [Itai, 1977].

[ND38] DIRECTED TWO-COMMODITY INTEGRAL FLOW

INSTANCE: Directed graph $G=(V,A)$, specified vertices s_1, s_2, t_1, and t_2, capacity $c(a) \in Z^+$ for each $a \in A$, requirements $R_1,R_2 \in Z^+$.
QUESTION: Are there two flow functions $f_1,f_2: A \rightarrow Z_0^+$ such that
(1) for each $a \in A$, $f_1(a)+f_2(a) \leqslant c(a)$,
(2) for each $v \in V-\{s,t\}$ and $i \in \{1,2\}$, flow f_i is conserved at v, and
(3) for $i \in \{1,2\}$, the net flow into t_i under flow f_i is at least R_i?

Reference: [Even, Itai, and Shamir, 1976]. Transformation from 3SAT.
Comment: Remains NP-complete even if $c(a)=1$ for all $a \in A$ and $R_1=1$. Variant in which $s_1=s_2$, $t_1=t_2$, and arcs can be restricted to carry only one specified commodity is also NP-complete (follows from [Even, Itai, and Shamir, 1976]). Corresponding M-commodity problem with non-integral flows allowed is polynomially equivalent to LINEAR PROGRAMMING for all $M \geqslant 2$ [Itai, 1977].

[ND39] UNDIRECTED TWO-COMMODITY INTEGRAL FLOW

INSTANCE: Graph $G=(V,E)$, specified vertices s_1, s_2, t_1, and t_2, a capacity $c(e) \in Z^+$ for each $e \in E$, requirements $R_1,R_2 \in Z^+$.

QUESTION: Are there two flow functions f_1,f_2: $\{(u,v),(v,u):\{u,v\} \in E\} \rightarrow Z_0^+$ such that

(1) for all $\{u,v\} \in E$ and $i \in \{1,2\}$, either $f_i((u,v))=0$ or $f_i((v,u))=0$,

(2) for each $\{u,v\} \in E$,

$$\max\{f_1((u,v)),f_1((v,u))\} + \max\{f_2((u,v),f_2(v,u)\} \leqslant c(\{u,v\}),$$

(3) for each $v \in V-\{s,t\}$ and $i \in \{1,2\}$, flow f_i is conserved at v, and

(4) for $i \in \{1,2\}$, the net flow into t_i under flow f_i is at least R_i?

Reference: [Even, Itai, and Shamir, 1976]. Transformation from DIRECTED TWO-COMMODITY INTEGRAL FLOW.

Comment: Remains NP-complete even if $c(e)=1$ for all $e \in E$. Solvable in polynomial time if $c(e)$ is even for all $e \in E$. Corresponding problem with non-integral flows allowed can be solved in polynomial time.

[ND40] DISJOINT CONNECTING PATHS

INSTANCE: Graph $G=(V,E)$, collection of disjoint vertex pairs $(s_1,t_1),(s_2,t_2),\ldots,(s_k,t_k)$.

QUESTION: Does G contain k mutually vertex-disjoint paths, one connecting s_i and t_i for each i, $1 \leqslant i \leqslant k$?

Reference: [Knuth, 1974c], [Karp, 1975a], [Lynch, 1974]. Transformation from 3SAT.

Comment: Remains NP-complete for planar graphs [Lynch, 1974], [Lynch, 1975]. Complexity is open for any fixed $k \geqslant 2$, but can be solved in polynomial time if $k=2$ and G is planar or chordal [Perl and Shiloach, 1978]. (A polynomial time algorithm for the general 2 path problem has been announced in [Shiloach, 1978]). The directed version of this problem is also NP-complete in general and solvable in polynomial time when $k=2$ and G is planar or acyclic [Perl and Shiloach, 1978].

[ND41] MAXIMUM LENGTH-BOUNDED DISJOINT PATHS

INSTANCE: Graph $G=(V,E)$, specified vertices s and t, positive integers $J,K \leqslant |V|$.

QUESTION: Does G contain J or more mutually vertex-disjoint paths from s to t, none involving more than K edges?

Reference: [Itai, Perl, and Shiloach, 1977]. Transformation from 3SAT.

Comment: Remains NP-complete for all fixed $K \geqslant 5$. Solvable in polynomial time for $K \leqslant 4$. Problem where paths need only be edge-disjoint is NP-complete for all fixed $K \geqslant 5$, polynomially solvable for $K \leqslant 3$, and open for $K=4$. The same results hold if G is a directed graph and the paths must be directed paths. The problem of finding the maximum number of disjoint paths from s to t, under no length constraint, is solvable in polynomial time by standard network flow techniques in both the vertex-disjoint and edge-disjoint cases.

[ND42] MAXIMUM FIXED-LENGTH DISJOINT PATHS

INSTANCE: Graph $G=(V,E)$, specified vertices s and t, positive integers $J,K \leqslant |V|$.

QUESTION: Does G contain J or more mutually vertex-disjoint paths from s to t, each involving exactly K edges?

Reference: [Itai, Perl, and Shiloach, 1977]. Transformation from 3SAT.

Comment: Remains NP-complete for fixed $K \geqslant 4$. Solvable in polynomial time for $K \leqslant 3$. Corresponding problem for edge-disjoint paths is NP-complete for fixed $K \geqslant 4$, polynomially solvable for $K \leqslant 2$, and open for $K=3$. The same results hold for directed graphs and directed paths, except that the arc-disjoint version is polynomially solvable for $K=3$ and open for $K=4$.

A2.5 MISCELLANEOUS

[ND43] QUADRATIC ASSIGNMENT PROBLEM

INSTANCE: Non-negative integer costs c_{ij}, $1 \leqslant i,j \leqslant n$, and distances d_{kl}, $1 \leqslant k,l \leqslant m$, bound $B \in Z^+$.

QUESTION: Is there a one-to-one function $f:\{1,2,\ldots,n\} \rightarrow \{1,2,\ldots,m\}$ such that

$$\sum_{i=1}^{n} \sum_{\substack{j=1 \\ j \neq i}}^{n} c_{ij} d_{f(i)f(j)} \leqslant B \ ?$$

Reference: [Sahni and Gonzalez, 1976]. Transformation from HAMILTONIAN CIRCUIT.

Comment: Special case in which each $d_{kl}=k-l$ and all $c_{ji}=c_{ij} \in \{0,1\}$ is the NP-complete OPTIMAL LINEAR ARRANGEMENT problem. The general problem is discussed, for example, in [Garfinkel and Nemhauser, 1972].

[ND44] MINIMIZING DUMMY ACTIVITIES IN PERT NETWORKS

INSTANCE: Directed acyclic graph $G=(V,A)$ where vertices represent tasks and the arcs represent precedence constraints, and a positive integer $K \leqslant |V|$.

QUESTION: Is there a PERT network corresponding to G with K or fewer dummy activities, i.e., a directed acyclic graph $G'=(V',A')$ where $V'=\{v_i^-,v_i^+: v_i \in V\}$ and $\{(v_i^-,v_i^+): v_i \in V\} \subseteq A'$, and such that $|A'| \leqslant |V|+K$ and there is a path from v_i^+ to v_j^- in G' if and only if there is a path from v_i to v_j in G?

Reference: [Krishnamoorthy and Deo, 1977b]. Transformation from VERTEX COVER.

[ND45] CONSTRAINED TRIANGULATION

INSTANCE: Graph $G=(V,E)$, coordinates $x(v)$, $y(v) \in Z$ for each $v \in V$.

QUESTION: Is there a subset $E' \subseteq E$, such that the set of line segments $\{[(x(u),y(u)),(x(v),y(v))]: \{u,v\} \in E'\}$ is a triangulation of the set of points $\{(x(v),y(v)): v \in V\}$ in the plane?

Reference: [Lloyd, 1977].

Comment: NP-complete in the strong sense.

[ND46] INTERSECTION GRAPH FOR SEGMENTS ON A GRID

INSTANCE: Graph $G=(V,E)$, positive integers M,N.

QUESTION: Is G the intersection graph for a set of line segments on an $M \times N$ grid, i.e., is there a one-to-one function f that maps each $v \in V$ to a line segment $f(v)=[(x,y),(z,w)]$, where $1 \leqslant x \leqslant z \leqslant M$, $1 \leqslant y \leqslant w \leqslant N$, and either $x=z$ or $y=w$, such that $\{u,v\} \in E$ if and only if the line segments $f(u)$ and $f(v)$ intersect?

Reference: [Gavril, 1977a]. Transformation from 3-PARTITION.

Comment: The analogous problem, which asks if G is the intersection graph for a set of rectangles on an $M \times N$ grid, is also NP-complete [Gavril, 1977a].

[ND47] EDGE EMBEDDING ON A GRID

INSTANCE: Graph $G=(V,E)$, positive integers M,N.

QUESTION: Is there a one-to-one function $f: V \rightarrow \{1,2,\ldots,M\} \times \{1,2,\ldots,N\}$ such that if $\{u,v\} \in E$, $f(u)=(x_1,y_1)$, and $f(v)=(x_2,y_2)$, then either $x_1=x_2$ or $y_1=y_2$, i.e., $f(u)$ and $f(v)$ are both on the same "line" of the grid?

Reference: [Gavril, 1977a]. Transformation from 3-PARTITION.

[ND48] GEOMETRIC CONNECTED DOMINATING SET

INSTANCE: Set $P \subseteq Z \times Z$ of points in the plane, positive integers B and K.

QUESTION: Is there a subset $P' \subseteq P$ with $|P'| \leqslant K$ such that all points in $P-P'$ are within Euclidean distance B of some point in P', and such that the graph $G=(P',E)$, with an edge between two points in P' if and only if they are within distance B of each other, is connected?

Reference: [Lichtenstein, 1977]. Transformation from PLANAR 3SAT.

Comment: Remains NP-complete if the Euclidean metric is replaced by the L_1 rectilinear metric or the L_∞ metric [Garey and Johnson, ——].

[ND49] MINIMUM BROADCAST TIME

INSTANCE: Graph $G=(V,E)$, subset $V_0 \subseteq V$, and a positive integer K.

QUESTION: Can a message be "broadcast" from the base set V_0 to all other vertices in time K, i.e., is there a sequence $V_0,E_1,V_1,E_2,\ldots,E_K,V_K$ such that each $V_i \subseteq V$, each $E_i \subseteq E$, $V_K=V$, and, for $1 \leqslant i \leqslant K$, (1) each edge in E_i has exactly one endpoint in V_{i-1}, (2) no two edges in E_i share a common endpoint, and (3) $V_i=V_{i-1} \cup \{v: \{u,v\} \in E_i\}$?

Reference: [Garey and Johnson, ——]. Transformation from 3DM. For more on this problem, see [Farley, Hedetniemi, Mitchell, and Proskurowski, 1977].

Comment: Remains NP-complete for any fixed $K \geqslant 4$, but is solvable in polynomial time by matching if $K=1$. The special case where $|V_0|=1$ remains NP-complete, but is solvable in polynomial time for trees [Cockayne, Hedetniemi, and Slater, 1978].

[ND50] MIN-MAX MULTICENTER

INSTANCE: Graph $G=(V,E)$, weight $w(v) \in Z_0^+$ for each $v \in V$, length $l(e) \in Z_0^+$ for each $e \in E$, positive integer $K \leqslant |V|$, positive rational number B.

QUESTION: Is there a set P of K "points on G" (where a point on G can be either a vertex in V or a point on an edge $e \in E$, with e regarded as a line segment of length $l(e)$) such that if $d(v)$ is the length of the shortest path from v to the closest point in P, then $\max\{d(v) \cdot w(v): v \in V\} \leqslant B$?

Reference: [Kariv and Hakimi, 1976a]. Transformation from DOMINATING SET.

Comment: Also known as the "p-center" problem. Remains NP-complete if $w(v)=1$ for all $v \in V$ and $l(e)=1$ for all $e \in E$. Solvable in polynomial time for any fixed K and for arbitrary K if G is a tree [Kariv and Hakimi, 1976a]. Variant in which we must choose a subset $P \subseteq V$ is also NP-complete but solvable for fixed K and for trees [Slater, 1976].

[ND51] **MIN-SUM MULTICENTER**

INSTANCE: Graph $G=(V,E)$, weight $w(v) \in Z_0^+$ for each $v \in V$, length $l(e) \in Z_0^+$ for each $e \in E$, positive integer $K \leqslant |V|$, positive rational number B.

QUESTION: Is there a set P of K "points on G" such that if $d(v)$ is the length of the shortest path from v to the closest point in P, then $\sum_{v \in V} d(v) \cdot w(v) \leqslant B$?

Reference: [Kariv and Hakimi, 1976b]. Transformation from DOMINATING SET.

Comment: Also known as the "p-median" problem. It can be shown that there is no loss of generality in restricting P to being a subset of V. Remains NP-complete if $w(v)=1$ for all $v \in V$ and $l(e)=1$ for all $e \in E$. Solvable in polynomial time for any fixed K and for arbitrary K if G is a tree.

A3 SETS AND PARTITIONS

A3.1 COVERING, HITTING, AND SPLITTING

[SP1] 3-DIMENSIONAL MATCHING (3DM)

INSTANCE: Set $M \subseteq W \times X \times Y$, where W, X, and Y are disjoint sets having the same number q of elements.
QUESTION: Does M contain a matching, i.e., a subset $M' \subseteq M$ such that $|M'| = q$ and no two elements of M' agree in any coordinate?

Reference: [Karp, 1972]. Transformation from 3SAT (see Section 3.1.2).
Comment: Remains NP-complete if M is "pairwise consistent," i.e., if for all elements a, b, c, whenever there exist elements w, z, and y such that $(a,b,w) \in M$, $(a,x,c) \in M$, and $(y,b,c) \in M$, then $(a,b,c) \in M$ (this follows from the proof of Theorem 3.1.2). Also remains NP-complete if no element occurs in more than three triples, but is solvable in polynomial time if no element occurs in more than two triples [Garey and Johnson, ——]. The related 2-DIMENSIONAL MATCHING problem (where $M \subseteq W \times X$) is also solvable in polynomial time (e.g., see [Lawler, 1976a]).

[SP2] EXACT COVER BY 3-SETS (X3C)

INSTANCE: Set X with $|X| = 3q$ and a collection C of 3-element subsets of X.
QUESTION: Does C contain an exact cover for X, i.e., a subcollection $C' \subseteq C$ such that every element of X occurs in exactly one member of C'?

Reference: [Karp, 1972]. Transformation from 3DM.
Comment: Remains NP-complete if no element occurs in more than three subsets, but is solvable in ploynomial time if no element occurs in more than two subsets [Garey and Johnson, ——]. Related EXACT COVER BY 2-SETS problem is also solvable in polynomial time by matching techniques.

[SP3] SET PACKING

INSTANCE: Collection C of finite sets, positive integer $K \leqslant |C|$.
QUESTION: Does C contain at least K mutually disjoint sets?

Reference: [Karp, 1972]. Transformation from X3C.
Comment: Remains NP-complete even if all $c \in C$ have $|c| \leqslant 3$. Solvable in polynomial time by matching techniques if all $c \in C$ have $|c| \leqslant 2$.

[SP4] SET SPLITTING

INSTANCE: Collection C of subsets of a finite set S.
QUESTION: Is there a partition of S into two subsets S_1 and S_2 such that no subset in C is entirely contained in either S_1 or S_2?

Reference: [Lovasz, 1973]. Transformation from NOT-ALL-EQUAL 3SAT. The problem is also known as HYPERGRAPH 2-COLORABILITY.
Comment: Remains NP-complete even if all $c \in C$ have $|c| \leqslant 3$. Solvable in polynomial time if all $c \in C$ have $|c| \leqslant 2$ (becomes GRAPH 2-COLORABILITY).

[SP5] MINIMUM COVER

INSTANCE: Collection C of subsets of a finite set S, positive integer $K \leqslant |C|$.
QUESTION: Does C contain a cover for S of size K or less, i.e., a subset $C' \subseteq C$ with $|C'| \leqslant K$ such that every element of S belongs to at least one member of C'?

Reference: [Karp, 1972]. Transformation from X3C.
Comment: Remains NP-complete even if all $c \in C$ have $|c| \leqslant 3$. Solvable in polynomial time by matching techniques if all $c \in C$ have $|c| \leqslant 2$.

[SP6] MINIMUM TEST SET

INSTANCE: Collection C of subsets of a finite set S, positive integer $K \leqslant |C|$.
QUESTION: Is there a subcollection $C' \subseteq C$ with $|C'| \leqslant K$ such that for each pair of distinct elements $u,v \in S$, there is some set $c \in C'$ that contains exactly one of u and v?

Reference: [Garey and Johnson, ——]. Transformation from 3DM.
Comment: Remains NP-complete if all $c \in C$ have $|c| \leqslant 3$, but is solvable in polynomial time if all $c \in C$ have $|c| \leqslant 2$. Variant in which C' can contain unions of subsets in C as well as subsets in C is also NP-complete [Ibaraki, Kameda, and Toida, 1977].

[SP7] SET BASIS

INSTANCE: Collection C of subsets of a finite set S, positive integer $K \leqslant |C|$.
QUESTION: Is there a collection B of subsets of S with $|B| = K$ such that, for each $c \in C$, there is a subcollection of B whose union is exactly c?

Reference: [Stockmeyer, 1975]. Transformation from VERTEX COVER.
Comment: Remains NP-complete if all $c \in C$ have $|c| \leqslant 3$, but is trivial if all $c \in C$ have $|c| \leqslant 2$.

[SP8] HITTING SET

INSTANCE: Collection C of subsets of a finite set S, positive integer $K \leqslant |S|$.
QUESTION: Is there a subset $S' \subseteq S$ with $|S'| \leqslant K$ such that S' contains at least one element from each subset in C?

Reference: [Karp, 1972]. Transformation from VERTEX COVER.
Comment: Remains NP-complete even if $|c| \leqslant 2$ for all $c \in C$.

[SP9] INTERSECTION PATTERN

INSTANCE: An $n \times n$ matrix $A = (a_{ij})$ with entries in Z_0^+.
QUESTION: Is there a collection $C = \{C_1, C_2, \ldots, C_n\}$ of sets such that for all i,j, $1 \leqslant i,j \leqslant n$, $a_{ij} = |C_i \cap C_j|$?

Reference: [Chvátal, 1978]. Transformation from GRAPH 3-COLORABILITY.
Comment: Remains NP-complete even if all $a_{ii} = 3$, $1 \leqslant i \leqslant m$ (and hence all C_i must have cardinality 3). If all $a_{ii} = 2$, it is equivalent to edge graph recognition and hence can be solved in polynomial time (e.g., see [Harary, 1969]).

[SP10] COMPARATIVE CONTAINMENT

INSTANCE: Two collections $R=\{R_1,R_2,\ldots,R_k\}$ and $S=\{S_1,S_2,\ldots,S_l\}$ of subsets of a finite set X, weights $w(R_i) \in Z^+$, $1 \leqslant i \leqslant k$, and $w(S_j) \in Z^+$, $1 \leqslant j \leqslant l$.
QUESTION: Is there a subset $Y \subseteq X$ such that

$$\sum_{Y \subseteq R_i} w(R_i) \geqslant \sum_{Y \subseteq S_j} w(S_j) \ ?$$

Reference: [Plaisted, 1976]. Transformation from VERTEX COVER.
Comment: Remains NP-complete even if all subsets in R and S have weight 1 [Garey and Johnson, ——].

[SP11] 3-MATROID INTERSECTION

INSTANCE: Three matroids $(E,F_1),(E,F_2),(E,F_3)$, positive integer $K \leqslant |E|$. (A matroid (E,F) consists of a set E of elements and a non-empty family F of subsets of E such that (1) $S \in F$ implies all subsets of S are in F and (2) if two sets $S,S' \in F$ satisfy $|S|=|S'|+1$, then there exists an element $e \in S-S'$ such that $(S' \cup \{e\}) \in F$.)
QUESTION: Is there a subset $E' \subseteq E$ such that $|E'|=K$ and $E' \in (F_1 \cap F_2 \cap F_3)$?

Reference: Transformation from 3DM.
Comment: The related 2-MATROID INTERSECTION problem can be solved in polynomial time, even if the matroids are described by giving polynomial time algorithms for recognizing their members, and even if each element $e \in E$ has a weight $w(e) \in Z^+$, with the goal being to find an $E' \in (F_1 \cap F_2)$ having maximum total weight (e.g., see [Lawler, 1976a]).

A3.2 WEIGHTED SET PROBLEMS

[SP12] PARTITION

INSTANCE: Finite set A and a size $s(a) \in Z^+$ for each $a \in A$.
QUESTION: Is there a subset $A' \subseteq A$ such that $\sum_{a \in A'} s(a) = \sum_{a \in A-A'} s(a)$?

Reference: [Karp, 1972]. Transformation from 3DM (see Section 3.1.5).
Comment: Remains NP-complete even if we require that $|A'|=|A|/2$, or if the elements in A are ordered as $a_1,a_2,\ldots,a_{2n}$ and we require that A' contain exactly one of a_{2i-1},a_{2i} for $1 \leqslant i \leqslant n$. However, all these problems can be solved in pseudo-polynomial time by dynamic programming (see Section 4.2).

[SP13] SUBSET SUM

INSTANCE: Finite set A, size $s(a) \in Z^+$ for each $a \in A$, positive integer B.
QUESTION: Is there a subset $A' \subseteq A$ such that the sum of the sizes of the elements in A' is exactly B?

Reference: [Karp, 1972]. Transformation from PARTITION.
Comment: Solvable in pseudo-polynomial time (see Section 4.2).

[SP14] SUBSET PRODUCT

INSTANCE: Finite set A, a size $s(a) \in Z^+$ for each $a \in A$, and a positive integer B.

QUESTION: Is there a subset $A' \subseteq A$ such that the product of the sizes of the elements in A' is exactly B?

Reference: [Yao, 1978b]. Transformation from X3C.

Comment: NP-complete in the strong sense.

[SP15] 3-PARTITION

INSTANCE: Set A of $3m$ elements, a bound $B \in Z^+$, and a size $s(a) \in Z^+$ for each $a \in A$ such that $B/4 < s(a) < B/2$ and such that $\sum_{a \in A} s(a) = mB$.

QUESTION: Can A be partitioned into m disjoint sets $A_1, A_2, \ldots, A_m$ such that, for $1 \leqslant i \leqslant m$, $\sum_{a \in A_i} s(a) = B$ (note that each A_i must therefore contain exactly three elements from A)?

Reference: [Garey and Johnson, 1975]. Transformation from 3DM (see Section 4.2).

Comment: NP-complete in the strong sense.

[SP16] NUMERICAL 3-DIMENSIONAL MATCHING

INSTANCE: Disjoint sets W, X, and Y, each containing m elements, a size $s(a) \in Z^+$ for each element $a \in W \cup X \cup Y$, and a bound $B \in Z^+$.

QUESTION: Can $W \cup X \cup Y$ be partitioned into m disjoint sets $A_1, A_2, \ldots, A_m$ such that each A_i contains exactly one element from each of W, X, and Y and such that, for $1 \leqslant i \leqslant m$, $\sum_{a \in A_i} s(a) = B$?

Reference: [Garey and Johnson, ——]. Transformation from 3DM (see proof of Theorem 4.4).

Comment: NP-complete in the strong sense.

[SP17] NUMERICAL MATCHING WITH TARGET SUMS

INSTANCE: Disjoint sets X and Y, each containing m elements, a size $s(a) \in Z^+$ for each element $a \in X \cup Y$, and a target vector $\langle B_1, B_2, \ldots, B_m \rangle$ with positive integer entries.

QUESTION: Can $X \cup Y$ be partitioned into m disjoint sets $A_1, A_2, \ldots, A_m$, each containing exactly one element from each of X and Y, such that, for $1 \leqslant i \leqslant m$, $\sum_{a \in A_i} s(a) = B_i$?

Reference: Transformation from NUMERICAL 3-DIMENSIONAL MATCHING.

Comment: NP-complete in the strong sense, but solvable in polynomial time if $B_1 = B_2 = \cdots = B_m$.

[SP18] EXPECTED COMPONENT SUM

INSTANCE: Collection C of m-dimensional vectors $v = (v_1, v_2, \ldots, v_m)$ with non-negative integer entries, positive integers K and B.

QUESTION: Is there a partition of C into disjoint sets $C_1, C_2, \ldots, C_K$ such that

$$\sum_{i=1}^{K} \max_{1 \leqslant j \leqslant m} \left(\sum_{v \in C_i} v_j \right) \geqslant B \,?$$

Reference: [Garey and Johnson, ——]. Transformation from X3C. The problem is due to [Witsenhausen, 1978] and corresponds to finding a partition that maximizes the expected value of the largest component sum, assuming all sets in the partition are equally likely.

Reference: NP-complete even if all entries are 0's and 1's. Solvable in polynomial time if K is fixed. The variant in which we ask for a partition with K non-empty sets that yields a sum of B or *less* is NP-complete even if K is fixed at 3 and all entries are 0's and 1's.

[SP19] MINIMUM SUM OF SQUARES

INSTANCE: Finite set A, a size $s(a) \in Z^+$ for each $a \in A$, positive integers $K \leqslant |A|$ and J.

QUESTION: Can A be partitioned into K disjoint sets $A_1, A_2, \ldots, A_K$ such that

$$\sum_{i=1}^{K} \left[\sum_{a \in A_i} s(a) \right]^2 \leqslant J \text{ ?}$$

Reference: Transformation from PARTITION or 3-PARTITION.

Comment: NP-complete in the strong sense. NP-complete in the ordinary sense and solvable in pseudo-polynomial time for any fixed K. Variants in which the bound K on the number of sets is replaced by a bound B on either the maximum set cardinality or the maximum total set size are also NP-complete in the strong sense [Wong and Yao, 1976]. In all these cases, NP-completeness is preserved if the exponent 2 is replaced by any fixed rational $\alpha > 1$.

[SP20] Kth LARGEST SUBSET (*)

INSTANCE: Finite set A, size $s(a) \in Z^+$ for each $a \in A$, positive integers K and B.

QUESTION: Are there K or more distinct subsets $A' \subseteq A$ for which the sum of the sizes of the elements in A' does not exceed B?

Reference: [Johnson and Kashdan, 1976]. Transformation from SUBSET SUM.

Comment: Not known to be in NP. Solvable in pseudo-polynomial time (polynomial in K, $|A|$, and $\log \sum s(a)$) [Lawler, 1972]. The corresponding enumeration problem is #P-complete.

[SP21] Kth LARGEST m-TUPLE (*)

INSTANCE: Sets $X_1, X_2, \ldots, X_m \subseteq Z^+$, a size $s(x) \in Z^+$ for each $x \in X_i$, $1 \leqslant i \leqslant m$, and positive integers K and B.

QUESTION: Are there K or more distinct m-tuples $(x_1, x_2, \ldots, x_m)$ in $X_1 \times X_2 \times \cdots \times X_m$ for which $\sum_{i=1}^{m} s(x_i) \geqslant B$?

Reference: [Johnson and Mizoguchi, 1978]. Transformation from PARTITION.

Comment: Not known to be in NP. Solvable in polynomial time for fixed m, and in pseudo-polynomial time in general (polynomial in K, $\sum |X_i|$, and $\log \sum s(x)$). The corresponding enumeration problem is #P-complete.

A4 STORAGE AND RETRIEVAL

A4.1 DATA STORAGE

[SR1] BIN PACKING

INSTANCE: Finite set U of items, a size $s(u) \in Z^+$ for each $u \in U$, a positive integer bin capacity B, and a positive integer K.
QUESTION: Is there a partition of U into disjoint sets $U_1, U_2, \ldots, U_K$ such that the sum of the sizes of the items in each U_i is B or less?

Reference: Transformation from PARTITION, 3-PARTITION.
Comment: NP-complete in the strong sense. NP-complete and solvable in pseudo-polynomial time for each fixed $K \geqslant 2$. Solvable in polynomial time for any fixed B by exhaustive search.

[SR2] DYNAMIC STORAGE ALLOCATION

INSTANCE: Set A of items to be stored, each $a \in A$ having a size $s(a) \in Z^+$, an arrival time $r(a) \in Z_0^+$, and a departure time $d(a) \in Z^+$, and a positive integer storage size D.
QUESTION: Is there a feasible allocation of storage for A, i.e., a function $\sigma: A \rightarrow \{1,2, \ldots, D\}$ such that for every $a \in A$ the allocated storage interval $I(a) = [\sigma(a), \sigma(a) + s(a) - 1]$ is contained in $[1,D]$ and such that, for all $a, a' \in A$, if $I(a) \cap I(a')$ is nonempty then either $d(a) \leqslant r(a')$ or $d(a') \leqslant r(a)$?

Reference: [Stockmeyer, 1976b]. Transformation from 3-PARTITION.
Comment: NP-complete in the strong sense, even if $s(a) \in \{1,2\}$ for all $a \in A$. Solvable in polynomial time if all item sizes are the same, by interval graph coloring algorithms (e.g., see [Gavril, 1972]).

[SR3] PRUNED TRIE SPACE MINIMIZATION

INSTANCE: Finite set S, collection F of functions $f: S \rightarrow Z^+$, and a positive integer K.
QUESTION: Is there a sequence $\langle f_1, f_2, \ldots, f_m \rangle$ of distinct functions from F such that for every two elements $a, b \in S$ there is some i, $1 \leqslant i \leqslant m$, for which $f_i(a) \neq f_i(b)$ and such that, if $N(i)$ denotes the number of distinct i-tuples $X = (x_1, x_2, \ldots, x_i)$ for which there is more than one $a \in S$ having $(f_1(a), f_2(a), \ldots, f_i(a)) = X$, then $\sum_{i=1}^{m} N(i) \leqslant K$?

Reference: [Comer and Sethi, 1976]. Transformation from 3DM.
Comment: Remains NP-complete even if all $f \in F$ have range $\{0,1\}$. Variants in which the "pruned trie" data structure abstracted above is replaced by "full trie," "collapsed trie," or "pruned 0-trie" are also NP-complete. The related "access time minimization" problem is also NP-complete for pruned tries, where we ask for a sequence $\langle f_1, f_2, \ldots, f_m \rangle$ of functions from F that distinguishes every two elements from S as above and such that, if the access time $L(a)$ for $a \in S$ is defined to be the least i for which no other $b \in S$ has $(f_1(b), f_2(b), \ldots, f_i(b))$ identical to $(f_1(a), f_2(a), \ldots, f_i(a))$, then $\sum_{a \in S} L(a) \leqslant K$.

[SR4] EXPECTED RETRIEVAL COST

INSTANCE: Set R of records, rational probability $p(r) \in [0,1]$ for each $r \in R$, with $\sum_{r \in R} p(r) = 1$, number m of sectors, and a positive integer K.
QUESTION: Is there a partition of R into disjoint subsets $R_1, R_2, \ldots, R_m$ such that, if $p(R_i) = \sum_{r \in R_i} p(r)$ and the "latency cost" $d(i,j)$ is defined to be $j-i-1$ if $1 \leqslant i < j \leqslant m$ and to be $m-i+j-1$ if $1 \leqslant j \leqslant i \leqslant m$, then the sum over all ordered pairs i,j, $1 \leqslant i,j \leqslant m$, of $p(R_i) \cdot p(R_j) \cdot d(i,j)$ is at most K?
Reference: [Cody and Coffman, 1976]. Transformation from PARTITION, 3-PARTITION.
Comment: NP-complete in the strong sense. NP-complete and solvable in pseudo-polynomial time for each fixed $m \geqslant 2$.

[SR5] ROOTED TREE STORAGE ASSIGNMENT

INSTANCE: Finite set X, collection $C = \{X_1, X_2, \ldots, X_n\}$ of subsets of X, positive integer K.
QUESTION: Is there a collection $C' = \{X_1', X_2', \ldots, X_n'\}$ of subsets of X such that $X_i \subseteq X_i'$ for $1 \leqslant i \leqslant n$, such that $\sum_{i=1}^{n} |X_i' - X_i| \leqslant K$, and such that there is a directed rooted tree $T = (X,A)$ in which the elements of each X_i', $1 \leqslant i \leqslant n$, form a directed path?
Reference: [Gavril, 1977a]. Transformation from ROOTED TREE ARRANGEMENT.

[SR6] MULTIPLE COPY FILE ALLOCATION

INSTANCE: Graph $G = (V,E)$, for each $v \in V$ a usage $u(v) \in Z^+$ and a storage cost $s(v) \in Z^+$, and a positive integer K.
QUESTION: Is there a subset $V' \subseteq V$ such that, if for each $v \in V$ we let $d(v)$ denote the number of edges in the shortest path in G from v to a member of V', we have

$$\sum_{v \in V'} s(v) + \sum_{v \in V} d(v) \cdot u(v) \leqslant K \ ?$$

Reference: [Van Sickle and Chandy, 1977]. Transformation from VERTEX COVER.
Comment: NP-complete in the strong sense, even if all $v \in V$ have the same value of $u(v)$ and the same value of $s(v)$.

[SR7] CAPACITY ASSIGNMENT

INSTANCE: Set C of communication links, set $M \subseteq Z^+$ of capacities, cost function $g: C \times M \rightarrow Z^+$, delay penalty function $d: C \times M \rightarrow Z^+$ such that, for all $c \in C$ and $i < j \in M$, $g(c,i) \leqslant g(c,j)$ and $d(c,i) \geqslant d(c,j)$, and positive integers K and J.
QUESTION: Is there an assignment $\sigma: C \rightarrow M$ such that the total cost $\sum_{c \in C} g(c,\sigma(c))$ does not exceed K and such that the total delay penalty $\sum_{c \in C} d(c,\sigma(c))$ does not exceed J?
Reference: [Van Sickle and Chandy, 1977]. Transformation from SUBSET SUM.
Comment: Solvable in pseudo-polynomial time.

A4.2 COMPRESSION AND REPRESENTATION

[SR8] SHORTEST COMMON SUPERSEQUENCE

INSTANCE: Finite alphabet Σ, finite set R of strings from Σ^*, and a positive integer K.

QUESTION: Is there a string $w \in \Sigma^*$ with $|w| \leqslant K$ such that each string $x \in R$ is a subsequence of w, i.e., $w = w_0 x_1 w_1 x_2 w_2 \cdots x_k w_k$ where each $w_i \in \Sigma^*$ and $x = x_1 x_2 \cdots x_k$?

Reference: [Maier, 1978]. Transformation from VERTEX COVER.

Comment: Remains NP-complete even if $|\Sigma| = 5$. Solvable in polynomial time if $|R| = 2$ (by first computing the largest common subsequence) or if all $x \in R$ have $|x| \leqslant 2$.

[SR9] SHORTEST COMMON SUPERSTRING

INSTANCE: Finite alphabet Σ, finite set R of strings from Σ^*, and a positive integer K.

QUESTION: Is there a string $w \in \Sigma^*$ with $|w| \leqslant K$ such that each string $x \in R$ is a substring of w, i.e., $w = w_0 x w_1$ where each $w_i \in \Sigma^*$?

Reference: [Maier and Storer, 1977]. Transformation from VERTEX COVER for cubic graphs.

Comment: Remains NP-complete even if $|\Sigma| = 2$ or if all $x \in R$ have $|x| \leqslant 8$ and contain no repeated symbols. Solvable in polynomial time if all $x \in R$ have $|x| \leqslant 2$.

[SR10] LONGEST COMMON SUBSEQUENCE

INSTANCE: Finite alphabet Σ, finite set R of strings from Σ^*, and a positive integer K.

QUESTION: Is there a string $w \in \Sigma^*$ with $|w| \geqslant K$ such that w is a subsequence of each $x \in R$?

Reference: [Maier, 1978]. Transformation from VERTEX COVER.

Comment: Remains NP-complete even if $|\Sigma| = 2$. Solvable in polynomial time for any fixed K or for fixed $|R|$ (by dynamic programming, e.g., see [Wagner and Fischer, 1974]). The analogous LONGEST COMMON SUBSTRING problem is trivially solvable in polynomial time.

[SR11] BOUNDED POST CORRESPONDENCE PROBLEM

INSTANCE: Finite alphabet Σ, two sequences $a = (a_1, a_2, \ldots, a_n)$ and $b = (b_1, b_2, \ldots, b_n)$ of strings from Σ^*, and a positive integer $K \leqslant n$.

QUESTION: Is there a sequence $i_1, i_2, \ldots, i_k$ of $k \leqslant K$ (not necessarily distinct) positive integers, each between 1 and n, such that the two strings $a_{i_1} a_{i_2} \cdots a_{i_k}$ and $b_{i_1} b_{i_2} \cdots b_{i_k}$ are identical?

Reference: [Constable, Hunt, and Sahni, 1974]. Generic transformation.

Comment: Problem is undecidable if no upper bound is placed on k, e.g., see [Hopcroft and Ullman, 1969].

[SR12] HITTING STRING

INSTANCE: Finite set A of strings over $\{0,1,*\}$, all having the same length n.
QUESTION: Is there a string $x \in \{0,1\}^*$ with $|x| = n$ such that for each string $a \in A$ there is some i, $1 \leqslant i \leqslant n$, for which the ith symbol of a and the ith symbol of x are identical?

Reference: [Fagin, 1974]. Transformation from 3SAT.

[SR13] SPARSE MATRIX COMPRESSION

INSTANCE: An $m \times n$ matrix A with entries $a_{ij} \in \{0,1\}$, $1 \leqslant i \leqslant m$, $1 \leqslant j \leqslant n$, and a positive integer $K \leqslant mn$.
QUESTION: Is there a sequence $(b_1, b_2, \ldots, b_{n+K})$ of integers b_i, each satisfying $0 \leqslant b_i \leqslant m$, and a function $s: \{1,2, \ldots, m\} \rightarrow \{1,2, \ldots, K\}$ such that, for $1 \leqslant i \leqslant m$ and $1 \leqslant j \leqslant n$, the entry $a_{ij} = 1$ if and only if $b_{s(i)+j-1} = i$?

Reference: [Even, Lichtenstein, and Shiloach, 1977]. Transformation from GRAPH 3-COLORABILITY.
Comment: Remains NP-complete for fixed $K = 3$.

[SR14] CONSECUTIVE ONES SUBMATRIX

INSTANCE: An $m \times n$ matrix A of 0's and 1's and a positive integer K.
QUESTION: Is there an $m \times K$ submatrix B of A that has the "consecutive ones" property, i.e., such that the columns of B can be permuted so that in each row all the 1's occur consecutively?

Reference: [Booth, 1975]. Transformation from HAMILTONIAN PATH.
Comment: The variant in which we ask instead that B have the "circular ones" property, i.e., that the columns of B can be permuted so that in each row either all the 1's or all the 0's occur consecutively, is also NP-complete. Both problems can be solved in polynomial time if $K = n$ (in which case we are asking if A has the desired property), e.g., see [Fulkerson and Gross, 1965], [Tucker, 1971], and [Booth and Lueker, 1976].

[SR15] CONSECUTIVE ONES MATRIX PARTITION

INSTANCE: An $m \times n$ matrix A of 0's and 1's.
QUESTION: Can the rows of A be partitioned into two groups such that the resulting $m_1 \times n$ and $m_2 \times n$ matrices $(m_1 + m_2 = m)$ each have the consecutive ones property?

Reference: [Lipsky, 1978]. Transformation from HAMILTONIAN PATH for cubic graphs.

[SR16] CONSECUTIVE ONES MATRIX AUGMENTATION

INSTANCE: An $m \times n$ matrix A of 0's and 1's and a positive integer K.
QUESTION: Is there a matrix A', obtained from A by changing K or fewer 0 entries to 1's, such that A' has the consecutive ones property?

Reference: [Booth, 1975], [Papadimitriou, 1976a]. Transformation from OPTIMAL LINEAR ARRANGEMENT.

Comment: Variant in which we ask instead that A' have the circular ones property is also NP-complete.

[SR17] CONSECUTIVE BLOCK MINIMIZATION

INSTANCE: An $m \times n$ matrix A of 0's and 1's and a positive integer K.
QUESTION: Is there a permutation of the columns of A that results in a matrix B having at most K blocks of consecutive 1's, i.e., having at most K entries b_{ij} such that $b_{ij}=1$ and either $b_{i,j+1}=0$ or $j=n$?

Reference: [Kou, 1977]. Transformation from HAMILTONIAN PATH.
Comment: Remains NP-complete if "$j=n$" is replaced by "$j=n$ and $b_{i1}=0$" [Booth, 1975]. If K equals the number of rows of A that are not all 0, then these problems are equivalent to testing A for the consecutive ones property or the circular ones property, respectively, and can be solved in polynomial time.

[SR18] CONSECUTIVE SETS

INSTANCE: Finite alphabet Σ, collection $C=\{\Sigma_1,\Sigma_2, \ldots, \Sigma_n\}$ of subsets of Σ, and a positive integer K.
QUESTION: Is there a string $w \in \Sigma^*$ with $|w| \leqslant K$ such that, for each i, the elements of Σ_i occur in a consecutive block of $|\Sigma_i|$ symbols of W?

Reference: [Kou, 1977]. Transformation from HAMILTONIAN PATH.
Comment: The variant in which we ask only that the elements of each Σ_i occur in a consecutive block of $|\Sigma_i|$ symbols of the string ww (i.e., we allow blocks that circulate from the end of w back to its beginning) is also NP-complete [Booth, 1975]. If K is the number of distinct symbols in the Σ_i, then these problems are equivalent to determining whether a matrix has the consecutive ones property or the circular ones property and are solvable in polynomial time.

[SR19] 2-DIMENSIONAL CONSECUTIVE SETS

INSTANCE: Finite alphabet Σ, collection $C=\{\Sigma_1,\Sigma_2, \ldots, \Sigma_n\}$ of subsets of Σ.
QUESTION: Is there a partition of Σ into disjoint sets $X_1,X_2, \ldots, X_k$ such that each X_i has at most one element in common with each Σ_j and such that, for each $\Sigma_j \in C$, there is an index $l(j)$ such that Σ_j is contained in

$$X_{l(j)} \cup X_{l(j)+1} \cup \cdots \cup X_{l(j)+|\Sigma_j|-1} \ ?$$

Reference: [Lipsky, 1977b]. Transformation from GRAPH 3-COLORABILITY.
Comment: Remains NP-complete if all $\Sigma_j \in C$ have $|\Sigma_j| \leqslant 5$, but is solvable in polynomial time if all $\Sigma_j \in C$ have $|\Sigma_j| \leqslant 2$.

[SR20] STRING-TO-STRING CORRECTION

INSTANCE: Finite alphabet Σ, two strings $x,y \in \Sigma^*$, and a positive integer K.
QUESTION: Is there a way to derive the string y from the string x by a sequence of K or fewer operations of single symbol deletion or adjacent symbol interchange?

Reference: [Wagner, 1975]. Transformation from SET COVERING.
Comment: Solvable in polynomial time if the operation set is expanded to include the operations of changing a single character and of inserting a single character,

even if interchanges are not allowed (e.g., see [Wagner and Fischer, 1974]), or if the *only* operation is adjacent symbol interchange [Wagner, 1975]. See reference for related results for cases in which different operations can have different costs.

[SR21] GROUPING BY SWAPPING

INSTANCE: Finite alphabet Σ, string $x \in \Sigma^*$, and a positive integer K.
QUESTION: Is there a sequence of K or fewer adjacent symbol interchanges that converts x into a string y in which all occurrences of each symbol $a \in \Sigma$ are in a single block, i.e., y has no subsequences of the form aba for $a,b \in \Sigma$ and $a \neq b$?

Reference: [Howell, 1977]. Transformation from FEEDBACK EDGE SET.

[SR22] EXTERNAL MACRO DATA COMPRESSION

INSTANCE: Alphabet Σ, string $s \in \Sigma^*$, pointer cost $h \in Z^+$, and a bound $B \in Z^+$.
QUESTION: Are there strings D (dictionary string) and C (compressed string) in $(\Sigma \cup \{p_i: 1 \leqslant i \leqslant |s|\})^*$, where the symbols p_i are "pointers," such that

$$|D| + |C| + (h-1)\cdot(\text{number of occurrences of pointers in } D \text{ and } C) \leqslant B$$

and such that there is a way of identifying pointers with substrings of D so that S can be obtained from C by repeatedly replacing pointers in C by their corresponding substrings in D?

Reference: [Storer, 1977], [Storer and Szymanski, 1978]. Transformation from VERTEX COVER.

Comment: Remains NP-complete even if h is any fixed integer 2 or greater. Many variants, including those in which D can contain no pointers and/or no pointers can refer to overlapping strings, are also NP-complete. If the alphabet size is fixed at 3 or greater, and the pointer cost is $\lceil h \cdot \log|s| \rceil$, the problem is also NP-complete. For further variants, including the case of "original pointers," see references.

[SR23] INTERNAL MACRO DATA COMPRESSION

INSTANCE: Alphabet Σ, string $s \in \Sigma^*$, pointer cost $h \in Z^+$, and a bound $B \in Z^+$.
QUESTION: Is there a single string $C \in (\Sigma \cup \{p_i: 1 \leqslant i \leqslant |s|\})^*$ such that

$$|C| + (h-1)\cdot\ (\text{number of occurences of pointers in } C) \leqslant B$$

and such that there is a way of identifying pointers with substrings of C so that s can be obtained from C by using C as both compressed string and dictionary string in the manner indicated in the previous problem?

Reference: [Storer, 1977], [Storer and Szymanski, 1978]. Transformation from VERTEX COVER.

Comment: Remains NP-complete even if h is any fixed integer 2 or greater. For other NP-complete variants (as in the previous problem), see references.

[SR24] REGULAR EXPRESSION SUBSTITUTION

INSTANCE: Two finite alphabets $X = \{x_1, x_2, \ldots, x_n\}$ and $Y = \{y_1, y_2, \ldots, y_m\}$, a regular expression R over $X \cup Y$, regular expressions $R_1, R_2, \ldots, R_n$ over Y, and a string $w \in Y^*$.
QUESTION: Is there a string z in the language determined by R and for each i,

$1 \leqslant i \leqslant n$, a string w_i in the language determined by R_i such that, if each string w_i is substituted for every occurrence of the symbol x_i in z, then the resulting string is identical to w?

Reference: [Aho and Ullman, 1977]. Transformation from X3C.

[SR25] RECTILINEAR PICTURE COMPRESSION

INSTANCE: An $n \times n$ matrix M of 0's and 1's, and a positive integer K.

QUESTION: Is there a collection of K or fewer rectangles that covers precisely those entries in M that are 1's, i.e., is there a sequence of quadruples (a_i,b_i,c_i,d_i), $1 \leqslant i \leqslant K$, where $a_i \leqslant b_i$, $c_i \leqslant d_i$, $1 \leqslant i \leqslant K$, such that for every pair (i,j), $1 \leqslant i,j \leqslant n$, $M_{ij}=1$ if and only if there exists a k, $1 \leqslant k \leqslant K$, such that $a_k \leqslant i \leqslant b_k$ and $c_k \leqslant j \leqslant d_k$?

Reference: [Masek, 1978]. Transformation from 3SAT.

A4.3 DATABASE PROBLEMS

[SR26] MINIMUM CARDINALITY KEY

INSTANCE: A set A of "attribute names," a collection F of ordered pairs of subsets of A (called "functional dependencies" on A), and a positive integer M.

QUESTION: Is there a key of cardinality M or less for the relational system $<A,F>$, i.e., a minimal subset $K \subseteq A$ with $|K| \leqslant M$ such that the ordered pair (K,A) belongs to the "closure" F^* of F defined by (1) $F \subseteq F^*$, (2) $B \subseteq C \subseteq A$ implies $(C,B) \in F^*$, (3) $(B,C),(C,D) \in F^*$ implies $(B,D) \in F^*$, and (4) $(B,C),(B,D) \in F^*$ implies $(B,C \cup D) \in F^*$?

Reference: [Lucchesi and Osborne, 1977], [Lipsky, 1977a]. Transformation from VERTEX COVER. See [Date, 1975] for general background on relational data bases.

[SR27] ADDITIONAL KEY

INSTANCE: A set A of attribute names, a collection F of functional dependencies on A, a subset $R \subseteq A$, and a set K of keys for the relational scheme $<R,F>$.

QUESTION: Does R have a key not already contained in K, i.e., is there an $R' \subseteq R$ such that $R' \notin K$, $(R',R) \in F^*$, and for no $R'' \subseteq R'$ is $(R'',R) \in F^*$?

Reference: [Beeri and Bernstein, 1978]. Transformation from HITTING SET.

[SR28] PRIME ATTRIBUTE NAME

INSTANCE: A set A of attribute names, a collection F of functional dependencies on A, and a specified name $x \in A$.

QUESTION: Is x a "prime attribute name" for $<A,F>$, i.e., is there a key K for $<A,F>$ such that $x \in K$?

Reference: [Lucchesi and Osborne, 1977]. Transformation from MINIMUM CARDINALITY KEY.

[SR29] BOYCE-CODD NORMAL FORM VIOLATION

INSTANCE: A set A of attribute names, a collection F of functional dependencies on A, and a subset $A' \subseteq A$.

QUESTION: Does A' violate Boyce-Codd normal form for the relational system $<A,F>$, i.e., is there a subset $X \subseteq A'$ and two attribute names $y,z \in A'-X$ such that $(X,\{y\}) \in F^*$ and $(X,\{z\}) \notin F^*$, where F^* is the closure of F?

Reference: [Bernstein and Beeri, 1976], [Beeri and Bernstein, 1978]. Transformation from HITTING SET.

Comment: Remains NP-complete even if A' is required to satisfy "third normal form," i.e., if $X \subseteq A'$ is a key for the system $<A',F>$ and if two names $y,z \in A'-X$ satisfy $(X,\{y\}) \in F^*$ and $(X,\{z\}) \notin F^*$, then z is a prime attribute for $<A',F>$.

[SR30] CONJUNCTIVE QUERY FOLDABILITY

INSTANCE: Finite domain set D, a collection $R = \{R_1, R_2, \ldots, R_m\}$ of relations, where each R_i consists of a set of d_i-tuples with entries from D, a set X of distinguished variables, a set Y of undistinguished variables, and two "queries" Q_1 and Q_2 over X,Y,D, and R, where a query Q has the form

$$(x_1,x_2, \ldots, x_k)(\exists y_1,y_2, \ldots, y_l)(A_1 \wedge A_2 \wedge \cdots \wedge A_r)$$

for some k,l, and r, with $X' = \{x_1,x_2, \ldots, x_k\} \subseteq X$, $Y' = \{y_1,y_2, \ldots, y_l\} \subseteq Y$, and each A_i of the form $R_j(u_1,u_2, \ldots, u_{d_j})$ with each $u \in D \cup X' \cup Y'$ (see reference for interpretation of such expressions in terms of data bases).

QUESTION: Is there a function $\sigma: Y \rightarrow X \cup Y \cup D$ such that, if for each $y \in Y$ the symbol $\sigma(y)$ is substituted for every occurrence of y in Q_1, then the result is query Q_2?

Reference: [Chandra and Merlin, 1977]. Transformation from GRAPH 3-COLORABILITY.

Comment: The isomorphism problem for conjunctive queries (with two queries being isomorphic if they are the same up to one-to-one renaming of the variables, reordering of conjuncts, and reordering within quantifications) is polynomially equivalent to graph isomorphism.

[SR31] CONJUNCTIVE BOOLEAN QUERY

INSTANCE: Finite domain set D, a collection $R = \{R_1, R_2, \ldots, R_m\}$ of relations, where each R_i consists of a set of d_i-tuples with entries from D, and a conjunctive Boolean query Q over R and D, where such a query Q is of the form

$$(\exists y_1,y_2, \ldots, y_l)(A_1 \wedge A_2 \wedge \cdots \wedge A_r)$$

with each A_i of the form $R_j(u_1,u_2, \ldots, u_{d_j})$ where each $u \in \{y_1,y_2, \ldots, y_l\} \cup D$.

QUESTION: Is Q, when interpreted as a statement about R and D, true?

Reference: [Chandra and Merlin, 1977]. Transformation from CLIQUE.

Comment: If we are allowed to replace the conjunctive query Q by an arbitrary first-order sentence involving the predicates in R, then the problem becomes *PSPACE*-complete, even for $D = \{0,1\}$.

[SR32] TABLEAU EQUIVALENCE

INSTANCE: A set A of attribute names, a collection F of ordered pairs of subsets of A, a set X of distinguished variables, a set Y of undistinguished variables, a set C_a of constants for each $a \in A$, and two "tableaux" T_1 and T_2 over X, Y, and the C_a. (A tableau is essentially a matrix with a column for each attribute and entries from X, Y, and the C_a, along with a blank symbol. For details and an interpretation in terms of relational expressions, see reference.)
QUESTION: Are T_1 and T_2 "weakly equivalent," i.e., do they represent identical relations under "universal interpretations"?

Reference: [Aho, Sagiv, and Ullman, 1978]. Transformation from 3SAT.
Comment: Remains NP-complete even if the tableaux come from "expressions" that have no "select" operations, or if the tableaux come from expressions that have select operations but F is empty, or if F is empty, the tableaux contain no constants, and the tableaux do not necessarily come from expressions at all. Problem is solvable in polynomial time for "simple" tableaux. The same results hold also for "strong equivalence," where the two tableaux must represent identical relations under *all* interpretations. The problem of tableau "containment," however, is NP-complete even for simple tableaux and for still further restricted tableaux [Sagiv and Yannakakis, 1978].

[SR33] SERIALIZABILITY OF DATABASE HISTORIES

INSTANCE: Set V of database variables, collection T of "transactions" (R_i, W_i), $1 \leq i \leq n$, where R_i and W_i are both subsets of V (called the "read set" and the "write set," respectively), and a "history" H for T, where a history is simply a permutation of all the R_i and the W_i in which each R_i occurs before the corresponding W_i.
QUESTION: Is there a serial history H' for T (i.e., a history in which each R_i occurs immediately before the corresponding W_i) that is equivalent to H in the sense that (1) both histories have the same set of "live" transactions (where a transaction (R_i, W_i) is live in a history if there is some $v \in V$ such that either W_i is the last write set to contain v or W_i is the last write set to contain v before v appears in the read set of some other live transaction), and (2) for any two live transactions (R_i, W_i) and (R_j, W_j) and any $v \in W_i \cap R_j$, W_i is the last write set to contain v before R_j in H if and only if W_i is the last write set to contain v before R_j in H'?

Reference: [Papadimitriou, Bernstein, and Rothnie, 1977], [Papadimitriou, 1978c]. Transformation from MONOTONE 3SAT.
Comment: For related polynomial time solvable subcases and variants, see [Papadimitriou, 1978c].

[SR34] SAFETY OF DATABASE TRANSACTION SYSTEMS (*)

INSTANCE: Set V of database variables, and a collection T of "transactions" (R_i, W_i), $1 \leq i \leq n$, where R_i and W_i are both subsets of V.
QUESTION: Is every history H for T equivalent to some serial history?

Reference: [Papadimitriou, Bernstein, and Rothnie, 1977]. Transformation from HITTING SET.
Comment: Not known either to be in NP or to be in co-NP. Testing whether every history H for T is "D-equivalent" to some serial history can be done in polynomi-

al time, where two histories are D-equivalent if one can be obtained from the other by a sequence of interchanges of adjacent sets in such a way that at each step the new history is equivalent to the previous one.

[SR35] CONSISTENCY OF DATABASE FREQUENCY TABLES

INSTANCE: Set A of attribute names, domain set D_a for each $a \in A$, set V of objects, collection F of frequency tables for some pairs $a,b \in A$ (where a frequency table for $a,b \in A$ is a function $f_{a,b}: D_a \times D_b \to Z^+$ with the sum, over all pairs $x \in D_a$ and $y \in D_b$, of $f_{a,b}(x,y)$ equal to $|V|$), and a set K of triples (v,a,x) with $v \in V$, $a \in A$, and $x \in D_a$, representing the known attribute values.
QUESTION: Are the frequency tables in F consistent with the known attribute values in K, i.e., is there a collection of functions $g_a: V \to D_a$, for each $a \in A$, such that $g_a(v) = x$ if $(v,a,x) \in K$ and such that, for each $f_{a,b} \in F$, $x \in D_a$, and $y \in D_b$, the number of $v \in V$ for which $g_a(v) = x$ and $g_b(v) = y$ is exactly $f_{a,b}(x,y)$?

Reference: [Reiss, 1977b]. Transformation from 3SAT.
Comment: Above result implies that no polynomial time algorithm can be given for "compromising" a data base from its frequency tables by deducing prespecified attribute values, unless P = NP (see reference for details).

[SR36] SAFETY OF FILE PROTECTION SYSTEMS (*)

INSTANCE: Set R of "rights," set O of objects, set $S \subseteq O$ of subjects, set $P(s,o) \subseteq R$ of rights for each ordered pair $s \in S$ and $o \in O$, a finite set C of commands, each having the form "if $r_1 \in P(X_1,Y_1)$ and $r_2 \in P(X_2,Y_2)$ and $\cdots$ and $r_m \in P(X_m,Y_m)$, then $\theta_1, \theta_2, \ldots, \theta_n$" for $m,n \geqslant 0$ and each θ_i of the form "enter r_i into $P(X_j,Y_k)$" or "delete r_i from $P(K_j,Y_k)$," and a specified right $r' \in R$.
QUESTION: Is there a sequence of commands from C and a way of identifying each r_i, X_j, and Y_k with a particular element of R,S, and O, respectively, such that at some point in the execution of the sequence, the right r' is entered into a set $P(s,o)$ that previously did not contain r' (see reference for details on the execution of such a sequence)?

Reference: [Harrison, Ruzzo, and Ullman, 1976]. Transformation from LINEAR BOUNDED AUTOMATON ACCEPTANCE.
Comment: PSPACE-complete. Undecidable if operations that create or delete "subjects" and "objects" are allowed, even for certain "fixed" systems in which only the initial values of the $P(s,o)$ are allowed to vary. If no command can contain more than one operation, then the problem is NP-complete in general and solvable in polynomial time for fixed systems.

A5 SEQUENCING AND SCHEDULING

A5.1 SEQUENCING ON ONE PROCESSOR

[SS1] SEQUENCING WITH RELEASE TIMES AND DEADLINES

INSTANCE: Set T of tasks and, for each task $t \in T$, a length $l(t) \in Z^+$, a release time $r(t) \in Z_0^+$, and a deadline $d(t) \in Z^+$.

QUESTION: Is there a one-processor schedule for T that satisfies the release time constraints and meets all the deadlines, i.e., a one-to-one function $\sigma: T \rightarrow Z_0^+$, with $\sigma(t) > \sigma(t')$ implying $\sigma(t) \geqslant \sigma(t') + l(t')$, such that, for all $t \in T$, $\sigma(t) \geqslant r(t)$ and $\sigma(t) + l(t) \leqslant d(t)$?

Reference: [Garey and Johnson, 1977b]. Transformation from 3-PARTITION (see Section 4.2).

Comment: NP-complete in the strong sense. Solvable in pseudo-polynomial time if the number of allowed values for $r(t)$ and $d(t)$ is bounded by a constant, but remains NP-complete (in the ordinary sense) even when each can take on only two values. If all task lengths are 1, or "preemptions" are allowed, or all release times are 0, the general problem can be solved in polynomial time, even under "precedence constraints" [Lawler, 1973], [Lageweg, Lenstra, and Rinnooy Kan, 1976]. Can also be solved in polynomial time even if release times and deadlines are allowed to be arbitrary rationals and there are precedence constraints, so long as all tasks have equal length [Carlier, 1978], [Simons, 1978], [Garey, Johnson, Simons, and Tarjan, 1978], or preemptions are allowed [Blazewicz, 1976].

[SS2] SEQUENCING TO MINIMIZE TARDY TASKS

INSTANCE: Set T of tasks, partial order $\lessdot$ on T, for each task $t \in T$ a length $l(t) \in Z^+$ and a deadline $d(t) \in Z^+$, and a positive integer $K \leqslant |T|$.

QUESTION: Is there a one-processor schedule σ for T that obeys the precedence constraints, i.e., such that $t \lessdot t'$ implies $\sigma(t) + l(t) < \sigma(t')$, and such that there are at most K tasks $t \in T$ for which $\sigma(t) + l(t) > d(t)$?

Reference: [Garey and Johnson, 1976c]. Transformation from CLIQUE (see Section 3.2.3).

Comment: Remains NP-complete even if all task lengths are 1 and $\lessdot$ consists only of "chains" (each task has at most one immediate predecessor and at most one immediate successor) [Lenstra, 1977]. The general problem can be solved in polynomial time if $K=0$ [Lawler, 1973], or if $\lessdot$ is empty [Moore, 1968] [Sidney, 1973]. The $\lessdot$ empty case remains polynomially solvable if "agreeable" release times (i.e., $r(t) < r(t')$ implies $d(t) \leqslant d(t')$) are added [Kise, Ibaraki, and Mine, 1978], but is NP-complete for arbitrary release times (see previous problem).

[SS3] SEQUENCING TO MINIMIZE TARDY TASK WEIGHT

INSTANCE: Set T of tasks, for each task $t \in T$ a length $l(t) \in Z^+$, a weight $w(t) \in Z^+$, and a deadline $d(t) \in Z^+$, and a positive integer K.

QUESTION: Is there a one-processor schedule σ for T such that the sum of $w(t)$, taken over all $t \in T$ for which $\sigma(t) + l(t) > d(t)$, does not exceed K?

Reference: [Karp, 1972]. Transformation from PARTITION.

Comment: Can be solved in pseudo-polynomial time (time polynomial in $|T|$,

$\sum l(t)$, and $\log \sum w(t)$) [Lawler and Moore, 1969]. Can be solved in polynomial time if weights are "agreeable" (i.e., $w(t) < w(t')$ implies $l(t) \geqslant l(t')$) [Lawler, 1976c].

[SS4] SEQUENCING TO MINIMIZE WEIGHTED COMPLETION TIME

INSTANCE: Set T of tasks, partial order $\lessdot$ on T, for each task $t \in T$ a length $l(t) \in Z^+$ and a weight $w(t) \in Z^+$, and a positive integer K.

QUESTION: Is there a one-processor schedule σ for T that obeys the precedence constraints and for which the sum, over all $t \in T$, of $(\sigma(t) + l(t)) \cdot w(t)$ is K or less?

Reference: [Lawler, 1978]. Transformation from OPTIMAL LINEAR ARRANGEMENT.

Comment: NP-complete in the strong sense and remains so even if all task lengths are 1 or all task weights are 1. Can be solved in polynomial time for $\lessdot$ a "forest" [Horn, 1972], [Adolphson and Hu, 1973], [Garey, 1973], [Sidney, 1975] or if $\lessdot$ is "series-parallel" or "generalized series-parallel" [Knuth, 1973], [Lawler, 1978], [Adolphson, 1977], [Monma and Sidney, 1977]. If the partial order $\lessdot$ is replaced by individual task deadlines, the resulting problem in NP-complete in the strong sense [Lenstra, 1977], but can be solved in polynomial time if all task weights are equal [Smith, 1956]. If there are individual task release times instead of deadline, the resulting problem is NP-complete in the strong sense, even if all task weights are 1 [Lenstra, Rinnooy Kan, and Brucker, 1977]. The "preemptive" version of this latter problem is NP-complete in the strong sense [Labetoulle, Lawler, Lenstra, and Rinnooy Kan, 1978], but is solvable in polynomial time if all weights are equal [Graham, Lawler, Lenstra, and Rinnooy Kan, 1978].

[SS5] SEQUENCING TO MINIMIZE WEIGHTED TARDINESS

INSTANCE: Set T of tasks, for each task $t \in T$ a length $l(t) \in Z^+$, a weight $w(t) \in Z^+$, and a deadline $d(t) \in Z^+$, and a positive integer K.

QUESTION: Is there a one-processor schedule σ for T such that the sum, taken over all $t \in T$ satisfying $\sigma(t) + l(t) > d(t)$, of $(\sigma(t) + l(t) - d(t)) \cdot w(t)$ is K or less?

Reference: [Lawler, 1977a]. Transformation from 3-PARTITION.

Comment: NP-complete in the strong sense. If all weights are equal, the problem can be solved in pseudo-polynomial time [Lawler, 1977a] and is open as to ordinary NP-completeness. If all lengths are equal (with weights arbitrary), it can be solved in polynomial time by bipartite matching. If precedence constraints are added, the problem is NP-complete even with equal lengths and equal weights [Lenstra and Rinnooy Kan, 1978a]. If release times are added instead, the problem is NP-complete in the strong sense for equal task weights (see SEQUENCING WITH RELEASE TIMES AND DEADLINES), but can be solved by bipartite matching for equal lengths and arbitrary weights [Graham, Lawler, Lenstra, and Rinnooy Kan, 1978].

[SS6] SEQUENCING WITH DEADLINES AND SET-UP TIMES

INSTANCE: Set C of "compilers," set T of tasks, for each $t \in T$ a length $l(t) \in Z^+$, a deadline $d(t) \in Z^+$, and a compiler $k(t) \in C$, and for each $c \in C$ a "set-up time" $l(c) \in Z_0^+$.

QUESTION: Is there a one-processor schedule σ for T that meets all the task deadlines and that satisfies the additional constraint that, whenever two tasks t and t' with $\sigma(t) < \sigma(t')$ are scheduled "consecutively" (i.e., no other task t'' has $\sigma(t) < \sigma(t'') < \sigma(t')$) and have different compilers (i.e., $k(t) \neq k(t')$), then $\sigma(t') \geqslant \sigma(t) + l(t) + l(k(t'))$?

Reference: [Bruno and Downey, 1978]. Transformation from PARTITION.

Comment: Remains NP-complete even if all set-up times are equal. The related problem in which set-up times are replaced by "changeover costs," and we want to know if there is a schedule that meets all the deadlines and has total changeover cost at most K, is NP-complete even if all changeover costs are equal. Both problems can be solved in pseudo-polynomial time when the number of distinct deadlines is bounded by a constant. If the number of deadlines is unbounded, it is open whether these problems are NP-complete in the strong sense.

[SS7] SEQUENCING TO MINIMIZE MAXIMUM CUMULATIVE COST

INSTANCE: Set T of tasks, partial order $\lessdot$ on T, a "cost" $c(t) \in Z$ for each $t \in T$ (if $c(t) < 0$, it can be viewed as a "profit"), and a constant $K \in Z$.

QUESTION: Is there a one-processor schedule σ for T that obeys the precedence constraints and which has the property that, for every task $t \in T$, the sum of the costs for all tasks t' with $\sigma(t') \leqslant \sigma(t)$ is at most K?

Reference: [Abdel-Wahab, 1976]. Transformation from REGISTER SUFFICIENCY.

Comment: Remains NP-complete even if $c(t) \in \{-1,0,1\}$ for all $t \in T$. Can be solved in polynomial time if $\lessdot$ is series-parallel [Abdel-Wahab and Kameda, 1978], [Monma and Sidney, 1977].

A5.2 MULTIPROCESSOR SCHEDULING

[SS8] MULTIPROCESSOR SCHEDULING

INSTANCE: Set T of tasks, number $m \in Z^+$ of processors, length $l(t) \in Z^+$ for each $t \in T$, and a deadline $D \in Z^+$.

QUESTION: Is there an m-processor schedule for T that meets the overall deadline D, i.e., a function $\sigma: T \rightarrow Z_0^+$ such that, for all $u \geqslant 0$, the number of tasks $t \in T$ for which $\sigma(t) \leqslant u < \sigma(t) + l(t)$ is no more than m and such that, for all $t \in T$, $\sigma(t) + l(t) \leqslant D$?

Reference: Transformation from PARTITION (see Section 3.2.1).

Comment: Remains NP-complete for $m=2$, but can be solved in pseudo-polynomial time for any fixed m. NP-complete in the strong sense for m arbitrary (3-PARTITION is a special case). If all tasks have the same length, then this problem is trivial to solve in polynomial time, even for "different speed" processors.

[SS9] PRECEDENCE CONSTRAINED SCHEDULING

INSTANCE: Set T of tasks, each having length $l(t)=1$, number $m \in Z^+$ of processors, partial order $\lessdot$ on T, and a deadline $D \in Z^+$.

QUESTION: Is there an m-processor schedule σ for T that meets the overall deadline D and obeys the precedence constraints, i.e., such that $t \lessdot t'$ implies $\sigma(t') \geqslant \sigma(t) + l(t) = \sigma(t) + 1$?

Reference: [Ullman, 1975]. Transformation from 3SAT.

Comment: Remains NP-complete for $D=3$ [Lenstra and Rinnooy Kan, 1978a]. Can be solved in polynomial time if $m=2$ (e.g., see [Coffman and Graham, 1972]) or if m is arbitrary and $\lessdot$ is a "forest" [Hu, 1961] or has a chordal graph as complement [Papadimitriou and Yannakakis, 1978b]. Complexity remains open for all fixed $m \geqslant 3$ when $\lessdot$ is arbitrary. The $m=2$ case becomes NP-complete if both task lengths 1 and 2 are allowed [Ullman, 1975]. If each task t can only be executed by a specified processor $p(t)$, the problem is NP-complete for $m=2$ and $\lessdot$ arbitrary, and for m arbitrary and $\lessdot$ a forest, but can be solved in polynomial time for m arbitrary if $\lessdot$ is a "cyclic forest" [Goyal, 1976].

[SS10] RESOURCE CONSTRAINED SCHEDULING

INSTANCE: Set T of tasks, each having length $l(t)=1$, number $m \in Z^+$ of processors, number $r \in Z^+$ of resources, resource bounds B_i, $1 \leqslant i \leqslant r$, resource requirement $R_i(t)$, $0 \leqslant R_i(t) \leqslant B_i$, for each task t and resource i, and an overall deadline $D \in Z^+$.

QUESTION: Is there an m-processor schedule σ for T that meets the overall deadline D and obeys the resource constraints, i.e., such that for all $u \geqslant 0$, if $S(u)$ is the set of all $t \in T$ for which $\sigma(t) \leqslant u < \sigma(t) + l(t)$, then for each resource i the sum of $R_i(t)$ over all $t \in S(u)$ is at most B_i?

Reference: [Garey and Johnson, 1975]. Transformation from 3-PARTITION.

Comment: NP-complete in the strong sense, even if $r=1$ and $m=3$. Can be solved in polynomial time by matching for $m=2$ and r arbitrary. If a partial order $\lessdot$ is added, the problem becomes NP-complete in the strong sense for $r=1$, $m=2$, and $\lessdot$ a "forest." If each resource requirement is restricted to be either 0 or B_i, the problem is NP-complete for $m=2$, $r=1$, and $\lessdot$ arbitrary [Ullman, 1976].

[SS11] SCHEDULING WITH INDIVIDUAL DEADLINES

INSTANCE: Set T of tasks, each having length $l(t)=1$, number $m \in Z^+$ of processors, partial order $\lessdot$ on T, and for each task $t \in T$ a deadline $d(t) \in Z^+$.

QUESTION: Is there an m-processor schedule σ for T that obeys the precedence constraints and meets all the deadlines, i.e., $\sigma(t) + l(t) \leqslant d(t)$ for all $t \in T$?

Reference: [Brucker, Garey, and Johnson, 1977]. Transformation from VERTEX COVER.

Comment: Remains NP-complete even if $\lessdot$ is an "out-tree" partial order (no task has more than one immediate predecessor), but can be solved in polynomial time if $\lessdot$ is an "in-tree" partial order (no task has more than one immediate successor). Solvable in polynomial time if $m=2$ and $\lessdot$ is arbitrary [Garey and Johnson, 1976c], even if individual release times are included [Garey and Johnson, 1977b]. For $\lessdot$ empty, can be solved in polynomial time by matching for m arbitrary, even with release times and with a single resource having 0-1 valued requirements [Blazewicz, 1977b], [Blazewicz, 1978].

[SS12] **PREEMPTIVE SCHEDULING**

INSTANCE: Set T of tasks, number $m \in Z^+$ of processors, partial order $\lessdot$ on T, length $l(t) \in Z^+$ for each $t \in T$, and an overall deadline $D \in Z^+$.

QUESTION: Is there an m-processor "preemptive" schedule for T that obeys the precedence constraints and meets the overall deadline? (Such a schedule σ is identical to an ordinary m-processor schedule, except that we are allowed to subdivide each task $t \in T$ into any number of subtasks $t_1, t_2, \ldots, t_k$ such that $\sum_{i=1}^{k} l(t_i) = l(t)$ and it is required that $\sigma(t_i+1) \geqslant \sigma(t_i)+l(t_i)$ for $1 \leqslant i < k$. The precedence constraints are extended to subtasks by requiring that every subtask of t precede every subtask of t' whenever $t \lessdot t'$.)

Reference: [Ullman, 1975]. Transformation from 3SAT.

Comment: Can be solved in polynomial time if $m=2$ [Muntz and Coffman, 1969], if $\lessdot$ is a "forest" [Muntz and Coffman, 1970], or if $\lessdot$ is empty and individual task deadlines are allowed [Horn, 1974]. If "(uniform) different speed" processors are allowed, the problem can be solved in polynomial time if $m=2$ or if $\lessdot$ is empty [Horvath, Lam, and Sethi, 1977], [Gonzalez and Sahni, 1978b] in the latter case even if individual task deadlines are allowed [Sahni and Cho, 1977a]; if both $m=2$ and $\lessdot$ is empty, it can be solved in polynomial time, even if both integer release times and deadlines are allowed [Labetoulle, Lawler, Lenstra, and Rinnooy Kan, 1977]. For "unrelated" processors, the case with m fixed and $\lessdot$ empty can be solved in polynomial time [Gonzalez, Lawler, and Sahni, 1978], and the case with m arbitrary and $\lessdot$ empty can be solved by linear programming [Lawler and Labetoulle, 1978].

[SS13] **SCHEDULING TO MINIMIZE WEIGHTED COMPLETION TIME**

INSTANCE: Set T of tasks, number $m \in Z^+$ of processors, for each task $t \in T$ a length $l(t) \in Z^+$ and a weight $w(t) \in Z^+$, and a positive integer K.

QUESTION: Is there an m-processor schedule σ for T such that the sum, over all $t \in T$, of $(\sigma(t)+l(t)) \cdot w(t)$ is no more than K?

Reference: [Lenstra, Rinnooy Kan, and Brucker, 1977]. Transformation from PARTITION.

Comment: Remains NP-complete for $m=2$, and is NP-complete in the strong sense for m arbitrary [Lageweg and Lenstra, 1977]. The problem is solvable in pseudo-polynomial time for fixed m. These results continue to hold if "preemptive" schedules are allowed [McNaughton, 1959]. Can be solved in polynomial time if all lengths are equal (by matching techniques). If instead all weights are equal, it can be solved in polynomial time even for "different speed" processors [Conway, Maxwell, and Miller, 1967] and for "unrelated" processors [Horn, 1973], [Bruno, Coffman, and Sethi, 1974]. The "preemptive" case for different speed processors also can be solved in polynomial time [Gonzalez, 1977]. If precedence constraints are allowed, the original problem is NP-complete in the strong sense even if all weights are equal, $m=2$, and the partial order is either an "in-tree" or an "out-tree" [Sethi, 1977a]. If resources are allowed, the same subcases mentioned under RESOURCE CONSTRAINED SCHEDULING are NP-complete, even for equal weights [Blazewicz, 1977a].

A5.3 SHOP SCHEDULING

[SS14] OPEN-SHOP SCHEDULING

INSTANCE: Number $m \in Z^+$ of processors, set J of jobs, each job $j \in J$ consisting of m tasks $t_1[j], t_2[j], \ldots, t_m[j]$ (with $t_i[j]$ to be executed by processor i), a length $l(t) \in Z_0^+$ for each such task t, and an overall deadline $D \in Z^+$.

QUESTION: Is there an open-shop schedule for J that meets the deadline, i.e., a collection of one-processor schedules $\sigma_i: J \rightarrow Z_0^+$, $1 \leqslant i \leqslant m$, such that $\sigma_i(j) > \sigma_i(k)$ implies $\sigma_i(j) \geqslant \sigma_i(k) + l(t_i[k])$, such that for each $j \in J$ the intervals $[\sigma_i(j), \sigma_i(j) + l(t_i[j]))$ are all disjoint, and such that $\sigma_i(j) + l(t_i[j]) \leqslant D$ for $1 \leqslant i \leqslant m$, $1 \leqslant j \leqslant |J|$?

Reference: [Gonzalez and Sahni, 1976]. Transformation from PARTITION.

Comment: Remains NP-complete if $m = 3$, but can be solved in polynomial time if $m = 2$. NP-complete in the strong sense for m arbitrary [Lenstra, 1977]. The general problem is solvable in polynomial time if "preemptive" schedules are allowed [Gonzalez and Sahni, 1976], even if two distinct release times are allowed [Cho and Sahni, 1978]. The $m = 2$ preemptive case can be solved in polynomial time even if arbitrary release times are allowed, and the general preemptive case with arbitrary release times and deadlines can be solved by linear programming [Cho and Sahni, 1978].

[SS15] FLOW-SHOP SCHEDULING

INSTANCE: Number $m \in Z^+$ of processors, set J of jobs, each job $j \in J$ consisting of m tasks $t_1[j], t_2[j], \ldots, t_m[j]$, a length $l(t) \in Z_0^+$ for each such task t, and an overall deadline $D \in Z^+$.

QUESTION: Is there a flow-shop schedule for J that meets the overall deadline, where such a schedule is identical to an open-shop schedule with the additional constraint that, for each $j \in J$ and $1 \leqslant i < m$, $\sigma_{i+1}(j) \geqslant \sigma_i(j) + l(t_i[j])$?

Reference: [Garey, Johnson, and Sethi, 1976]. Transformation from 3-PARTITION.

Comment: NP-complete in the strong sense for $m = 3$. Solvable in polynomial time for $m = 2$ [Johnson, 1954]. The same results hold if "preemptive" schedules are allowed [Gonzalez and Sahni, 1978a], although if release times are added in this case, the problem is NP-complete in the strong sense, even for $m = 2$ [Cho and Sahni, 1978]. If the goal is to meet a bound K on the sum, over all $j \in J$, of $\sigma_m(j) + l(t_m[j])$, then the non-preemptive problem is NP-complete in the strong sense even if $m = 2$ [Garey, Johnson, and Sethi, 1976].

[SS16] NO-WAIT FLOW-SHOP SCHEDULING

INSTANCE: (Same as for FLOW-SHOP SCHEDULING).

QUESTION: Is there a flow-shop schedule for J that meets the overall deadline and has the property that, for each $j \in J$ and $1 \leqslant i < m$, $\sigma_{i+1}(j) = \sigma_i(j) + l(t_i[j])$?

Reference: [Lenstra, Rinnooy Kan, and Brucker, 1977]. Transformation from DIRECTED HAMILTONIAN PATH.
Comment: NP-complete in the strong sense for any fixed $m \geqslant 4$ [Papadimitriou and Kanellakis, 1978]. Solvable in polynomial time for $m=2$ [Gilmore and Gomory, 1964]. (However, NP-complete in the strong sense for $m=2$ if jobs with no tasks on the first processor are allowed [Sahni and Cho, 1977b].) Open for fixed $m=3$. If the goal is to meet a bound K on the sum, over all $j \in J$, of $\sigma_m(j)+l(t_m[j])$, then the problem is NP-complete in the strong sense for m arbitrary [Lenstra, Rinnooy Kan, and Brucker, 1977] and open for fixed $m \geqslant 2$. The analogous "no-wait" versions of OPEN-SHOP SCHEDULING and JOB-SHOP SCHEDULING are NP-complete in the strong sense for $m=2$ [Sahni and Cho, 1977b].

[SS17] TWO-PROCESSOR FLOW-SHOP WITH BOUNDED BUFFER

INSTANCE: (Same as for FLOW-SHOP SCHEDULING with $m=2$, with the addition of a "buffer bound" $B \in Z_0^+$.)
QUESTION: Is there a flow-shop schedule for J that meets the overall deadline and such that, for all $u \geqslant 0$, the number of jobs $j \in J$ for which both $\sigma_1(j)+l(t_1[j]) \leqslant u$ and $\sigma_2(j) > u$ does not exceed B?
Reference: [Papadimitriou and Kanellakis, 1978]. Transformation from NUMERICAL 3-DIMENSIONAL MATCHING.
Comment: NP-complete in the strong sense for any fixed B, $1 \leqslant B < \infty$. Solvable in polynomial time if $B=0$ [Gilmore and Gomory, 1964] or if $B \geqslant |J|-1$ [Johnson, 1954].

[SS18] JOB-SHOP SCHEDULING

INSTANCE: Number $m \in Z^+$ of processors, set J of jobs, each $j \in J$ consisting of an ordered collection of tasks $t_k[j]$, $1 \leqslant k \leqslant n_j$, for each such task t a length $l(t) \in Z_0^+$ and a processor $p(t) \in \{1,2,\ldots,m\}$, where $p(t_k[j]) \neq p(t_{k+1}[j])$ for all $j \in J$ and $1 \leqslant k < n_j$, and a deadline $D \in Z^+$.
QUESTION: Is there a job-shop schedule for J that meets the overall deadline, i.e., a collection of one-processor schedules σ_i mapping $\{t: p(t)=i\}$ into Z_0^+, $1 \leqslant i \leqslant m$, such that $\sigma_i(t) > \sigma_i(t')$ implies $\sigma_i(t) \geqslant \sigma_i(t')+l(t)$, such that $\sigma(t_{k+1}[j]) \geqslant \sigma(t_k[j])+l(t_k[j])$ (where the appropriate subscripts are to be assumed on σ) for all $j \in J$ and $1 \leqslant k < n_j$, and such that for all $j \in J$ $\sigma(t_{n_j}[j])+l(t_{n_j}[j]) \leqslant D$ (again assuming the appropriate subscript on σ)?
Reference: [Garey, Johnson, and Sethi, 1976]. Transformation from 3-PARTITION.
Comment: NP-complete in the strong sense for $m=2$. Can be solved in polynomial time if $m=2$ and $n_j \leqslant 2$ for all $j \in J$ [Jackson, 1956]. NP-complete (in the ordinary sense) if $m=2$ and $n_j \leqslant 3$ for all $j \in J$, or if $m=3$ and $n_j \leqslant 2$ for all $j \in J$ [Gonzalez and Sahni, 1978a]. All the above results continue to hold if "preemptive" schedules are allowed [Gonzalez and Sahni, 1978a]. If in the nonpreemptive case all tasks have the same length, the problem is NP-complete for $m=3$ and open for $m=2$ [Lenstra and Rinnooy Kan, 1978b].

A5.4 MISCELLANEOUS

[SS19] TIMETABLE DESIGN

INSTANCE: Set H of "work periods," set C of "craftsmen," set T of "tasks," a subset $A(c) \subseteq H$ of "available hours" for each craftsman $c \in C$, a subset $A(t) \subseteq H$ of "available hours" for each task $t \in T$, and, for each pair $(c,t) \in C \times T$, a number $R(c,t) \in Z_0^+$ of "required work periods."
QUESTION: Is there a timetable for completing all the tasks, i.e., a function $f: C \times T \times H \rightarrow \{0,1\}$ (where $f(c,t,h)=1$ means that craftsman c works on task t during period h) such that (1) $f(c,t,h)=1$ only if $h \in A(c) \cap A(t)$, (2) for each $h \in H$ and $c \in C$ there is at most one $t \in T$ for which $f(c,t,h)=1$, (3) for each $h \in H$ and $t \in T$ there is at most one $c \in C$ for which $f(c,t,h)=1$, and (4) for each pair $(c,t) \in C \times T$ there are exactly $R(c,t)$ values of h for which $f(c,t,h)=1$?
Reference: [Even, Itai, and Shamir, 1976]. Transformation from 3SAT.
Comment: Remains NP-complete even if $|H|=3$, $A(t)=H$ for all $t \in T$, and each $R(c,t) \in \{0,1\}$. The general problem can be solved in polynomial time if $|A(c)| \leqslant 2$ for all $c \in C$ or if $A(c)=A(t)=H$ for all $c \in C$ and $t \in T$.

[SS20] STAFF SCHEDULING

INSTANCE: Positive integers m and k, a collection C of m-tuples, each having k 1's and $m-k$ 0's (representing possible worker schedules), a "requirement" m-tuple $\bar{R}$ of non-negative integers, and a number n of workers.
QUESTION: Is there a schedule $f: C \rightarrow Z_0^+$ such that $\sum_{\bar{c} \in C} f(\bar{c}) \leqslant n$ and such that $\sum_{\bar{c} \in C} f(\bar{c}) \cdot \bar{c} \geqslant \bar{R}$?
Reference: [Garey and Johnson, ——] Transformation from X3C.
Comment: Solvable in polynomial time if every $\bar{c} \in C$ has the cyclic one's property, i.e., has all its 1's occuring in consecutive positions with position 1 regarded as following position m [Bartholdi, Orlin, and Ratliff, 1977]. (This corresponds to workers who are available only for consecutive hours of the day, or days of the week.)

[SS21] PRODUCTION PLANNING

INSTANCE: Number $n \in Z^+$ of periods, for each period i, $1 \leqslant i \leqslant n$, a demand $r_i \in Z_0^+$, a production capacity $c_i \in Z_0^+$, a production set-up cost $b_i \in Z_0^+$, an incremental production cost coefficient $p_i \in Z_0^+$, and an inventory cost coefficient $h_i \in Z_0^+$, and an overall bound $B \in Z^+$.
QUESTION: Do there exist production amounts $x_i \in Z_0^+$ and associated inventory levels $I_i = \sum_{j=1}^{i}(x_j - r_j)$, $1 \leqslant i \leqslant n$, such that all $x_i \leqslant c_i$, all $I_i \geqslant 0$, and

$$\sum_{i=1}^{n}(p_i x_i + h_i I_i) + \sum_{x_i > 0} b_i \leqslant B \ ?$$

Reference: [Lenstra, Rinnooy Kan, and Florian, 1978]. Transformation from PARTITION.
Comment: Solvable in pseudo-polynomial time, but remains NP-complete even if all demands are equal, all set-up costs are equal, and all inventory costs are 0. If all capacities are equal, the problem can be solved in polynomial time [Florian and Klein, 1971]. The cited algorithms can be generalized to allow for arbitrary mono-

tone non-decreasing concave cost functions, if these can be computed in polynomial time.

[SS22] DEADLOCK AVOIDANCE

INSTANCE: Set $\{P_1, P_2, \ldots, P_m\}$ of "process flow diagrams" (directed acyclic graphs), set Q of "resources," state S of system giving current "active" vertex in each process and "allocation" of resources (see references for details).
QUESTION: Is S "unsafe," i.e., are there control flows for the various processes from state S such that no sequence of resource allocations and deallocations can enable the system to reach a "final" state?

Reference: [Araki, Sugiyama, Kasami, and Okui, 1977], [Sugiyama, Araki, Okui, and Kasami, 1977]. Transformation from 3SAT.
Comment: Remains NP-complete even if allocation calls are "properly nested" and no allocation call involves more than two resources. See references for additional complexity results. See also [Gold, 1978] for results and algorithms for a related model of the deadlock problem.

A6 MATHEMATICAL PROGRAMMING

[MP1] INTEGER PROGRAMMING

INSTANCE: Finite set X of pairs $(\bar{x},b)$, where $\bar{x}$ is an m-tuple of integers and b is an integer, an m-tuple $\bar{c}$ of integers, and an integer B.
QUESTION: Is there an m-tuple $\bar{y}$ of integers such that $\bar{x}\cdot\bar{y} \leqslant b$ for all $(\bar{x},b) \in X$ and such that $\bar{c}\cdot\bar{y} \geqslant B$ (where the dot-product $\bar{u}\cdot\bar{v}$ of two m-tuples $\bar{u}=(u_1,u_2,\ldots,u_m)$ and $\bar{v}=(v_1,v_2,\ldots,v_m)$ is given by $\sum_{i=1}^{m} u_i v_i$)?

Reference: [Karp, 1972], [Borosh and Treybig, 1976]. Transformation from 3SAT. The second reference proves membership in NP.
Comment: NP-complete in the strong sense. Variant in which all components of $\bar{y}$ are required to belong to $\{0,1\}$ (ZERO-ONE INTEGER PROGRAMMING) is also NP-complete, even if each b, all components of each $\bar{x}$, and all components of $\bar{c}$ are required to belong to $\{0,1\}$. Also NP-complete are the questions of whether a $\bar{y}$ with non-negative integer entries exists such that $\bar{x}\cdot\bar{y}=b$ for all $(\bar{x},b) \in X$, and the question of whether there exists any $\bar{y}$ with integer entries such that $\bar{x}\cdot\bar{y} \geqslant 0$ for all $(\bar{x},b) \in X$ [Sahni, 1974].

[MP2] QUADRATIC PROGRAMMING (*)

INSTANCE: Finite set X of pairs $(\bar{x},b)$, where $\bar{x}$ is an m-tuple of rational numbers and b is a rational number, two m-tuples $\bar{c}$ and $\bar{d}$ of rational numbers, and a rational number B.
QUESTION: Is there an m-tuple $\bar{y}$ of rational numbers such that $\bar{x}\cdot\bar{y} \leqslant b$ for all $(\bar{x},b) \in X$ and such that $\sum_{i=1}^{m} (c_i y_i^2 + d_i y_i) \geqslant B$, where c_i, y_i, and d_i denote the ith components of $\bar{c}$, $\bar{y}$, and $\bar{d}$ respectively?

Reference: [Sahni, 1974]. Transformation from PARTITION.
Comment: Not known to be in NP, unless the c_i's are all non-negative [Klee, 1978]. If the constraints are quadratic and the objective function is linear (the reverse of the situation above), then the problem is also NP-hard [Sahni, 1974]. If we add to this last problem the requirement that all entries of $\bar{y}$ be integers, then the problem becomes undecidable [Jeroslow, 1973].

[MP3] COST-PARAMETRIC LINEAR PROGRAMMING

INSTANCE: Finite set X of pairs $(\bar{x},b)$, where $\bar{x}$ is an m-tuple of integers and b is an integer, a set $J \subseteq \{1,2,\ldots,m\}$, and a positive rational number q.
QUESTION: Is there an m-tuple $\bar{c}$ with rational entries such that $(\bar{c}\cdot\bar{c})^{1/2} \leqslant q$ and such that, if Y is the set of all m-tuples $\bar{y}$ with non-negative rational entries satisfying $\bar{x}\cdot\bar{y} \geqslant b$ for all $(\bar{x},b) \in X$, then the minimum of $\sum_{j \in J} c_j y_j$ over all $\bar{y} \in Y$ exceeds

$$\tfrac{1}{2}\max\{|c_j| : j \in J\} + \sum_{j \in J} \min\{0,c_j\} \ ?$$

Reference: [Jeroslow, 1976]. Transformation from 3SAT.
Comment: Remains NP-complete for any fixed $q > 0$. The problem arises from first order error analysis for linear programming.

[MP4] FEASIBLE BASIS EXTENSION

INSTANCE: An $m \times n$ integer matrix A, $m < n$, a column vector $\bar{a}$ of length m, and a subset S of the columns of A with $|S| < m$.
QUESTION: Is there a *feasible basis* B for $A\bar{x} = \bar{a}$, $\bar{x} \geqslant 0$, i.e., a nonsingular $m \times m$ submatrix B of A such that $B^{-1}\bar{a} \geqslant 0$, and such that B contains all the columns in S?

Reference: [Murty, 1972]. Transformation from HAMILTONIAN CIRCUIT.

[MP5] MINIMUM WEIGHT SOLUTION TO LINEAR EQUATIONS

INSTANCE: Finite set X of pairs $(\bar{x}, b)$, where $\bar{x}$ is an m-tuple of integers and b is an integer, and a positive integer $K \leqslant m$.
QUESTION: Is there an m-tuple $\bar{y}$ with rational entries such that $\bar{y}$ has at most K non-zero entries and such that $\bar{x} \cdot \bar{y} = b$ for all $(\bar{x}, b) \in X$?

Reference: [Garey and Johnson, ——]. Transformation from X3C.
Comment: NP-complete in the strong sense. Solvable in polynomial time if $K = m$.

[MP6] OPEN HEMISPHERE

INSTANCE: Finite set X of m-tuples of integers, and a positive integer $K \leqslant |X|$.
QUESTION: Is there an m-tuple $\bar{y}$ of rational numbers such that $\bar{x} \cdot \bar{y} > 0$ for at least K m-tuples $\bar{x} \in X$?

Reference: [Johnson and Preparata, 1978]. Transformation from MAXIMUM 2-SATISFIABILITY.
Comment: NP-complete in the strong sense, but solvable in polynomial time for any fixed m, even in a "weighted" version of the problem. The same results hold for the related CLOSED HEMISPHERE problem in which we ask that $\bar{y}$ satisfy $\bar{x} \cdot \bar{y} \geqslant 0$ for at least K m-tuples $\bar{x} \in X$ [Johnson and Preparata, 1978]. If $K = 0$ or $K = |X|$, both problems are polynomially equivalent to linear programming [Reiss and Dobkin, 1976].

[MP7] K-RELEVANCY

INSTANCE: Finite set X of pairs $(\bar{x}, b)$, where $\bar{x}$ is an m-tuple of integers and b is an integer, and a positive integer $K \leqslant |X|$.
QUESTION: Is there a subset $X' \subseteq X$ with $|X'| \leqslant K$ such that, for all m-tuples $\bar{y}$ of rational numbers, if $\bar{x} \cdot \bar{y} \leqslant b$ for all $(\bar{x}, b) \in X'$, then $\bar{x} \cdot \bar{y} \leqslant b$ for all $(\bar{x}, b) \in X$?

Reference: [Reiss and Dobkin, 1976]. Transformation from X3C.
Comment: NP-complete in the strong sense. Equivalent to linear programming if $K = |X| - 1$ [Reiss and Dobkin, 1976]. Other NP-complete problems of this form, where a standard linear programming problem is modified by asking that the desired property hold for some subset of K constraints, can be found in the reference.

[MP8] TRAVELING SALESMAN POLYTOPE NON-ADJACENCY

INSTANCE: Graph $G = (V, E)$, two Hamiltonian circuits C and C' for G.
QUESTION: Do C and C' correspond to non-adjacent vertices of the "traveling salesman polytope" for G?

Reference: [Papadimitriou, 1978a]. Transformation from 3SAT.
Comment: Result also holds for the "non-symmetric" case where G is a directed graph and C and C' are directed Hamiltonian circuits. Analogous polytope non-adjacency problems for graph matching and CLIQUE can be solved in polynomial time [Chvátal, 1975].

[MP9] KNAPSACK

INSTANCE: Finite set U, for each $u \in U$ a size $s(u) \in Z^+$ and a value $v(u) \in Z^+$, and positive integers B and K.
QUESTION: Is there a subset $U' \subseteq U$ such that $\sum_{u \in U'} s(u) \leqslant B$ and such that $\sum_{u \in U'} v(u) \geqslant K$?
Reference: [Karp, 1972]. Transformation from PARTITION.
Comment: Remains NP-complete if $s(u) = v(u)$ for all $u \in U$ (SUBSET SUM). Can be solved in pseudo-polynomial time by dynamic programming (e.g., see [Dantzig, 1957] or [Lawler, 1976a]).

[MP10] INTEGER KNAPSACK

INSTANCE: Finite set U, for each $u \in U$ a size $s(u) \in Z^+$ and a value $v(u) \in Z^+$, and positive integers B and K.
QUESTION: Is there an assignment of a non-negative integer $c(u)$ to each $u \in U$ such that $\sum_{u \in U} c(u) \cdot s(u) \leqslant B$ and such that $\sum_{u \in U} c(u) \cdot v(u) \geqslant K$?
Reference: [Lueker, 1975]. Transformation from SUBSET SUM.
Comment: Remains NP-complete if $s(u) = v(u)$ for all $u \in U$. Solvable in pseudo-polynomial time by dynamic programming. Solvable in polynomial time if $|U| = 2$ [Hirschberg and Wong, 1976].

[MP11] CONTINUOUS MULTIPLE CHOICE KNAPSACK

INSTANCE: Finite set U, for each $u \in U$ a size $s(u) \in Z^+$ and a value $v(u) \in Z^+$, a partition of U into disjoint sets $U_1, U_2, \ldots, U_m$, and positive integers B and K.
QUESTION: Is there a choice of a unique element $u_i \in U_i$, $1 \leqslant i \leqslant m$, and an assignment of rational numbers r_i, $0 \leqslant r_i \leqslant 1$, to these elements, such that $\sum_{i=1}^{m} r_i \cdot s(u_i) \leqslant B$ and $\sum_{i=1}^{m} r_i \cdot v(u_i) \geqslant K$?
Reference: [Ibaraki, 1978]. Transformation from PARTITION.
Comment: Solvable in pseudo-polynomial time, but remains NP-complete even if $|U_i| \leqslant 2$, $1 \leqslant i \leqslant m$. Solvable in polynomial time by "greedy" algorithms if $|U_i| = 1$, $1 \leqslant i \leqslant m$, or if we only require that the $r_i \geqslant 0$ but place no upper bound on them. [Ibaraki, Hasegawa, Teranaka, and Iwase, 1978].

[MP12] PARTIALLY ORDERED KNAPSACK

INSTANCE: Finite set U, partial order $<$ on U, for each $u \in U$ a size $s(u) \in Z^+$ and a value $v(u) \in Z^+$, positive integers B and K.
QUESTION: Is there a subset $U' \subseteq U$ such that if $u \in U'$ and $u' < u$, then $u' \in U'$, and such that $\sum_{u \in U'} s(u) \leqslant B$ and $\sum_{u \in U'} v(u) \geqslant K$?
Reference: [Garey and Johnson, ——]. Transformation from CLIQUE. Problem is discussed in [Ibarra and Kim, 1975b].

Comment: NP-complete in the strong sense, even if $s(u)=v(u)$ for all $u \in U$. General problem is solvable in pseudo-polynomial time if $<$ is a "tree" partial order [Garey and Johnson, ——].

[MP13] COMPARATIVE VECTOR INEQUALITIES

INSTANCE: Sets $X=\{\bar{x}_1,\bar{x}_2, \ldots, \bar{x}_k\}$ and $Y=\{\bar{y}_1,\bar{y}_2, \ldots, \bar{y}_l\}$ of m-tuples of integers.

QUESTION: Is there an m-tuple $\bar{z}$ of integers such that the number of m-tuples $\bar{x}_i$ satisfying $\bar{x}_i \geqslant \bar{z}$ is at least as large as the number of m-tuples $\bar{y}_j$ satisfying $\bar{y}_j \geqslant \bar{z}$, where two m-tuples $\bar{u}$ and $\bar{v}$ satisfy $\bar{u} \geqslant \bar{v}$ if and only if no component of $\bar{u}$ is less than the corresponding component of $\bar{v}$?

Reference: [Plaisted, 1976]. Transformation from COMPARATIVE CONTAINMENT (with equal weights).

Comment: Remains NP-complete even if all components of the $\bar{x}_i$ and $\bar{y}_j$ are required to belong to $\{0,1\}$.

A7 ALGEBRA AND NUMBER THEORY

A7.1 DIVISIBILITY PROBLEMS

[AN1] QUADRATIC CONGRUENCES

INSTANCE: Positive integers a, b, and c.
QUESTION: Is there a positive integer $x < c$ such that $x^2 \equiv a \pmod{b}$?

Reference: [Manders and Adleman, 1978]. Transformation from 3SAT.
Comment: Remains NP-complete even if the instance includes a prime factorization of b and solutions to the congruence modulo all prime powers occurring in the factorization. Solvable in polynomial time if $c = \infty$ (i.e., there is no upper bound on x) and the prime factorization of b is given. Assuming the Extended Riemann Hypothesis, the problem is solvable in polynomial time when b is prime. The general problem is trivially solvable in pseudo-polynomial time.

[AN2] SIMULTANEOUS INCONGRUENCES

INSTANCE: Collection $\{(a_1,b_1), \ldots, (a_n,b_n)\}$ of ordered pairs of positive integers, with $a_i \leqslant b_i$ for $1 \leqslant i \leqslant n$.
QUESTION: Is there an integer x such that, for $1 \leqslant i \leqslant n$, $x \not\equiv a_i \pmod{b_i}$?

Reference: [Stockmeyer and Meyer, 1973]. Transformation from 3SAT.

[AN3] SIMULTANEOUS DIVISIBILITY OF LINEAR POLYNOMIALS (*)

INSTANCE: Vectors $a_i = (a_i[0], \ldots, a_i[m])$ and $b_i = (b_i[0], \ldots, b_i[m])$, $1 \leqslant i \leqslant n$, with positive integer entries.
QUESTION: Do there exist positive integers $x_1, x_2, \ldots, x_m$ such that, for $1 \leqslant i \leqslant n$, $a_i[0] + \sum_{j=1}^{m} (a_i[j] \cdot x_j)$ divides $b_i[0] + \sum_{j=1}^{m} (b_i[j] \cdot x_j)$?

Reference: [Lipshitz, 1977], [Lipshitz, 1978]. Transformation from QUADRATIC DIOPHANTINE EQUATIONS.
Comment: Not known to be in NP, but belongs to NP for any fixed n. NP-complete for any fixed $n \geqslant 5$. General problem is undecidable if the vector entries and the x_j are allowed to range over the ring of "integers" in a real quadratic extension of the rationals. See reference for related decidability and undecidability results.

[AN4] COMPARATIVE DIVISIBILITY

INSTANCE: Sequences $a_1, a_2, \ldots, a_n$ and $b_1, b_2, \ldots, b_m$ of positive integers.
QUESTION: Is there a positive integer c such that the number of i for which c divides a_i is more than the number of j for which c divides b_j?

Reference: [Plaisted, 1976]. Transformation from 3SAT.
Comment: Remains NP-complete even if all a_i are different and all b_j are different [Garey and Johnson, ——].

[AN5] EXPONENTIAL EXPRESSION DIVISIBILITY (*)

INSTANCE: Sequences $a_1, a_2, \ldots, a_n$ and $b_1, b_2, \ldots, b_m$ of positive integers, and an integer q.

QUESTION: Does $\prod_{i=1}^{n} (q^{a_i}-1)$ divide $\prod_{j=1}^{m} (q^{b_j}-1)$?

Reference: [Plaisted, 1976]. Transformation from 3SAT.

Comment: Not known to be in NP or co-NP, but solvable in pseudo-polynomial time using standard greatest common divisor algorithms. Remains NP-hard for any fixed value of q with $|q|>1$, even if the a_i and b_j are restricted to being products of distinct primes.

[AN6] NON-DIVISIBILITY OF A PRODUCT POLYNOMIAL

INSTANCE: Sequences $A_i = <(a_i[1],b_i[1]), \ldots, (a_i[k],b_i[k])>$, $1 \leq i \leq m$, of pairs of integers, with each $b_i[j] \geq 0$, and an integer N.

QUESTION: Is $\prod_{i=1}^{m} (\sum_{j=1}^{k} a_i[j] \cdot z^{b_i[j]})$ *not* divisible by $z^N - 1$?

Reference: [Plaisted, 1977a], [Plaisted, 1977b]. Transformation from 3SAT. Proof of membership in NP is non-trivial and appears in the second reference.

Comment: The related problem in which we are given two sequences $<a_1,a_2, \ldots, a_m>$ and $<b_1,b_2, \ldots, b_n>$ of positive integers and are asked whether $\prod_{i=1}^{m} (z^{a_i}-1)$ does not divide $\prod_{j=1}^{n} (z^{b_j}-1)$ is also NP-complete [Plaisted, 1976].

[AN7] NON-TRIVIAL GREATEST COMMON DIVISOR (*)

INSTANCE: Sequences $A_i = <(a_i[1],b_i[1]), \ldots, (a_i[k],b_i[k])>$, $1 \leq i \leq m$, of pairs of integers, with each $b_i[j] \geq 0$.

QUESTION: Does the greatest common divisor of the polynomials $\sum_{j=1}^{k} a_i[j] \cdot z^{b_i[j]}$, $1 \leq i \leq m$, have degree greater than zero?

Reference: [Plaisted, 1977a]. Transformation from 3SAT.

Comment: Not known to be in NP or co-NP. Remains NP-hard if each $a_i[j]$ is either -1 or $+1$ [Plaisted, 1976] or if $m=2$ [Plaisted, 1977b]. The analogous problem in which the instance also includes a positive integer K, and we are asked if the least common multiple of the given polynomials has degree less than K, is NP-hard under the same restrictions. Both problems can be solved in pseudo-polynomial time using standard algorithms.

A7.2 SOLVABILITY OF EQUATIONS

[AN8] QUADRATIC DIOPHANTINE EQUATIONS

INSTANCE: Positive integers a, b, and c.

QUESTION: Are there positive integers x and y such that $ax^2 + by = c$?

Reference: [Manders and Adleman, 1978]. Transformation from 3SAT.

Comment: Diophantine equations of the forms $ax^k = c$ and $\sum_{i=1}^{k} a_i \cdot x_i = c$ are solvable in polynomial time for arbitrary values of k. The general Diophantine problem, "Given a polynomial with integer coefficients in k variables, does it have an integer solution?" is undecidable, even for $k=13$ [Matijasevic and Robinson, 1975]. However, the given problem can be generalized considerably (to simultaneous equations in many variables) while remaining in NP, so long as only one variable enters into the equations in a non-linear way (see [Gurari and Ibarra, 1978]).

[AN9] ALGEBRAIC EQUATIONS OVER GF[2]

INSTANCE: Polynomials $P_i(x_1,x_2, \ldots, x_n), 1 \leq i \leq m$, over GF[2], i.e., each polynomial is a sum of terms, where each term is either the integer 1 or a product of distinct x_i.

QUESTION: Do there exist $u_1,u_2, \ldots, u_n \in \{0,1\}$ such that, for $1 \leq i \leq m$, $P_i(u_1,u_2, \ldots, u_n) = 0$, where arithmetic operations are as defined in GF[2], with $1+1=0$ and $1 \cdot 1 = 1$?

Reference: [Fraenkel and Yesha, 1977]. Transformation from X3C.

Comment: Remains NP-complete even if none of the polynomials has a term involving more than two variables [Valiant, 1977c]. Easily solved in polynomial time if no term involves more than one variable or if there is just one polynomial. Variant in which the u_j are allowed to range over the algebraic closure of GF[2] is NP-hard, even if no term involves more than two variables [Fraenkel and Yesha, 1977].

[AN10] ROOT OF MODULUS 1 (*)

INSTANCE: Ordered pairs $(a[i], b[i])$, $1 \leq i \leq n$, of integers, with each $b[i] \geq 0$.

QUESTION: Does the polynomial $\sum_{i=1}^{n} a[i] \cdot z^{b[i]}$ have a root on the complex unit circle, i.e., is there a complex number q with $|q| = 1$ such that $\sum_{i=1}^{n} a[i] \cdot q^{b[i]} = 0$?

Reference: [Plaisted, 1977b]. Transformation from 3SAT.

Comment: Not known to be in NP or co-NP.

[AN11] NUMBER OF ROOTS FOR A PRODUCT POLYNOMIAL (*)

INSTANCE: Sequences $A_i = <(a_i[1],b_i[1]), \ldots, (a_i[k],b_i[k])>$, $1 \leq i \leq m$, of pairs of integers, with each $b_i[j] \geq 0$, and a positive integer K.

QUESTION: Does the polynomial $\prod_{i=1}^{m} (\sum_{j=1}^{k} a_i[j] \cdot z^{b_i[j]})$ have fewer than K distinct complex roots?

Reference: [Plaisted, 1977a]. Transformation from 3SAT.

Comment: Not known to be in NP or co-NP. Remains NP-hard if each $a_i[j]$ is either -1 or $+1$, as does the variant in which the instance also includes an integer M and we are asked whether the product polynomial has fewer than K complex roots of multiplicity M [Plaisted, 1976].

[AN12] PERIODIC SOLUTION RECURRENCE RELATION (*)

INSTANCE: Ordered pairs (c_i,b_i), $1 \leq i \leq m$, of integers, with all b_i positive.

QUESTION: Is there a sequence $a_0,a_1, \ldots, a_{n-1}$ of integers, with $n \geq \max\{b_i\}$, such that the infinite sequence $a_0,a_1,\ldots$ defined by the recurrence relation

$$a_i = \sum_{j=1}^{m} c_j \cdot a_{(i-b_j)}$$

satisfies $a_i \equiv a_{i(\text{mod } n)}$, for all $i \geq n$?

Reference: [Plaisted, 1977b]. Tranformation from 3SAT

Comment: Not known to be in NP or co-NP. See reference for related results.

A7.3 MISCELLANEOUS

[AN13] PERMANENT EVALUATION (*)

INSTANCE: An $n \times n$ matrix M of 0's and 1's, and a positive integer $K \leqslant n!$.
QUESTION: Is the value of the permanent of M equal to K?

Reference: [Valiant, 1977a]. Transformation from 3SAT.
Comment: The problem is NP-hard but not known to be in NP, as is the case for the variants in which we ask whether the value of the permanent is "K or less" or "K or more." The problem of computing the value of the permanent of M is #P-complete.

[AN14] COSINE PRODUCT INTEGRATION

INSTANCE: Sequence $(a_1,a_2, \ldots, a_n)$ of integers.
QUESTION: Does $\int_0^{2\pi} (\prod_{i=1}^{n} \cos(a_i\theta))\,d\theta = 0$?

Reference: [Plaisted, 1976]. Transformation from PARTITION.
Comment: Solvable in pseudo-polynomial time. See reference for related complexity results concerning integration.

[AN15] EQUILIBRIUM POINT

INSTANCE: Set $x=\{x_1,x_2, \ldots, x_n\}$ of variables, collection $\{F_i\colon 1 \leqslant i \leqslant n\}$ of product polynomials over X and the integers, and a finite "range-set" $M_i \subseteq Z$ for $1 \leqslant i \leqslant n$.
QUESTION: Does there exist a sequence $y_1,y_2, \ldots, y_n$ of integers, with $y_i \in M_i$, such that for $1 \leqslant i \leqslant n$ and all $y \in M_i$,

$$F_i(y_1,y_2, \ldots, y_{i-1},y_i,y_{i+1}, \ldots, y_n) \geqslant F_i(y_1,y_2, \ldots, y_{i-1},y,y_{i+1}, \ldots, y_n)\ ?$$

Reference: [Sahni, 1974]. Transformation from 3SAT.
Comment: Remains NP-complete even if $M_i=\{0,1\}$ for $1 \leqslant i \leqslant n$.

[AN16] UNIFICATION WITH COMMUTATIVE OPERATORS

INSTANCE: Set V of variables, set C of constants, ordered pairs (e_i,f_i), $1 \leqslant i \leqslant n$, of "expressions," where an expression is either a variable from V, a constant from C, or $(e+f)$ where e and f are expressions.
QUESTION: Is there an assignment to each $v \in V$ of a variable-free expression $I(v)$ such that, if $I(e)$ denotes the expression obtained by replacing each occurrence of each variable v in e by $I(v)$, then $I(e_i) \equiv I(f_i)$ for $1 \leqslant i \leqslant n$, where $e \equiv f$ if $e=f$ or if $e=(a+b)$, $f=(c+d)$, and either $a \equiv c$ and $b \equiv d$ or $a \equiv d$ and $b \equiv c$?

Reference: [Sethi, 1977b]. Transformation from 3SAT.
Comment: Remains NP-complete even if no e_j or f_i contains more than 7 occurrences of constants and variables. The variant in which the operator is non-commutative (and hence $e \equiv f$ only if $e=f$) is solvable in polynomial time [Paterson and Wegman, 1976].

[AN17] UNIFICATION FOR FINITELY PRESENTED ALGEBRAS

INSTANCE: Finite presentation of an algebra A in terms of a set G of generators, a collection O of operators of various finite dimensions, and a collection Γ of defining relations on well-formed formulas over G and O; two well-formed expressions e and f over G,O, and a variable set V (see reference for details).
QUESTION: Is there an assignment to each $v \in V$ of a unique "term" $I(v)$ over G and O such that, if $I(e)$ and $I(f)$ denote the expressions obtained by replacing all variables in e and f by their corresponding terms, then $I(e)$ and $I(f)$ represent the same element in A?

Reference: [Kozen, 1977a], [Kozen, 1976]. Transformation from 3SAT. Proof of membership in NP is non-trivial and appears in the second reference.
Comment: Remains NP-complete if only one of e and f contains variable symbols, but is solvable in polynomial time if neither contains variable symbols. See [Kozen, 1977b] for quantified versions of this problem that are complete for PSPACE and for the various levels of the polynomial hierarchy.

[AN18] INTEGER EXPRESSION MEMBERSHIP

INSTANCE: Integer expression e over the operations $\cup$ and $+$, where if $n \in Z^+$, the binary representation of n is an integer expression representing n, and if f and g are integer expressions representing the sets F and G, then $f \cup g$ is an integer expression representing the set $F \cup G$ and $f+g$ is an integer expression representing the set $\{m+n: m \in F \text{ and } n \in G\}$, and a positive integer K.
QUESTION: Is K in the set represented by e?

Reference: [Stockmeyer and Meyer, 1973]. Transformation from SUBSET SUM.
Comment: The related INTEGER EXPRESSION INEQUIVALENCE problem, "given two integer expressions e and f, do they represent different sets?" is NP-hard and in fact complete for Σ_2^p in the polynomial hierarchy ([Stockmeyer and Meyer, 1973], [Stockmeyer, 1976a], see also Section 7.2). If the operator "$\neg$" is allowed, with $\neg e$ representing the set of all positive integers not represented by e, then both the membership and inequivalence problems become PSPACE-complete [Stockmeyer and Meyer, 1973].

A8 GAMES AND PUZZLES

[GP1] GENERALIZED HEX (*)

INSTANCE: Graph $G=(V,E)$ and two specified vertices $s,t \in V$.
QUESTION: Does player 1 have a forced win in the following game played on G? The players alternate choosing a vertex from $V-\{s,t\}$, with those chosen by player 1 being colored "blue" and those chosen by player 2 being colored "red." Play continues until all such vertices have been colored, and player 1 wins if and only if there is a path from s to t in G that passes through only blue vertices.

Reference: [Even and Tarjan, 1976]. Transformation from QBF.
Comment: PSPACE-complete. The variant in which players alternate choosing an edge instead of a vertex, known as "the Shannon switching game on edges," can be solved in polynomial time [Bruno and Weinberg, 1970]. If G is a directed graph and player 1 wants a "blue" directed path from s to t, both the vertex selection game and the arc selection game are PSPACE-complete [Even and Tarjan, 1976].

[GP2] GENERALIZED GEOGRAPHY (*)

INSTANCE: Directed graph $G=(V,A)$ and a specified vertex $v_0 \in V$.
QUESTION: Does player 1 have a forced win in the following game played on G? Players alternate choosing a new arc from A. The first arc chosen must have its tail at v_0 and each subsequently chosen arc must have its tail at the vertex that was the head of the previous arc. The first player unable to choose such a new arc loses.

Reference: [Schaefer, 1978a]. Transformation from QBF.
Comment: PSPACE-complete, even if G is bipartite, planar, and has no in- or out-degree exceeding 2 and no degree exceeding 3 (PLANAR GEOGRAPHY) [Lichtenstein and Sipser, 1978]. This game is a generalization of the "Geography" game in which players alternate choosing countries, each name beginning with the same letter that ends the previous country's name.

[GP3] GENERALIZED KAYLES (*)

INSTANCE: Graph $G=(V,E)$.
QUESTION: Does player 1 have a forced win in the following game played on G? Players alternate choosing a vertex in the graph, removing that vertex and all vertices adjacent to it from the graph. Player 1 wins if and only if player 2 is the first player left with no vertices to choose from.

Reference: [Schaefer, 1978a]. Transformation from QBF.
Comment: PSPACE-complete. The variant in which $G=(V_1 \cup V_2,E)$ is bipartite, with each edge involving one vertex from V_1 and one from V_2, and player i can only choose vertices from the set V_i (but still removes all adjacent vertices as before) is also PSPACE-complete. For a description of the game Kayles upon which this generalization is based, see [Conway, 1976].

[GP4] SEQUENTIAL TRUTH ASSIGNMENT (*)

INSTANCE: A sequence $U=<u_1,u_2, \ldots, u_n>$ of variables and a collection C of clauses over U (as in an instance of SATISFIABILITY).
QUESTION: Does player 1 have a forced win in the following game played on U

and C? Players alternate assigning truth values to the variables in U, with player 1 assigning a value to u_{2i-1} and player 2 assigning a value to u_{2i} on their i^{th} turns. Player 1 wins if and only if the resulting truth assignment satisfies all clauses in C.

Reference: [Stockmeyer and Meyer, 1973]. Transformation from QBF.
Comment: PSPACE-complete, even if each clause in C has only three literals. Solvable in polynomial time if no clause has more than two literals [Schaefer, 1978b].

[GP5] VARIABLE PARTITION TRUTH ASSIGNMENT (*)

INSTANCE: A set U of variables and a collection C of clauses over U.
QUESTION: Does player 1 have a forced win in the following game played on U and C? Players alternate choosing a variable from U until all variables have been chosen. Player 1 wins if and only if a satisfying truth assignment for C is obtained by setting "true" all variables chosen by player 1 and setting "false" all variables chosen by player 2.

Reference: [Schaefer, 1978a]. Transformation from QBF.
Comment: PSPACE-complete, even if each clause consists only of un-negated literals (i.e., contains no literals of the form $\bar{u}$ for $u \in U$). Analogous results for several other games played on logical expressions can be found in the reference.

[GP6] SIFT (*)

INSTANCE: Two collections A and B of subsets of a finite set X, with A and B having no subsets in common.
QUESTION: Does player 1 have a forced win in the following game played on A, B, and X? Players alternate choosing an element from X until the set X' of all elements chosen so far either intersects all the subsets in A or intersects all the subsets in B. Player 1 wins if and only if the final set X' of chosen elements intersects all the subsets in B and, if player 1 made the last move, does *not* intersect all subsets in A.

Reference: [Schaefer, 1978a]. Transformation from QBF.
Comment: PSPACE-complete.

[GP7] ALTERNATING HITTING SET (*)

INSTANCE: A collection C of subsets of a basic set B.
QUESTION: Does player 1 have a forced win in the following game played on C and B? Players alternate choosing a new element of B until, for each $c \in C$, some member of c has been chosen. The player whose choice causes this to happen loses.

Reference: [Schaefer, 1978a]. Transformation from QBF.
Comment: PSPACE-complete even if no set in C contains more than two elements, a subcase of the original HITTING SET problem that can be solved in polynomial time. If the roles of winner and loser are reversed, the problem is PSPACE-complete even if no set in C contains more than three elements.

[GP8] ALTERNATING MAXIMUM WEIGHTED MATCHING (*)

INSTANCE: Graph $G=(V,E)$, a weight $w(e) \in Z^+$ for each $e \in E$, and a bound $B \in Z^+$.

QUESTION: Does player 1 have a forced win in the following game played on G? Players alternate choosing a new edge from E, subject to the constraint that no edge can share an endpoint with any of the already chosen edges. If the sum of the weights of the edges chosen ever exceeds B, player 1 wins.

Reference: [Dobkin and Ladner, 1978]. Transformation from QBF.

Comment: PSPACE-complete, even though the corresponding weighted matching problem can be solved in polynomial time (e.g., see [Lawler, 1976a]).

[GP9] ANNIHILATION (*)

INSTANCE: Directed acyclic graph $G=(V,A)$, collection $\{A_i: 1 \leqslant i \leqslant r\}$ of (not necessarily disjoint) subsets of A, function f_0 mapping V into $\{0,1,2,\ldots,r\}$, where $f_0(v)=i>0$ means that a "token" of type i is "on" vertex v and $f_0(v)=0$ means that v is unoccupied.

QUESTION: Does player 1 have a forced win in the following game played on G? A position is a function $f: V \rightarrow \{0,1,\ldots,r\}$ with f_0 being the initial position and players alternating moves. A player moves by selecting a vertex $v \in V$ with $f(v)>0$ and an arc $(v,w) \in A_{f(v)}$, and the move corresponds to moving the token on vertex v to vertex w. The new position f' is the same as f except that $f'(v)=0$ and $f'(w)$ is either 0 or $f(v)$, depending, respectively, on whether $f(w)>0$ or $f(w)=0$. (If $f(w)>0$, then both the token moved to w and the token already there are "annihilated.") Player 1 wins if and only if player 2 is the first player unable to move.

Reference: [Fraenkel and Yesha, 1977]. Transformation from VERTEX COVER.

Comment: NP-hard and in PSPACE, but not known to be PSPACE-complete. Remains NP-hard even if $r=2$ and $A_1 \cap A_2$ is empty. Problem can be solved in polynomial time if $r=1$ [Fraenkel and Yesha, 1976]. Related NP-hardness results for other token-moving games on directed graphs (REMOVE, CONTRAJUNCTIVE, CAPTURE, BLOCKING, TARGET) can be found in [Fraenkel and Yesha, 1977].

[GP10] N×N CHECKERS (*)

INSTANCE: Positive integer N, a partition of the black squares of an $N \times N$ Checkerboard into those that are empty, those that are occupied by "Black kings," and those that are occupied by "Red kings," and the identity of the player (Red or Black) whose turn it is.

QUESTION: Does Black have a forced win from the given position in a game of Checkers played according to the standard rules, modified only to take into account the expanded board and number of pieces?

Reference: [Fraenkel, Garey, Johnson, Schaefer, and Yesha, 1978]. Transformation from PLANAR GEOGRAPHY.

Comment: PSPACE-hard, and PSPACE-complete for certain drawing rules. The related problem in which we ask whether Black can jump all of Red's pieces in one turn is solvable in polynomial time.

[GP11] N×N GO (*)

INSTANCE: Positive integer N, a partition of the "points" on an $N \times N$ Go board into those that are empty, those that are occupied by White stones and those that are occupied by Black stones, and the name (Black or White) of the player whose turn it is.

QUESTION: Does White have a forced win from the given position in a game of Go played according to the standard rules, modified only to take into account the expanded board?

Reference: [Lichtenstein and Sipser, 1978]. Transformation from PLANAR GEOGRAPHY.

Comment: PSPACE-hard.

[GP12] LEFT-RIGHT HACKENBUSH FOR REDWOOD FURNITURE

INSTANCE: A piece of "redwood furniture," i.e., a connected graph $G=(V,E)$ with a specified "ground" vertex $v \in V$ and a partition of the edges into sets L and R, where L is the set of all edges containing v (the set of "feet"), $R=E-L$, and each "foot" in L shares a vertex with at most one edge in R, which is its corresponding "leg" (not all edges in R need to be legs however), and a positive integer K.

QUESTION: Is the "value" of the Left-Right Hackenbush game played on G less than or equal to 2^{-K} (see [Conway, 1976] for the definition of the game, there called Hackenbush Restrained, and for the definition of "value")?

Reference: [Berlekamp, 1976]. Transformation from SET COVERING.

Comment: Remains NP-complete even for "bipartite" redwood furniture, but can be solved in polynomial time for the subclass of redwood furniture known as "redwood trees." As a consequence of this result, the problem of determining if player 1 has a win in an arbitrary game of Left-Right Hackenbush is NP-hard.

[GP13] SQUARE-TILING

INSTANCE: Set C of "colors," collection $T \subseteq C^4$ of "tiles" (where $<a,b,c,d>$ denotes a tile whose top, right, bottom, and left sides are colored a,b,c, and d, respectively), and a positive integer $N \leqslant |C|$.

QUESTION: Is there a tiling of an $N \times N$ square using the tiles in T, i.e., an assignment of a tile $A(i,j) \in T$ to each ordered pair i,j, $1 \leqslant i \leqslant N$, $1 \leqslant j \leqslant N$, such that (1) if $f(i,j)=<a,b,c,d>$ and $f(i+1,j)=<a',b',c',d'>$, then $a=c'$, and (2) if $f(i,j)=<a,b,c,d>$ and $f(i,j+1)=<a',b',c',d'>$, then $b=d'$?

Reference: [Garey, Johnson, and Papadimitriou, 1977]. Transformation from DIRECTED HAMILTONIAN PATH.

Comment: Variant in which we ask if T can be used to tile the entire plane ($Z \times Z$) "periodically" with period less than N is also NP-complete. In general, the problem of whether a set of tiles can be used to tile the plane is undecidable [Berger, 1966], as is the problem of whether a set of tiles can be used to tile the plane periodically.

[GP14] CROSSWORD PUZZLE CONSTRUCTION

INSTANCE: A finite set $W \subseteq \Sigma^*$ of words and an $n \times n$ matrix A of 0's and 1's.
QUESTION: Can an $n \times n$ crossword puzzle be built up from the words in W and blank squares corresponding to the 0's of A, i.e., if E is the set of pairs (i,j) such that $A_{ij}=0$, is there an assignment $f: E \rightarrow \Sigma$ such that the letters assigned to any maximal horizontal or vertical contiguous sequence of members of E form, in order, a word of W?

Reference: [Lewis and Papadimitriou, 1978]. Transformation from X3C.
Comment: Remains NP-complete even if all entries in A are 0.

[GP15] GENERALIZED INSTANT INSANITY

INSTANCE: Finite set C of "colors" and a set Q of cubes, with $|Q|=|C|$ and with each side of each cube in Q having some assigned color from C.
QUESTION: Can the cubes in Q be stacked in one vertical column such that each of the colors in C appears exactly once on each of the four sides of the column?

Reference: [Robertson and Munro, 1978]. Transformation from EXACT COVER.
Comment: The associated two-person game, in which players alternate placing a new cube on the stack, with player 1 trying to construct a stack as specified above and player 2 trying to prevent this, is PSPACE-complete with respect to whether the first player has a forced win. INSTANT INSANITY is a trade name of Parker Brothers, Inc.

A9 LOGIC

A9.1 PROPOSITIONAL LOGIC

[LO1] SATISFIABILITY

INSTANCE: Set U of variables, collection C of clauses over U (see Section 2.6 for definitions).
QUESTION: Is there a satisfying truth assignment for C?

Reference: [Cook, 1971a]. Generic transformation.
Comment: Remains NP-complete even if each $c \in C$ satisfies $|c|=3$ (3SAT), or if each $c \in C$ satisfies $|c| \leqslant 3$ and, for each $u \in U$, there are at most 3 clauses in C that contain either u or $\bar{u}$. Also remains NP-complete if each $c \in C$ has $|c| \leqslant 3$ and the bipartite graph $G=(V,E)$, where $V=U \cup C$ and E contains exactly those pairs $\{u,c\}$ such that either u or $\bar{u}$ belongs to the clause c, is planar (PLANAR 3SAT) [Lichtenstein, 1977]. The general problem is solvable in polynomial time if each $c \in C$ has $|c| \leqslant 2$ (e.g., see [Even, Itai, and Shamir, 1976]).

[LO2] 3-SATISFIABILITY (3SAT)

INSTANCE: Set U of variables, collection C of clauses over U such that each clause $c \in C$ has $|c|=3$.
QUESTION: Is there a satisfying truth assignment for C?

Reference: [Cook, 1971a]. Transformation from SATISFIABILITY.
Comment: Remains NP-complete even if each clause contains either only negated variables or only un-negated variables (MONOTONE 3SAT) [Gold, 1974], or if for each $u \in U$ there are at most 5 clauses in C that contain either u or $\bar{u}$.

[LO3] NOT-ALL-EQUAL 3SAT

INSTANCE: Set U of variables, collection C of clauses over U such that each clause $c \in C$ has $|c|=3$.
QUESTION: Is there a truth assignment for U such that each clause in C has at least one true literal and at least one false literal?

Reference: [Schaefer, 1978b]. Transformation from 3SAT. .

[LO4] ONE-IN-THREE 3SAT

INSTANCE: Set U of variables, collection C of clauses over U such that each clause $c \in C$ has $|c|=3$.
QUESTION: Is there a truth assignment for U such that each clause in C has exactly one true literal?

Reference: [Schaefer, 1978b]. Transformation from 3SAT.
Comment: Remains NP-complete even if no $c \in C$ contains a negated literal.

[LO5] MAXIMUM 2-SATISFIABILITY

INSTANCE: Set U of variables, collection C of clauses over U such that each clause $c \in C$ has $|c|=2$, positive integer $K \leqslant |C|$.

QUESTION: Is there a truth assignment for U that simultaneously satisfies at least K of the clauses in C?

Reference: [Garey, Johnson, and Stockmeyer, 1976]. Transformation from 3SAT.
Comment: Solvable in polynomial time if $K=|C|$ (e.g.,see [Even, Itai, and Shamir, 1976]).

[LO6] GENERALIZED SATISFIABILITY

INSTANCE: Positive integers $k_1,k_2,\ldots,k_m$, sequence $S=<R_1,R_2,\ldots,R_m>$ of subsets $R_i \subseteq \{T,F\}^{k_i}$, set U of variables, and, for $1 \leqslant i \leqslant m$, a collection C_i of k_i-tuples of variables from U.
QUESTION: Is there a truth assignment $t: U \rightarrow \{T,F\}$ such that for all i, $1 \leqslant i \leqslant m$, and for all k_i-tuples $(u[1],u[2],\ldots,u[k_i])$ in C_i, we have

$$(t(u[1]),t(u[2]),\ldots,t(u[k_i])) \in R_i \ ?$$

Reference: [Schaefer, 1978b]. Transformation from 3SAT.
Comment: For any fixed sequence S, the problem is NP-complete unless one of the following six alternatives holds, in which case the problem with that S is solvable in polynomial time:

(1) Each R_i contains $\{T\}^{k_i}$,
(2) each R_i contains $\{F\}^{k_i}$,
(3) each R_i is logically "equivalent" to some conjunctive normal form expression having at most one negated literal per clause,
(4) each R_i is logically "equivalent" to some conjunctive normal form expression having at most one un-negated literal per clause,
(5) each R_i is logically "equivalent" to some conjunctive normal form expression having at most 2 literals per clause, or
(6) each R_i is the "solution set" for some system of linear equations over GF[2].

The NP-completeness of 3SAT, ONE-IN-THREE 3SAT, and NOT-ALL-EQUAL 3SAT all follow from this classification. If the tuples in each C_i are allowed to be in $(U \cup \{T,F\})^{k_i}$ ("formulas with constants"), the problem is NP-complete even if (1) or (2) holds, but is still polynomially solvable if (3), (4), (5), or (6) holds. The quantified version of the problem "with constants," where we are also given a sequence $Q_1,Q_2,\ldots,Q_n$ of quantifiers (each Q_i being either $\forall$ or $\exists$) and ask if

$$(Q_1u_1)(Q_2u_2)\cdots(Q_nu_n)\,[c \in R_i \text{ for all } c \in C_i,\ 1 \leqslant i \leqslant m]$$

is PSPACE-complete, even for fixed S, so long as S does not meet any of (3), (4), (5), or (6), and is solvable in polynomial time for any fixed S that does meet one of (3), (4), (5), or (6).

[LO7] SATISFIABILITY OF BOOLEAN EXPRESSIONS

INSTANCE: Variable set U, a subset B of the set of 16 possible binary Boolean connectives, and a well-formed Boolean expression E over U and B.
QUESTION: Is there a truth assignment for U that satisfies E?

Reference: [Cook, 1971a]. Generic transformation.
Comment: Remains NP-complete if B is restricted to $\{\wedge,\vee,\rightarrow,\neg\}$, or any other

truth-functionally complete set of connectives. Also NP-complete for any truth-functionally incomplete set of connectives containing $\{\nrightarrow\}$, $\{\nleftarrow\}$, $\{\not\equiv,\vee\}$, or $\{\not\equiv,\wedge\}$ as a subset [Lewis, 1978]. Problem is solvable in polynomial time for any truth-functionally incomplete set of connectives *not* containing one of these four sets as a subset.

[LO8] NON-TAUTOLOGY

INSTANCE: Boolean expression E over a set U of variables, using the connectives "$\neg$" (not), "$\vee$" (or), "$\wedge$" (and), and "$\rightarrow$" (implies).
QUESTION: Is E *not* a tautology, i.e., is there a truth assignment for U that makes E false?

Reference: [Cook, 1971a]. Transformation from SATISFIABILITY.
Comment: Remains NP-complete even if E is in "disjunctive normal form" with at most 3 literals per disjunct.

[LO9] MINIMUM DISJUNCTIVE NORMAL FORM

INSTANCE: Set $U=\{u_1,u_2, \ldots, u_n\}$ of variables, set $A \subseteq \{T,F\}^n$ of "truth assignments," and a positive integer K.
QUESTION: Is there a disjunctive normal form expression E over U, having no more than K disjuncts, such that E is true for precisely those truth assignments in A, and no others?

Reference: [Gimpel, 1965]. Transformation from MINIMUM COVER.
Comment: Variant in which the instance contains a *complete* truth table, i.e., disjoint sets A and $B \subseteq \{T,F\}^n$ such that $A \cup B = \{T,F\}^n$, and E must be true for all truth assignments in A and false for all those in B, is also NP-complete, despite the possibly much larger instance size [Masek, 1978].

[LO10] TRUTH-FUNCTIONALLY COMPLETE CONNECTIVES

INSTANCE: Set U of variables, collection C of well-formed Boolean expressions over U.
QUESTION: Is C truth-functionally complete, i.e., is there a truth-functionally complete set of logical connectives (unary and binary operators) $D=\{\theta_1,\theta_2, \ldots, \theta_k\}$ such that for each $\theta_i \in D$ there is an expression $E \in C$ and a substitution $s: U \rightarrow \{a,b\}$ for which $s(E) \equiv a\theta_i b$ or $s(E) \equiv \theta_i a$ (depending on whether θ_i is binary or unary)?

Reference: [Statman, 1976]. Transformation from 3SAT.
Comment: Remains NP-complete even if $|C|=2$.

A9.2 MISCELLANEOUS

[LO11] QUANTIFIED BOOLEAN FORMULAS (QBF) (*)

INSTANCE: Set $U=\{u_1,u_2, \ldots, u_n\}$ of variables, well-formed quantified Boolean formula $F=(Q_1u_1)(Q_2u_2) \cdots (Q_nu_n)E$, where E is a Boolean expression and each Q_i is either $\forall$ or $\exists$.
QUESTION: Is F true?

Reference: [Stockmeyer and Meyer, 1973]. Generic transformation.
Comment: PSPACE-complete, even if E is in conjunctive normal form with three literals per clause (QUANTIFIED 3SAT), but solvable in polynomial time when there are at most two literals per clause [Schaefer, 1978b]. If F is restricted to at most k alternations of quantifiers (i.e., there are at most k indices i such that $Q_i \neq Q_{i+1}$), then the restricted problem is complete for some class in the polynomial hierarchy, depending on k and the allowed values for Q_1 (see Section 7.2).

[LO12] **FIRST ORDER THEORY OF EQUALITY (*)**
INSTANCE: Finite set $U=\{u_1,u_2, \ldots, u_n\}$ of variables, sentence S over U in the first order theory of equality. (Such sentences can be defined inductively as follows: An "expression" is of the form "$u=v$" where $u,v \in U$, or of the form "$\neg E$," "$(E \vee F)$," "$(E \wedge F)$," or "$(E \Rightarrow F)$" where E and F are expressions. A sentence is of the form $(Q_1 u_1)(Q_2 u_2) \cdots (Q_n u_n) E$ where E is an expression and each Q_i is either $\forall$ or $\exists$.)
QUESTION: Is S true in all models of the theory?

Reference: [Stockmeyer and Meyer, 1973]. Generic transformation.
Comment: PSPACE-complete. The analogous problem for any fixed first order theory that has a model in which some predicate symbol is interpreted as a relation that holds sometimes but not always is PSPACE-hard [Hunt, 1977].

[LO13] **MODAL LOGIC S5-SATISFIABILITY**
INSTANCE: Well-formed modal formula A over a finite set U of variables, where a modal formula is either a variable $u \in U$ or is of the form "$(A \wedge B)$," "$\neg A$," or "$\Box A$," where A and B are modal formulas.
QUESTION: Is A "S5-satisfiable," i.e., is there a model (W,R,V), where W is a set, R is a reflexive, transitive, and symmetric binary relation on W, and V is a mapping from $U \times W$ into $\{T,F\}$ such that, for some $w \in W$, $V(A,w)=T$, where V is extended to formulas by $V(A \wedge B,w)=T$ if and only if $V(A,w)=V(B,w)=T$, $V(\neg A,w)=T$ if and only if $V(A,w)=F$, and $V(\Box A,w)=T$ if and only if $V(A,w')=T$ for all $w' \in W$ satisfying $(w,w') \in R$?

Reference: [Ladner, 1977]. Transformation from 3SAT. Nontrivial part is proving membership in NP.

[LO14] **MODAL LOGIC PROVABILITY (*)**
INSTANCE: Well-formed modal formula A, modal system $S \in \{K,T,S4\}$ (see reference or [Hughes and Cresswell, 1968] for details of K, T, and $S4$).
QUESTION: Is A provable in system S?

Reference: [Ladner, 1977]. Transformation from QBF.
Comment: PSPACE-complete for fixed $S \in \{K,T,S4\}$ or for any fixed modal system S in which everything provable in K, but nothing not provable in $S4$, can be proved.

[LO15] PREDICATE LOGIC WITHOUT NEGATION

INSTANCE: Sets $U = \{u_1, u_2, \ldots, u_n\}$ of variables, $F = \{f_1^{m_1}, f_2^{m_2}, \ldots, f_k^{m_k}\}$ of function symbols, and $R = \{R_1^{r_1}, R_2^{r_2}, \ldots, R_j^{r_j}\}$ of relation symbols ($m_i \geqslant 0$ and $r_i \geqslant 0$ being the dimensions of the corresponding functions and relations), and a well-formed predicate logic sentence A without negations over U, F, and R. (Such a sentence can be defined inductively as follows: A term is a variable $u \in U$ or of the form "$f_i^{m_i}(t_1, t_2, \ldots, t_{m_i})$" where each t_j is a term. A formula is of the form "$t_1 = t_2$" where t_1 and t_2 are terms, "$R_i^{r_i}(t_1, t_2, \ldots, t_{r_i})$" where each t_j is a term, or "$(A \wedge B)$," "$(A \vee B)$," "$\forall u_i(A)$," or "$\exists u_i(A)$" where A and B are formulas and $u_i \in U$. A sentence is a formula in which all variables are quantified before they occur.)

QUESTION: Is A true under all interpretations of F and R?

Reference: [Kozen, 1977c]. Transformation from 3SAT. Nontrivial part is proving membership in NP.

Comment: Remains NP-complete even if there are no universal quantifiers, no relation symbols, and only two functions, both with dimension 0 (and hence constants).

[LO16] CONJUNCTIVE SATISFIABILITY WITH FUNCTIONS AND INEQUALITIES

INSTANCE: Set U of variables, set F of univariate function symbols, and a collection C of "clauses" of the form $U * V$ where $*$ is either "$\leqslant$," "$>$," "$=$," or "$\neq$," and U and V are either "0," "1," "u," "$f(0)$," "$f(1)$," or "$f(u)$," for some $f \in F$ and $u \in U$.

QUESTION: Is there an assignment of integer values to all the variables $u \in U$ and to all $f(u)$, for $u \in U$ and $f \in F$, such that all the clauses in C are satisfied under the usual interpretations of $\leqslant$, $>$, $=$, and $\neq$?

Reference: [Pratt, 1977]. Transformation from 3SAT.

Comment: Remains NP-complete even if $=$ and $\neq$ are not used. Solvable in polynomial time if $\leqslant$ and $>$ are not used [Nelson and Oppen, 1977], or if $=$ and $\neq$ are not used and no function symbols are allowed [Litvintchouk and Pratt, 1977]. Variant in which W and V are either of the form "u" or "$u+c$" for some $u \in U$ and $c \in Z$ is NP-complete if all four relations are allowed, but solvable in polynomial time if only $\leqslant$ and $>$ or only $=$ and $\neq$ are allowed [Chan, 1977].

[LO17] MINIMUM AXIOM SET

INSTANCE: Finite set S of "sentences," subset $T \subseteq S$ of "true sentences," an "implication relation" R consisting of pairs (A, s) where $A \subseteq S$ and $s \in S$, and a positive integer $K \leqslant |S|$.

QUESTION: Is there a subset $S_0 \subseteq T$ with $|S_0| \leqslant K$ and a positive integer n such that, if we define S_i, $1 \leqslant i \leqslant n$, to consist of exactly those $s \in S$ for which either $s \in S_{i-1}$ or there exists a $U \subseteq S_{i-1}$ such that $(U, s) \in R$, then $S_n = T$?

Reference: [Pudlák, 1975]. Transformation from X3C.

Comment: Remains NP-complete even if $T = S$.

[LO18] FIRST ORDER SUBSUMPTION

INSTANCE: Finite set U of "variable symbols," finite set C of "function symbols," collection $E=\{E_1,E_2,\ldots,E_m\}$ of expressions over $U \cup C$, collection $F=\{F_1,F_2,\ldots,F_n\}$ of expressions over C.

QUESTION: Is there a substitution mapping s that assigns to each $u \in U$ an expression $s(u)$ over C such that, if $s(E_i)$ denotes the result of substituting for each occurrence in E_i of each $u \in U$ the corresponding expression $s(u)$, then $\{s(E_1),s(E_2),\ldots,s(E_m)\}$ is a subset of $\{F_1,F_2,\ldots,F_n\}$?

Reference: [Baxter, 1976], [Baxter, 1977]. Transformation from 3SAT.

Comment: Remains NP-complete for any fixed $n \geqslant 3$, but is solvable in polynomial time for any fixed m.

[LO19] SECOND ORDER INSTANTIATION

INSTANCE: Two "second order logic expressions" E_1 and E_2, the second of which contains no variables (in a second order expression, functions can be variables; see references for details).

QUESTION: Is there a substitution for the variables of E_1 that yields an expression identical to E_2?

Reference: [Baxter, 1976]. Transformation from 3SAT. Proof of membership in NP is nontrivial.

Comment: The more general SECOND ORDER UNIFICATION problem, where both E_1 and E_2 can contain variables and we ask if there is a substitution for the variables in E_1 and E_2 that results in identical expressions, is not known to be decidable. THIRD ORDER UNIFICATION is undecidable [Huet, 1973], whereas FIRST ORDER UNIFICATION can be solved in polynomial time [Baxter, 1975], [Paterson and Wegman, 1978].

A10 AUTOMATA AND LANGUAGE THEORY

A10.1 AUTOMATA THEORY

[AL1] FINITE STATE AUTOMATON INEQUIVALENCE (*)

INSTANCE: Two nondeterministic finite state automata A_1 and A_2 having the same input alphabet Σ (where such an automaton $A = (Q,\Sigma,\delta,q_0,F)$ consists of a finite set Q of states, input alphabet Σ, transition function δ mapping $Q \times \Sigma$ into subsets of Q, initial state q_0, and a set $F \subseteq K$ of "accept" states, e.g., see [Hopcroft and Ullman, 1969]).
QUESTION: Do A_1 and A_2 recognize different languages?

Reference: [Kleene, 1956]. Transformation from REGULAR EXPRESSION NON-UNIVERSALITY.
Comment: PSPACE-complete, even if $|\Sigma| = 2$ and A_2 is the trivial automaton recognizing Σ^*. The general problem is NP-complete if $|\Sigma| = 1$, or if A_1 and A_2 both recognize finite languages (a property that can be checked in polynomial time, e.g., see [Hopcroft and Ullman, 1969]). Problem is solvable in polynomial time if A_1 and A_2 are deterministic finite state automata, e.g., see [Hopcroft and Ullman, 1969].

[AL2] TWO-WAY FINITE STATE AUTOMATON NON-EMPTINESS (*)

INSTANCE: A two-way nondeterministic finite state automaton $A = (Q,\Sigma,\delta,q_0,F)$ (where Q, Σ, q_0, and F are the same as for a one-way nondeterministic finite state automaton, but the transition function δ maps $Q \times \Sigma$ into subsets of $Q \times \{-1,0,1\}$, e.g., see [Hopcroft and Ullman, 1969]).
QUESTION: Is there an $x \in \Sigma^*$ such that A accepts x?

Reference: [Hunt, 1973b]. Transformation from LINEAR BOUNDED AUTOMATON ACCEPTANCE.
Comment: PSPACE-complete, even if $|\Sigma| = 2$ and A is deterministic. If $|\Sigma| = 1$ the general problem is NP-complete [Galil, 1976]. If A is a one-way nondeterministic finite state automaton, the general problem can be solved in polynomial time (e.g., see [Hopcroft and Ullman, 1969]). Analogous results for the question of whether A recognizes an infinite language can be found in the above references.

[AL3] LINEAR BOUNDED AUTOMATON ACCEPTANCE (*)

INSTANCE: A "linear bounded automaton" A with input alphabet Σ (see [Hopcroft and Ullman, 1969] for definition), and a string $x \in \Sigma^*$.
QUESTION: Does A accept x?

Reference: [Karp, 1972]. Generic transformation.
Comment: PSPACE-complete, even if A is deterministic (the LINEAR SPACE ACCEPTANCE problem of Section 7.4). Moreover, there exist fixed deterministic linear bounded automata for which the problem is PSPACE-complete.

[AL4] QUASI-REALTIME AUTOMATON ACCEPTANCE

INSTANCE: A multi-tape nondeterministic Turing machine M (Turing machine ***program***, in our terminology), whose input tape read-head must move right at each

step, and which must halt whenever the read-head sees a blank, and a string x over the input alphabet Σ of M. (For a more complete description of this type of machine and its equivalent formulations, see [Book and Greibach, 1970].)
QUESTION: Does M accept x?

Reference: [Book, 1972]. Generic transformation.
Comment: Remains NP-complete even if M has only a single work tape in addition to its input tape. See also QUASI-REALTIME LANGUAGE MEMBERSHIP (the languages accepted by quasi-realtime automata are the same as the quasi-realtime languages defined in that entry).

[AL5] NON-ERASING STACK AUTOMATON ACCEPTANCE (*)

INSTANCE: A "one-way nondeterministic non-erasing stack automaton" (a 1NESA) A with input alphabet Σ (see [Hopcroft and Ullman, 1969] for definition), and a string $x \in \Sigma^*$.
QUESTION: Does A accept x?

Reference: [Galil, 1976], [Hopcroft and Ullman, 1967]. Transformation from LINEAR BOUNDED AUTOMATON ACCEPTANCE. The second reference proves membership in PSPACE.
Comment: PSPACE-complete, even if $x \in \Sigma^*$ is fixed and A is restricted to be a "checking stack automaton" (as defined in [Greibach, 1969]). If x is the empty string and A is further restricted to be a checking stack automaton with a single stack symbol, the problem becomes NP-complete [Galil, 1976]. If instead x is allowed to vary and A is fixed, the problem is in NP for each 1NESA and remains so if A is allowed to be a general "nested stack automaton" [Rounds, 1973]. There exist particular 1NESAs for which the problem is NP-complete [Rounds, 1973], and these particular 1NESAs can be chosen to be checking stack automata [Shamir and Beeri, 1974] that are also "reading pushdown automata" [Hunt, 1976]. However, if A is restricted to be a "one-way nondeterministic pushdown automaton," then the problem can be solved in polynomial time (even with A allowed to vary), as indeed is the case for "two-way nondeterministic pushdown automata" [Aho, Hopcroft, and Ullman, 1968].

[AL6] FINITE STATE AUTOMATA INTERSECTION (*)

INSTANCE: Sequence $A_1, A_2, \ldots, A_n$ of deterministic finite state automata having the same input alphabet Σ.
QUESTION: Is there a string $x \in \Sigma^*$ accepted by each of the A_i, $1 \leqslant i \leqslant n$?

Reference: [Kozen, 1977d]. Transformation from LINEAR SPACE ACCEPTANCE.
Comment: PSPACE-complete. Solvable in polynomial time for any fixed n (e.g., see [Hopcroft and Ullman, 1969]).

[AL7] REDUCTION OF INCOMPLETELY SPECIFIED AUTOMATA

INSTANCE: An incompletely specified deterministic finite state automaton $A = (Q, \Sigma, \delta, q_0, F)$, where Q is the set of states, Σ is the input alphabet, δ is a "partial" transition function mapping a subset of $Q \times \Sigma$ into Q, $q_0 \in Q$ is the initial state, and $F \subseteq Q$ is the set of "accept" states, and a positive integer K.

QUESTION: Can the transition function δ be extended to a total function from $Q \times \Sigma$ into Q in such a way that the resulting completely specified automaton has an equivalent "reduced automaton" with K or fewer states?

Reference: [Pfleeger, 1973]. Transformation from GRAPH 3-COLORABILITY.
Comment: Remains NP-complete for any fixed $K \geqslant 6$. Related question in which "state-splitting" (as used in [Paull and Unger, 1959]) is allowed is also NP-complete for any fixed $K \geqslant 6$ [Pfleeger, 1973]. If both "state-splitting" and "symbol-splitting" (as used in [Grasselli and Luccio, 1966]) are allowed, the analogous problem in which the corresponding reduced automaton is to have the sum of the number of states and the number of symbols be no more than K is also NP-complete [Pfleeger, 1974]. The problem of determining the minimum state deterministic finite state automaton equivalent to a given *completely* specified one can be solved in polynomial time (e.g., see [Hopcroft, 1971] or [Aho and Ullman, 1972]). The corresponding problem for completely specified nondeterministic finite state automata is PSPACE-complete (see FINITE STATE AUTOMATA INEQUIVALENCE).

[AL8] MINIMUM INFERRED FINITE STATE AUTOMATON

INSTANCE: Finite alphabet Σ, two finite subsets $S, T \subseteq \Sigma^*$, positive integer K.
QUESTION: Is there a K-state deterministic finite automaton A that recognizes a language $L \subseteq \Sigma^*$ such that $S \subseteq L$ and $T \subseteq \Sigma^* - L$?

Reference: [Gold, 1974]. Transformation from MONOTONE 3SAT.
Comment: Can be solved in polynomial time if $S \cup T = \Sigma^{(n)}$ for some n, where $\Sigma^{(n)}$ is the set of all strings of length n or less over Σ [Trakhtenbrot and Barzdin, 1973]. However, for any fixed $\epsilon > 0$, the problem remains NP-complete if restricted to instances for which $(S \cup T) \subseteq \Sigma^{(n)}$ and $|\Sigma^{(n)} - (S \cup T)| \leqslant |\Sigma^{(n)}|^{\epsilon}$ [Angluin, 1977].

A10.2 FORMAL LANGUAGES

[AL9] REGULAR EXPRESSION INEQUIVALENCE (*)

INSTANCE: Regular expressions E_1 and E_2 over the operators $\{\cup, \cdot, *\}$ and the alphabet Σ (see Section 7.4 for definition).
QUESTION: Do E_1 and E_2 represent different languages?

Reference: [Stockmeyer and Meyer, 1973], [Stockmeyer, 1974a]. Generic transformation. The second reference proves membership in PSPACE.
Comment: PSPACE-complete, even if $|\Sigma| = 2$ and $E_2 = \Sigma^*$ (REGULAR EXPRESSION NON-UNIVERSALITY, see Section 7.4). In fact, PSPACE-complete if E_2 is any fixed expression representing an "unbounded" language [Hunt, Rosenkrantz, and Szymanski, 1976a]. NP-complete for fixed E_2 representing any infinite "bounded" language, but solvable in polynomial time for fixed E_2 representing any finite language. The general problem remains PSPACE-complete if E_1 and E_2 both have "star height" k for a fixed $k \geqslant 1$ [Hunt, Rosenkrantz, and Szymanski, 1976a], but is NP-complete for $k = 0$ ("star free") [Stockmeyer and Meyer, 1973], [Hunt, 1973a]. Also NP-complete if one or both of E_1 and E_2 represent bounded languages (a property that can be checked in polynomial time) [Hunt, Rosenkrantz, and Szymanski, 1976a] or if $|\Sigma| = 1$ [Stockmeyer and Meyer, 1973]. For related

results and intractable generalizations, see cited references, [Hunt, 1973b], and [Hunt and Rosenkrantz, 1978].

[AL10] MINIMUM INFERRED REGULAR EXPRESSION

INSTANCE: Finite alphabet Σ, two finite subsets $S,T \subseteq \Sigma^*$, positive integer K.
QUESTION: Is there a regular expression E over Σ that has K or fewer occurrences of symbols from Σ and such that, if $L \subseteq \Sigma^*$ is the language represented by E, then $S \subseteq L$ and $T \subseteq \Sigma^* - L$?

Reference: [Angluin, 1977]. Transformation from 3SAT.
Comment: Remains NP-complete even if E is required to contain no "$\cup$" operations or to be "star-free" (contain no "*" operations) [Angluin, 1976].

[AL11] REYNOLDS COVERING FOR CONTEXT-FREE GRAMMARS

INSTANCE: Context-free grammars $G_1=(N_1,\Sigma,\Pi_1,S_1)$ and $G_2=(N_2,\Sigma,\Pi_2,S_2)$, where Σ is a finite set of "terminal" symbols, N_i is a finite set of "nonterminal" symbols, $S_i \in N_i$ is the "initial" symbol, and Π_i is a set of "productions" of the form "$A \rightarrow w$," where $A \in N_i$ and $w \in (N_i \cup \Sigma)^*$.
QUESTION: Does G_2 "Reynolds cover" G_1, i.e., is there a function f mapping $N_1 \cup \Sigma$ into $N_2 \cup \Sigma$ such that $f(x)=x$ for all $x \in \Sigma$, $f(A) \in N_2$ for all $A \in N_1$, $f(S_1)=S_2$, and for each production $A \rightarrow x_1 x_2 \cdots x_n$ in Π_1, the image $f(A) \rightarrow f(x_1) f(x_2) \cdots f(x_n)$ of that production is in Π_2?

Reference: [Hunt and Rosenkrantz, 1977]. Transformation from 3SAT.
Comment: Remains NP-complete even if G_1 and G_2 are restricted to "regular" grammars. The same results hold for the related questions of whether G_2 "weakly Reynolds covers" G_1 or whether G_2 is a "homomorphic image" of G_1. The problem "Given G is there an $LL(k)$ context-free grammar H such that H Reynolds covers G?" is solvable in polynomial time, as are the related problems where $LL(k)$ is replaced by $LR(k)$ or one of a number of other grammar classes (see [Hunt and Rosenkrantz, 1977]).

[AL12] COVERING FOR LINEAR GRAMMARS (*)

INSTANCE: Two linear context-free grammars $G_1=(N_1,\Sigma,\Pi_1,S_1)$ and $G_2=(N_2,\Sigma,\Pi_2,S_2)$, where no production in such a grammar is allowed to have more than one nonterminal symbol on its right hand side.
QUESTION: Is there a function $h: P_1 \rightarrow P_2 \cup \{\lambda\}$ (where λ denotes the empty production) such that G_1 covers G_2 under h, i.e., such that for all strings $w \in \Sigma^*$ (1) if w is derivable from S_1 under the sequence of productions $p_1,p_2,\ldots,p_n$, then w is derivable from S_2 under the sequence $h(p_1),h(p_2),\ldots,h(p_n)$, and (2) if w is derivable from S_2 under the sequence of productions $q_1,q_2,\ldots,q_n$ from Π_2, then there exists a sequence of productions $p_1,p_2,\ldots,p_m$ that is a derivation of w in G_1 such that $h(p_1),h(p_2),\ldots,h(p_m)$ equals $q_1,q_2,\ldots,q_n$?

Reference: [Hunt, Rosenkrantz, and Szymanski, 1976a], [Hunt, Rosenkrantz, and Szymanski, 1976b]. Transformation from REGULAR EXPRESSION NON-UNIVERSALITY. The second reference proves membership in PSPACE.
Comment: PSPACE-complete, even for "regular" grammars. Undecidable for ar-

bitrary context-free grammars. See [Hunt and Rosenkrantz, 1977] for related results.

[AL13] STRUCTURAL INEQUIVALENCE FOR LINEAR GRAMMARS (*)

INSTANCE: Two linear context-free grammars $G_1 = (N_1, \Sigma, \Pi_1, S_1)$ and $G_2 = (N_2, \Sigma, \Pi_2, S_2)$.

QUESTION: Are G_1 and G_2 "structurally inequivalent," i.e., do the parenthesized grammars obtained from G_1 and G_2 by replacing each production $A \rightarrow w$ by $A \rightarrow (w)$ (where "(" and ")" are new terminal symbols) generate different languages?

Reference: [Hunt, Rosenkrantz, and Szymanski, 1976a]. Transformation from REGULAR EXPRESSION NON-UNIVERSALITY.

Comment: PSPACE-complete, even if G_1 and G_2 are regular and $|\Sigma| = 2$. NP-complete if $|\Sigma| = 1$. For arbitrary context-free grammars, problem is decidable but not known to be in PSPACE.

[AL14] REGULAR GRAMMAR INEQUIVALENCE (*)

INSTANCE: Regular grammars $G_1 = (N_1, \Sigma, \Pi_1, S_1)$ and $G_2 = (N_2, \Sigma, \Pi_2, S_2)$, where a regular grammar is a context-free grammar in which each production has the form $A \rightarrow aB$ or $A \rightarrow a$ with $A, B \in N$ and $a \in \Sigma$.

QUESTION: Do G_1 and G_2 generate different languages?

Reference: [Chomsky and Miller, 1958]. Transformation from FINITE STATE AUTOMATON INEQUIVALENCE.

Comment: PSPACE-complete, even if $|\Sigma| = 2$ and G_2 is a fixed grammar generating Σ^* (REGULAR GRAMMAR NON-UNIVERSALITY). The general problem is NP-complete if $|\Sigma| = 1$ or if both grammars generate finite languages (a property that can be checked in polynomial time, e.g., see [Hopcroft and Ullman, 1969]). If G_1 is allowed to be an arbitrary linear grammar and G_2 is a fixed grammar generating Σ^* (LINEAR GRAMMAR NON-UNIVERSALITY), the problem is undecidable [Hunt, Rosenkrantz, and Szymanski, 1976a].

[AL15] NON-LR(K) CONTEXT-FREE GRAMMAR

INSTANCE: Context-free grammar G, positive integer K written in *unary* notation.

QUESTION: Is G *not* an $LR(K)$ grammar (see reference for definition)?

Reference: [Hunt, Szymanski, and Ullman, 1975]. Generic transformation.

Comment: Solvable in polynomial time for any fixed K. If K is written in binary (as in our standard encodings), then the problem is complete for NEXP-TIME and hence intractable. Determining whether there exists an integer K such that G is an $LR(K)$ grammar is undecidable [Hunt and Szymanski, 1976a]. The same results hold if "$LR(K)$" is replaced by "$LL(K)$," "$LC(K)$," "$SLR(K)$," or any one of a number of other properties (see above references). However, in the case of $LL(K)$, if it is known that there is some K' for which G is $LR(K')$, then one can decide whether there exists a K for which G is $LL(K)$ in polynomial time [Hunt and Szymanski, 1978].

[AL16] ETOL GRAMMAR NON-EMPTINESS (*)

INSTANCE: An ETOL grammar $G=(N,\Sigma,\{\delta_1,\delta_2,\ldots,\delta_n\},S)$, where N is a finite set of "nonterminal" symbols, Σ is a finite set of "terminal" symbols, $S \in N$ is the "initial" symbol, and each δ_i is a "table" of productions that take symbols in $N \cup \Sigma$ to strings in $(N \cup \Sigma)^*$ (at each step of a derivation, every symbol in the current string is replaced according to some production from a particular chosen table, e.g., see [Herman and Rozenberg, 1975]).

QUESTION: Is the language generated by G non-empty?

Reference: [Jones and Skyum, 1976]. Transformation from REGULAR EXPRESSION NON-UNIVERSALITY.

Comment: PSPACE-complete, even if G is an "ϵ-free EDTOL grammar"; NP-complete if G contains just one table and exactly one production for each symbol (G is an "EDOL grammar"). The same results hold for the question of whether G generates an infinite language. These problems are solvable in polynomial time for context-free grammars, but undecidable for context-sensitive grammars, e.g., see [Hopcroft and Ullman, 1969].

[AL17] CONTEXT-FREE PROGRAMMED LANGUAGE MEMBERSHIP

INSTANCE: An ϵ-free, context-free programmed grammar $G=(N,\Sigma,\Pi,S)$ and a string $x \in \Sigma^*$. (In such a grammar, the productions in Π are of the form $A \rightarrow w(T)(F)$, where $A \in N$, $w \in (N \cup \Sigma)^* - \epsilon$, and T and F are subsets of Π indicating where the next production to be applied must be chosen from, depending on whether the last production chosen was applicable or not; see [Rosenkrantz, 1969].)

QUESTION: Is x in the language generated by G?

Reference: [Shamir and Beeri, 1974]. Transformation from 3SAT. The ϵ-free property ensures membership in NP.

Comment: Remains NP-complete even if all productions $A \rightarrow w(T)(F)$ in Π are required to have $T=F$ [van Leeuwen, 1975]. If $T=F=\Pi$ for all productions in Π and if productions to the empty string ϵ are permitted, we obtain the membership problem for context-free languages, which can be solved in polynomial time, e.g., see [Hopcroft and Ullman, 1969].

[AL18] QUASI-REAL-TIME LANGUAGE MEMBERSHIP

INSTANCE: Context-free grammars G_1, G_2, and G_3 having the same terminal alphabet Σ, a second finite alphabet Γ, a function $f: \Sigma \rightarrow \Gamma$, and a string $w \in \Gamma^*$.

QUESTION: Is w in the "quasi-real-time language" determined by G_1, G_2, G_3, and h, i.e., the language $L = h(L_1 \cap L_2 \cap L_3)$ consisting of all strings from Γ^* of the form $h(x_1)h(x_2) \cdots h(x_k)$ such that $x_1x_2 \cdots x_k \in (L_1 \cap L_2 \cap L_3)$, where L_1, L_2, and L_3 are the languages generated by G_1, G_2, and G_3 respectively?

Reference: [Hunt, 1973b], [Greibach, 1973b]. Transformation from 3SAT.

Comment: Solvable in polynomial time if h is one-to-one (by standard context-free parsing techniques, e.g., see [Hopcroft and Ullman, 1969]). The problem remains NP-complete if $L_3=\Sigma^*$, i.e., $L = h(L_1 \cap L_2)$ [Greibach, 1973a], but is polynomially solvable if both L_2 and L_3 equal Σ^*, i.e., $L = h(L_1)$.

[AL19] ETOL LANGUAGE MEMBERSHIP (*)

INSTANCE: An ETOL grammar $G=(N,\Sigma,\{\delta_1,\delta_2, \ldots, \delta_n\},S)$ and a string $w \in \Sigma^*$.

QUESTION: Is w in the language generated by G?

Reference: [van Leeuwen, 1975]. Transformation from 3SAT.

Comment: PSPACE-complete, even if G is an "ϵ-free EDTOL grammar" [Jones and Skyum, 1976]. If G is fixed, the problem is in NP and there exist particular grammars for which it is NP-complete, even if G is a "TOL grammar" (has no nonterminals) and is ϵ-free [Van Leeuwen, 1975]. The problem is solvable in polynomial time for fixed G if G is an "EDTOL grammar" [Jones and Skyum, 1977] or if G is an "EOL grammar" (has only one table) [Opatrńy and Culik, 1975].

[AL20] CONTEXT-SENSITIVE LANGUAGE MEMBERSHIP (*)

INSTANCE: Context-sensitive grammar $G=(N,\Sigma,\Pi,S)$ and a string $w \in \Sigma^*$. (In a context-sensitive grammar, each production has the form $x \rightarrow y$ where x and y are nonempty strings over $N \cup \Sigma$ and $|y| \geqslant |x|$).

QUESTION: Is w in the language generated by G?

Reference: [Kuroda, 1964]. Transformation from LINEAR BOUNDED AUTOMATON ACCEPTANCE.

Comment: PSPACE-complete, even for deterministic context-sensitive grammars. Moreover, there exist fixed context-sensitive grammars for which the problem is PSPACE-complete, and a fixed "linear time" context-sensitive grammar for which the problem is NP-complete [Book, 1978]. (For any fixed linear time context-sensitive grammar the problem is in NP.)

[AL21] TREE TRANSDUCER LANGUAGE MEMBERSHIP (*)

INSTANCE: A "top-down finite-state tree transducer" M with output alphabet Γ, a context-free grammar G, and a string $w \in \Gamma^*$ (see references for detailed definitions).

QUESTION: Is w in the "yield" of the "surface set" determined by M and G?

Reference: [Reiss, 1977a]. Generic transformation.

Comment: PSPACE-complete. Problem is in NP for fixed M and G, and there exist particular choices for M and G for which the problem is NP-complete [Rounds, 1973]. The general problem is solvable in polynomial time if M is required to be "linear", while for fixed M the problem is solvable in polynomial time if M is "deterministic" [Reiss, 1977b].

A11 PROGRAM OPTIMIZATION

A11.1 CODE GENERATION

[PO1] REGISTER SUFFICIENCY

INSTANCE: Directed acyclic graph $G=(V,A)$, positive integer K.
QUESTION: Is there a computation for G that uses K or fewer registers, i.e., an ordering $v_1,v_2,\ldots,v_n$ of the vertices in V, where $n=|V|$, and a sequence $S_0,S_1,\ldots,S_n$ of subsets of V, each satisfying $|S_i|\leqslant K$, such that S_0 is empty, S_n contains all vertices with in-degree 0 in G, and, for $1\leqslant i\leqslant n$, $v_i\in S_i$, $S_i-\{v_i\}\subseteq S_{i-1}$, and S_{i-1} contains all vertices u for which $(v_i,u)\in A$?

Reference: [Sethi, 1975]. Transformation from 3SAT.
Comment: Remains NP-complete even if all vertices of G have out-degree 2 or less. The variant in which "recomputation" is allowed (i.e., we ask for sequences $v_1,v_2,\ldots,v_m$ and $S_0,S_1,\ldots,S_m$, where no a priori bound is placed on m and the vertex sequence can contain repeated vertices, but all other properties stated above must hold) is NP-hard and is not known to be in NP.

[PO2] FEASIBLE REGISTER ASSIGNMENT

INSTANCE: Directed acyclic graph $G=(V,A)$, positive integer K, and a register assignment $f\colon V\rightarrow\{R_1,R_2,\ldots,R_k\}$.
QUESTION: Is there a computation for G using the given register assignment, i.e., an ordering $v_1,v_2,\ldots,v_n$ of V and a sequence $S_0,S_1,\ldots,S_n$ of subsets of V that satisfies all the properties given in REGISTER SUFFICIENCY and that in addition satisfies, for $1\leqslant j\leqslant K$ and $1\leqslant i\leqslant n$, there is at most one vertex $u\in S_i$ for which $f(u)=R_j$?

Reference: [Sethi, 1975]. Transformation from 3SAT.
Comment: Remains NP-complete even if all vertices of G have out-degree 2 or less.

[PO3] REGISTER SUFFICIENCY FOR LOOPS

INSTANCE: Set V of loop variables, a loop length $N\in Z^+$, for each variable $v\in V$ a start time $s(v)\in Z_0^+$ and a duration $l(v)\in Z^+$, and a positive integer K.
QUESTION: Can the loop variables be safely stored in K registers, i.e., is their an assignment $f\colon V\rightarrow\{1,2,\ldots,K\}$ such that if $f(v)=f(u)$ for some $u\neq v\in V$, then $s(u)\leqslant s(v)$ implies $s(u)+l(u)\leqslant s(v)$ and $s(v)+l(v)(\mathrm{mod}\,N)\leqslant s(u)$?

Reference: [Garey, Johnson, Miller, and Papadimitriou, 1978]. Transformation from permutation generation.
Comment: Solvable in polynomial time for any fixed K.

[PO4] CODE GENERATION ON A ONE-REGISTER MACHINE

INSTANCE: Directed acyclic graph $G=(V,A)$ in which no vertex has out-degree larger than 2, and a positive integer K.
QUESTION: Is there a program with K or fewer instructions for computing all the root vertices of G (i.e., those with in-degree 0) on a one-register machine, starting with all the leaves of G (i.e., those with out-degree 0) in memory and using only

LOAD, STORE, and OP instructions? (A LOAD instruction copies a specified vertex into the register. A STORE instruction copies the vertex in the register into memory. A new vertex v can be computed by an OP instruction if the vertex u in the register is such that $(v,u) \in A$ and, if there is another vertex u' such that $(v,u') \in A$, then u' is in memory. Execution of the OP instruction replaces u by v in the register. The computation of a new vertex is not completed until it is copied into memory by a STORE instruction.)

Reference: [Bruno and Sethi, 1976]. Transformation from 3SAT.
Comment: Remains NP-complete even if all vertices having in-degree larger than one have arcs only to leaves of G [Aho, Johnson, and Ullman, 1977a]. Solvable in polynomial time if G is a directed forest [Sethi and Ullman, 1970].

[PO5] CODE GENERATION WITH UNLIMITED REGISTERS

INSTANCE: Directed acyclic graph $G=(V,A)$ in which no vertex has out-degree larger than 2, partition of A into disjoints sets L and R such that two arcs leaving the same vertex always belong to different sets, and a positive integer K.
QUESTION: Is there a program with K or fewer instructions for computing all the root vertices of G, starting with all the leaves of G stored in registers and using only instructions of the form "$r_i \leftarrow r_j$" or "$r_i \leftarrow r_i \ op \ r_j$," $i,j \in Z^+$, where a vertex v with out-degree 2 and outgoing arcs $(v,u) \in L$ and $(v,w) \in R$ can be computed only by an instruction $r_i \leftarrow r_i \ op \ r_j$ when r_i contains u and r_j contains w?

Reference: [Aho, Johnson, and Ullman, 1977a]. Transformation from FEEDBACK VERTEX SET.
Comment: Remains NP-complete even if only leaves of G have in-degree exceeding 1. The "commutative" variant in which instructions of the form "$r_i \leftarrow r_j \ op \ r_i$" are also allowed is NP-complete [Aho, Johnson, and Ullman, 1977b]. Both problems can be solved in polynomial time if G is a forest or if 3-address instructions "$r_i \leftarrow r_j \ op \ r_k$" are allowed [Aho, Johnson, and Ullman, 1977a].

[PO6] CODE GENERATION FOR PARALLEL ASSIGNMENTS

INSTANCE: Set $V=\{v_1,v_2, \ldots, v_n\}$ of variables, set $A=\{A_1,A_2, \ldots, A_n\}$ of assignments, each A_i of the form "$v_i \leftarrow op(B_i)$" for some subset $B_i \subseteq V$, and a positive integer K.
QUESTION: Is there an ordering $v_{\pi(1)}, v_{\pi(2)}, \ldots, v_{\pi(n)}$ of V such that there are at most K values of i, $1 \leqslant i \leqslant n$, for which $v_{\pi(i)} \in B_{\pi(j)}$ for some $j > i$?

Reference: [Sethi, 1973]. Transformation from FEEDBACK VERTEX SET.
Comment: Remains NP-complete even if each B_i satisfies $|B_i| \leqslant 2$.

[PO7] CODE GENERATION WITH ADDRESS EXPRESSIONS

INSTANCE: Sequence $I=(I_1,I_2, \ldots, I_n)$ of instructions, for each $I_i \in I$ an expression $g(I_i)$ of the form "I_j," "I_j+k," "I_j-k," or "k" where $I_j \in I$ and $k \in Z^+$, and positive integers B,C, and M.
QUESTION: Can the instructions in I be stored as one- and two-byte instructions so that the total memory required is at most M, i.e., is there a one-to-one function $f:I \rightarrow \{1,2, \ldots, M\}$ such that $f(I_i) < f(I_j)$ whenever $i<j$ and such that, if $h(I_i)$ is defined to be $f(I_j)$, $f(I_j) \pm k$, or k depending on whether $g(I_i)$ is I_j, $I_j \pm k$, or

k, then for each i, $1 \leqslant i \leqslant n$, either $-C < f(I_i) - h(I_i) < B$ or $f(I_i)+1$ is not in the range of f?

Reference: [Szymanski, 1978]. Transformation from 3SAT.

Comment: Remains NP-complete for certain fixed values of B and C, e.g., 128 and 127 (much smaller values also are possible). Solvable in polynomial time if no "pathological" expressions occur (see reference for details).

[PO8] CODE GENERATION WITH UNFIXED VARIABLE LOCATIONS

INSTANCE: Sequence $I = (I_1, I_2, \ldots, I_n)$ of instructions, finite set V of variables, assignment $g: I \to I \cup V$, positive integers B and M.

QUESTION: Can the instructions in I be stored as one- and two-byte instructions and the variables stored among them so that the total memory required is at most M, i.e., is there a one-to-one function $f: I \cup V \to \{1, 2, \ldots, M\}$ such that $f(I_i) < f(I_j)$ whenever $i < j$ and such that, for $1 \leqslant i \leqslant n$, either $|f(I_i) - f(g(I_i))| < B$ or $f(I_i)+1$ is not in the range of f?

Reference: [Robertson, 1977]. Transformation from 3SAT.

Comment: Remains NP-complete even for certain fixed values of B, e.g., $B = 31$. Solvable in polynomial time if V is empty.

[PO9] ENSEMBLE COMPUTATION

INSTANCE: Collection C of subsets of a finite set A, positive integer J.

QUESTION: Is there a sequence $S = (z_1 \leftarrow x_1 \cup y_1, z_2 \leftarrow x_2 \cup y_2, \ldots, z_j \leftarrow x_j \cup y_j)$ of $j \leqslant J$ union operations, where each x_i and y_i is either $\{a\}$ for some $a \in A$ or z_k for some $k < i$, such that x_i and y_i are disjoint, $1 \leqslant i \leqslant j$, and such that for every subset $c \in C$ there exists some z_i, $1 \leqslant i \leqslant j$, that is identical to c?

Reference: [Garey and Johnson, ——]. Transformation from VERTEX COVER (see Section 3.2.2).

Comment: Remains NP-complete even if each $c \in C$ satisfies $|c| \leqslant 3$. The analogous problem in which x_i and y_i need not be disjoint for $1 \leqslant i \leqslant j$ is also NP-complete under the same restriction.

[PO10] MICROCODE BIT OPTIMIZATION

INSTANCE: Finite set A of "micro-commands," collection $C = \{C_1, C_2, \ldots, C_m\}$ of subsets of A called "micro-instructions," and a positive integer K.

QUESTION: Is there a K-bit instruction format for the given micro-instructions, i.e., is there a partition of A into disjoint subsets $A_1, A_2, \ldots, A_n$ such that no pair A_i, C_j have more than one element in common and such that $\sum_{i=1}^{n} \lceil \log_2(|A_i|+1) \rceil \leqslant K$?

Reference: [Robertson, 1978]. Transformation from 3DM.

A11.2 PROGRAMS AND SCHEMES

[PO11] INEQUIVALENCE OF PROGRAMS WITH ARRAYS

INSTANCE: Finite sets X, Θ, and R of variables, operators, and array variables, two programs P_1 and P_2 made up of "operate" ($x_0 \leftarrow \theta x_1 x_2 \cdots x_r$), "update" ($\alpha[x_i] \leftarrow x_j$), and "select" ($x_i \leftarrow \alpha[x_j]$) commands, where each $x_i \in X$, $\theta \in \Theta$, r is the "arity" of θ, and $\alpha \in R$, a finite value set V, and an interpretation of each operator $\theta \in \Theta$ as a specific function from V^r to V.
QUESTION: Is there an initial assignment of a value from V to each variable in X such that the two programs yield different final values for some variable in X (see reference for details on the execution of such programs)?

Reference: [Downey and Sethi, 1976]. Transformation from 3SAT.
Comment: Remains NP-complete even if there are no operate commands and only one array variable. Solvable in polynomial time if there are no update commands, or no select commands, or no array variables.

[PO12] INEQUIVALENCE OF PROGRAMS WITH ASSIGNMENTS

INSTANCE: Finite set X of variables, two programs P_1 and P_2, each a sequence of assignments of the form "$x_0 \leftarrow$ *if* $x_1{=}x_2$ *then* x_3 *else* x_4" where the x_i are in X, and a value set V.
QUESTION: Is there an initial assignment of a value from V to each variable in X such that the two programs yield different final values for some variable in X (see reference for details on the execution of such programs)?

Reference: [Downey and Sethi, 1976]. Transformation from 3SAT.
Comment: Remains NP-complete for $V=\{0,1\}$. This problem can be embedded in many inequivalence problems for simple programs, thus rendering them NP-hard [Downey and Sethi, 1976], [van Leeuwen, 1977].

[PO13] INEQUIVALENCE OF FINITE MEMORY PROGRAMS (*)

INSTANCE: Finite set X of variables, finite alphabet Σ, two programs P_1 and P_2, each a sequence $I_1, I_2, \ldots, I_m$ of instructions (not necessarily of the same length m) of the form "*read* x_i," "*write* v_j," "$x_i \leftarrow v_j$," "*if* $v_j = v_k$ *goto* I_l," "*accept*," or "*halt*," where each $x_i \in X$, each $v_j \in X \cup \Sigma \cup \{\$\}$, and I_m is either "*halt*" or "*accept*."
QUESTION: Is there a string $w \in \Sigma^*$ such that the two programs yield different outputs for input w (see reference for details on the execution of such programs)?

Reference: [Jones and Muchnik, 1977]. Transformation from LINEAR BOUNDED AUTOMATON ACCEPTANCE.
Comment: PSPACE-complete, even if P_2 is a fixed program with no *write* instructions and hence no output. See reference for a number of other special cases and variants that are PSPACE-complete or harder.

[PO14] INEQUIVALENCE OF LOOP PROGRAMS WITHOUT NESTING

INSTANCE: Finite set X of variables, subset $Y \subseteq X$ of input variables, specified output variable x_0, two loop programs P_1 and P_2 without nested loops, i.e., sequences of instructions of the form "$x \leftarrow y$," "$x \leftarrow x+1$," "$x \leftarrow 0$," "*loop* x," and

"*end*," where $x,y \in X$ and each *loop* instruction is followed by a corresponding *end* instruction before any further *loop* instructions occur.
QUESTION: Is there an initial assignment $f: Y \rightarrow Z^+$ of integers to the input variables such that the two programs halt with different values for the output variable x_0 (see references for details on the execution of such programs)?

Reference: [Constable, Hunt, and Sahni, 1974], [Tsichritzis, 1970]. Transformation from 3SAT. The second reference proves membership in NP.
Comment: Problem becomes undecidable if nested loops are allowed (even for nesting of only depth 2) [Meyer and Ritchie, 1967]. Solvable in polynomial time if *loop* statements are not allowed [Tsichritzis, 1970]. See [Hunt, 1977] for a generalization of the main result.

[PO15] INEQUIVALENCE OF SIMPLE FUNCTIONS

INSTANCE: Finite set X of variables, two expressions f and g over X, each being a composition of functions from the collection "$s(x)=x+1$," "$p(x)=\max\{x-1,0\}$," "$plus(x,y)=x+y$," "$div(x,t)=\lfloor x/t \rfloor$," "$mod(x,t)=x-t \cdot \lfloor x/t \rfloor$," "$w(x,y)=$ *if* $y=0$ *then* x *else* 0," and "$select_i^n(x_1,x_2,\ldots,x_n)=x_i$" where $x,y,x_i \in X$, $i,n,t \in Z^+$, and $i \leqslant n$.
QUESTION: Is there an assignment of non-negative integer values to the variables in X for which the values of f and g differ?

Reference: [Tsichritzis, 1970]. Transformation from INEQUIVALENCE OF LOOP PROGRAMS WITHOUT NESTING.
Comment: Remains NP-complete even if f and g are defined only in terms of $w(x,y)$, in terms of *plus* and *mod*, or in terms of *plus* and p [Lieberherr, 1977]. Variants in which f and g are defined in terms of *plus* and "$sub1(x)=\max\{0,1-x\}$," or solely in terms of "$minus(x,y)=\max\{0,x-y\}$," (where in both cases $x,y \in X \cup Z^+$) are also NP-complete [Constable, Hunt, and Sahni, 1974].

[PO16] STRONG INEQUIVALENCE OF IANOV SCHEMES

INSTANCE: Finite sets F and P of function and predicate symbols, single variable x, and two Ianov schemes over F,P, and x, each a sequence $I_1,I_2,\ldots,I_m$ of instructions of the form "$x \leftarrow f(x)$," "*if* $p(x)$ *then goto* I_j *else goto* I_k," and "*halt*," where $f \in F$ and $p \in P$.
QUESTION: Are the two given Ianov schemes not strongly equivalent, i.e., is there a domain set D, an interpretation of each $f \in F$ as a function $f: D \rightarrow D$, an interpretation of each $p \in P$ as a function $p: D \rightarrow \{T,F\}$, and an initial value $x_0 \in D$ for x, such that either both schemes halt with different final values for x or one halts and the other doesn't?

Reference: [Constable, Hunt, and Sahni, 1974], [Rutledge, 1964]. Transformation from 3SAT. Membership in NP follows from the second reference.
Comment: Remains NP-complete even if neither program contains any loops and P_2 is the trivial program that leaves the value of x unchanged. The strong inequivalence problem for Ianov schemes with two variables is undecidable, even if $|F|=|P|=1$ [Luckham, Park, and Paterson, 1970]. See references, [Hunt, 1978], and [Hunt and Szymanski, 1976b] for analogous results for other properties, such as "weak equivalence," "divergence," "halting," etc. Strong equivalence can be

tested in polynomial time for Ianov schemes that are "strongly free," i.e., in which at least one function application occurs between every two successive predicate tests [Constable, Hunt, and Sahni, 1974]. Strong equivalence is open for "free" Ianov schemes.

[PO17] STRONG INEQUIVALENCE FOR MONADIC RECURSION SCHEMES

INSTANCE: Finite sets F and P of function and predicate symbols, set G of "defined" function symbols disjoint from F, specified symbol $f_0 \in G$, and two linear monadic recursion schemes S_1 and S_2, each consisting of a defining statement for each $f \in G$ of the form "$fx =$ *if* px *then* αx *else* βx" where $p \in P$, $\alpha,\beta \in (F \cup G)^*$, and α and β each contain at most one occurrence of a symbol from G.

QUESTION: Is there a domain set D, an interpretation of each $f \in F$ as a function $f: D \to D$, an interpretation of each $p \in P$ as a function $P: D \to \{T,F\}$, and an initial value $x_0 \in D$ such that, as defined by the recursion schemes S_1 and S_2, either the two values for $f_0(x_0)$ differ or one is defined and the other isn't?

Reference: [Constable, Hunt, and Sahni, 1974]. Transformation from STRONG INEQUIVALENCE OF IANOV SCHEMES. Proof of membership in NP is nontrivial.

Comment: Remains NP-complete even if one scheme trivially sets $f_0(x) = x$ and the other is "right linear," i.e., each α and β only contains a defined symbol as its rightmost character. See reference for other NP-completeness and NP-hardness results concerning linear monadic recursion schemes.

[PO18] NON-CONTAINMENT FOR FREE B-SCHEMES

INSTANCE: Two free B-schemes S_1 and S_2, where a free B-scheme is a rooted, directed acyclic graph $G = (V,A)$, all of whose vertices have out-degree 0 (leaves) or 2 (tests), with the two arcs leaving a test vertex labeled L and R respectively, together with a set B of Boolean variable symbols and a label $l(v) \in B$ for each test vertex, such that no two test vertices on the same directed path get the same label, and a set F of function symbols along with a label $l(v) \in F \cup \{\Omega\}$ for each leaf in V.

QUESTION: Is S_1 not "contained" in S_2, i.e., is there an assignment $t: B_1 \cup B_2 \to \{L,R\}$ such that if the paths from the roots of G_1 and G_2 to leaf vertices determined by always leaving a test vertex v by the arc labeled $t(l(v))$ terminate at leaves labeled f_1 and f_2 respectively, then $f_1 \neq f_2$ and $f_1 \neq \Omega$?

Reference: [Fortune, Hopcroft, and Schmidt, 1977]. Transformation from 3SAT.

Comment: The "strong inequivalence" problem for free B-schemes (same as above, only all that we now require is that $f_1 \neq f_2$) is open, but can be solved in polynomial time if one of S_1 and S_2 is an "ordered" B-scheme. The open version is Turing equivalent to the strong inequivalence problem for free Ianov schemes (see STRONG INEQUIVALENCE OF IANOV SCHEMES).

[PO19] NON-FREEDOM FOR LOOP-FREE PROGRAM SCHEMES

INSTANCE: Finite sets F and P of function and predicate symbols, set X of variables, and a loop-free monadic program scheme S over F,P, and X, where such a scheme consists of a sequence $I_1, I_2, \ldots, I_m$ of instructions of the form "$x \leftarrow f(y)$," "*if* $p(x)$ *then goto* I_j *else goto* I_k," and "*halt*," with $x \in X$, $f \in F$, and $p \in P$, and must be such that no directed cycles occur in the corresponding "flow graph."

QUESTION: Is S non-free, i.e., is there a directed path in the flow graph for S that can never be followed in any computation, no matter what the interpretation of the functions and predicates in F and P and the initial values for the variables in X?

Reference: [Constable, Hunt, and Sahni, 1974]. Transformation from 3SAT.

Comment: Remains NP-complete for $|X|=2$. If $|X|=1$, the problem is solvable in polynomial time. If loops are allowed and $|X|$ is arbitrary, the problem is undecidable [Paterson, 1967].

[PO20] PROGRAMS WITH FORMALLY RECURSIVE PROCEDURES

INSTANCE: Finite set A of procedure identifiers, ALGOL-like program P involving procedure declarations and procedure calls for procedures in A (see reference for details).

QUESTION: Is any of the procedures in A "formally recursive" in program P (in the sense of [Langmaack, 1973])?

Reference: [Winklmann, 1977]. Transformation from 3SAT.

Comment: See reference for related results concerning deciding whether P has the "formal most-recent property," "formal parameter correctness," the "formal macro-property," and others.

A12 MISCELLANEOUS

[MS1] BETWEENNESS

INSTANCE: Finite set A, collection C of ordered triples (a,b,c) of distinct elements from A.
QUESTION: Is there a one-to-one function $f: A \rightarrow \{1,2,\ldots,|A|\}$ such that for each $(a,b,c) \in C$, we have either $f(a)<f(b)<f(c)$ or $f(c)<f(b)<f(a)$?

Reference: [Opatrný, 1978]. Transformation from SET SPLITTING.

[MS2] CYCLIC ORDERING

INSTANCE: Finite set A, collection C of ordered triples (a,b,c) of distinct elements from A.
QUESTION: Is there a one-to-one function $f: A \rightarrow \{1,2,\ldots,|A|\}$ such that, for each $(a,b,c) \in A$, we have either $f(a)<f(b)<f(c)$ or $f(b)<f(c)<f(a)$ or $f(c)<f(a)<f(b)$?

Reference: [Galil and Megiddo, 1977]. Transformation from 3SAT.

[MS3] NON-LIVENESS OF FREE CHOICE PETRI NETS

INSTANCE: Petri net $P=(n,M_0,T)$, where $n \in Z^+$, M_0 is an n-tuple of nonnegative integers, and T is a set of transitions $<a,b>$ in which both a and b are n-tuples of 0's and 1's, such that P has the "free choice" property, i.e., for each $<a,b> \in T$, either a contains exactly one 1 or in every other transition $<c,d> \in T$, c has a 0 in every position where a has a 1.
QUESTION: Is P not "live," i.e., is there a transition $t \in T$ and a sequence σ of transitions from T such that, for every sequence τ of transitions from T, the sequence $\sigma\tau t$ is not "fireable" at M_0, where $<a_1,b_1><a_2,b_2> \cdots <a_m,b_m>$ is fireable at M_0 if and only if the sequence $M_0,M_1,\ldots,M_{2m}$ in which $M_{2i+1}=M_{2i}-a_i$ and $M_{2i+2}=M_{2i+1}+b_i$, $0 \leqslant i < m$, contains no vector with a negative component?

Reference: [Jones, Landweber, and Lien, 1977]. Transformation from 3SAT. Proof of membership in NP is nontrivial and is based on a result of [Hack, 1972].

[MS4] REACHABILITY FOR 1-CONSERVATIVE PETRI NETS (*)

INSTANCE: Petri net $P=(n,M_0,T)$ that is "1-conservative," i.e., for each $<a,b> \in T$, a and b have the same number of 1's, and an n-tuple M of nonnegative integers.
QUESTION: Is M reachable from M_0 in P, i.e., is there a sequence $<a_1,b_1>$ $<a_2,b_2> \cdots <a_m,b_m>$ of transitions from T such that the sequence $M_0,M_1,\ldots,M_{2m}$ obtained as in the preceding problem contains no vector with a negative component and satisfies $M_{2m}=M$?

Reference: [Jones, Landweber, and Lien, 1977]. Transformation from LINEAR BOUNDED AUTOMATON ACCEPTANCE.

Comment: PSPACE-complete, even if P is also a free choice Petri net. Problem is not known to be decidable for arbitrary Petri nets, but is known to require at least exponential space [Lipton, 1975]. Analogous results hold for the "coverability" problem: Is there an M' having each of its components no smaller than the corresponding component of M such that M' is reachable from M_0? The related "K-boundedness" problem (given P and an integer K, is there no vector that exceeds K in every component that is reachable from M_0?) is PSPACE-complete for arbitrary Petri nets, as well as for 1-conservative free choice Petri nets. See [Jones, Landweber, and Lien, 1977] and [Hunt, 1977] for additional details and related results.

[MS5] FINITE FUNCTION GENERATION (*)

INSTANCE: Finite set A, a collection F of functions $f: A \rightarrow A$, and a specified function $h: A \rightarrow A$.
QUESTION: Can h be generated from the functions in F by composition?

Reference: [Kozen, 1977d]. Transformation from FINITE STATE AUTOMATA INTERSECTION.
Comment: PSPACE-complete.

[MS6] PERMUTATION GENERATION

INSTANCE: Permutation σ of the integers $\{1,2, \ldots, N\}$, and a sequence $S_1, S_2, \ldots, S_m$ of subsets of $\{1,2, \ldots, N\}$.
QUESTION: Can σ be expressed as a composition $\sigma = \sigma_1 \sigma_2 \cdots \sigma_m$, where for each i, $1 \leqslant i \leqslant m$, σ_i is a permuation of $\{1,2, \ldots, N\}$ that leaves all elements in $\{1,2, \ldots, N\} - S_i$ fixed?

Reference: [Garey, Johnson, Miller, Papadimitriou, 1978]. Transformation from X3C.
Comment: Solvable in polynomial time for any fixed N.

[MS7] DECODING OF LINEAR CODES

INSTANCE: An $n \times m$ matrix $A = (a_{ij})$ of 0's and 1's, a vector $\bar{y} = (y_1, y_2, \ldots, y_m)$ of 0's and 1's, and a positive integer K.
QUESTION: Is there a 0-1 vector $\bar{x} = (x_1, x_2, \ldots, x_n)$ with no more than K 1's such that, for $1 \leqslant j \leqslant m$, $\sum_{i=1}^{n} x_i \cdot a_{ij} \equiv y_j (\text{mod } 2)$?

Reference: [Berlekamp, McEliece, and van Tilborg, 1978]. Transformation from 3DM.
Comment: If $\bar{y}$ is the all zero vector, and hence we are asking for a "codeword" of Hamming weight K or less, the problem is open. The variant in which we ask for an $\bar{x}$ with *exactly* K 1's is NP-complete, even for fixed $\bar{y} = (0,0, \ldots, 0)$.

[MS8] SHAPLEY-SHUBIK VOTING POWER

INSTANCE: Ordered set $V = \{v_1, v_2, \ldots, v_n\}$ of voters, number of votes $w_i \in Z^+$ for each $v_i \in V$, and a quota $q \in Z^+$.
QUESTION: Does voter v_1 have non-zero "Shapley-Shubik voting power," where the voting power $p(v)$ for a voter $v \in V$ is defined to be $(1/n!)$ times the number of permutations π of $\{1,2, \ldots, n\}$ for which $\sum_{i=1}^{j-1} w_{\pi(i)} < q$, $\sum_{i=1}^{j} w_{\pi(i)} \geqslant q$, and $v = v_{\pi(j)}$?

Reference: [Garey and Johnson, ——]. Transformation from PARTITION. The definition of voting power is from [Shapley and Shubik, 1954].

Comment: Determining the value of the Shapley-Shubik voting power for a given voter is #P-complete, but that value can be computed in pseudo-polynomial time by dynamic programming.

[MS9] CLUSTERING

INSTANCE: Finite set X, a distance $d(x,y) \in Z_0^+$ for each pair $x,y \in X$, and two positive integers K and B.
QUESTION: Is there a partition of X into disjoint sets $X_1, X_2, \ldots, X_k$ such that, for $1 \leqslant i \leqslant k$ and all pairs $x,y \in X_i$, $d(x,y) \leqslant B$?

Reference: [Brucker, 1978]. Transformation from GRAPH 3-COLORABILITY.
Comment: Remains NP-complete even for fixed $K=3$ and all distances in $\{0,1\}$. Solvable in polynomial time for $K=2$. Variants in which we ask that the sum, over all X_i, of $\max\{d(x,y): x,y \in X_i\}$ or of $\sum_{x,y \in X_i} d(x,y)$ be at most B, are similarly NP-complete (with the last one NP-complete even for $K=2$).

[MS10] RANDOMIZATION TEST FOR MATCHED PAIRS (*)

INSTANCE: Sequence $(x_1,y_1),(x_2,y_2), \ldots, (x_n,y_n)$ of ordered pairs of integers, nonnegative integer K.
QUESTION: Are there at least K subsets $S \subseteq \{1,2, \ldots, n\}$ for which

$$\sum_{i \in S} |x_i - y_i| \leqslant \sum_{x_i > y_i} (x_i - y_i) \ ?$$

Reference: [Shamos, 1976]. Transformation from PARTITION.
Comment: Not known to be in NP. The corresponding enumeration problem is $\#P$-complete, but solvable in pseudo-polynomial time by dynamic programming.

[MS11] MAXIMUM LIKELIHOOD RANKING

INSTANCE: An $n \times n$ matrix $A = (a_{ij})$ with integer entries satisfying $a_{ij} + a_{ji} = 0$ for all $i,j \in \{1,2, \ldots, n\}$, positive integer B.
QUESTION: Is there a matrix $B = (b_{ij})$ obtained from A by simultaneous row and column permutations such that

$$\sum_{1 \leqslant i < j \leqslant n} \min\{b_{ij}, 0\} \geqslant -B \ ?$$

Reference: [Rafsky, 1977]. Transformation from FEEDBACK ARC SET.
Comment: NP-complete in the strong sense.

[MS12] MATRIX DOMINATION

INSTANCE: An $n \times n$ matrix M with entries from $\{0,1\}$, and a positive integer K.
QUESTION: Is there a set of K or fewer non-zero entries in M that dominate all others, i.e., s subset $C \subseteq \{1,2, \ldots, n\} \times \{1,2, \ldots, n\}$ with $|C| \leqslant K$ such that $M_{ij} = 1$ for all $(i,j) \in C$ and such that, whenever $M_{ij} = 1$, there exists an $(i',j') \in C$ for which either $i = i'$ or $j = j'$?

Reference: [Yannakakis and Gavril, 1978]. Transformation from MINIMUM MAXIMAL MATCHING.
Comment: Remains NP-complete even if M is upper triangular.

[MS13] **MATRIX COVER**

INSTANCE: An $n \times n$ matrix $A = (a_{ij})$ with nonnegative integer entries, and an integer K.

QUESTION: Is there a function $f: \{1,2, \ldots, n\} \to \{-1,+1\}$ such that

$$\sum_{1 \leqslant i,j \leqslant n} a_{ij} \cdot f(i) \cdot f(j) \geqslant K \ ?$$

Reference: [Garey and Johnson, ——]. Transformation from MAX CUT.

Comment: NP-complete in the strong sense and remains so if A is required to be positive definite.

[MS14] **SIMPLY DEVIATED DISJUNCTION**

INSTANCE: Collection M of m-tuples $(M_i[1], M_i[2], \ldots, M_i[m])$, $1 \leqslant i \leqslant n$, with each $M_i[j]$ being either 0,1, or x.

QUESTION: Is there a partition of $\{1,2, \ldots, m\}$ into disjoint sets I,J and an assignment $f: \{1,2, \ldots, m\} \to \{0,1\}$ such that, if Φ is the formula $\bigvee_{j \in I} (M[j] = f(j))$ and Ψ is the formula $\bigvee_{j \in J} (M[j] = f(j))$, then Φ and Ψ are simply deviated in M, i.e., the number of $M_i \in M$ such that Φ and Ψ are both true for M_i times the number of $M_i \in M$ such that Φ and Ψ are both false for M_i is larger than the number of $M_i \in M$ such that Φ is true and Ψ is false for M_i times the number of $M_i \in M$ such that Φ is false and Ψ is true for M_i? (The definition of "simply deviated" is from [Havránek, 1975].)

Reference: [Pudlák and Springsteel, 1975]. Transformation from MAX CUT.

Comment: Remains NP-complete even if $f(j) = 1$ for $1 \leqslant j \leqslant m$. Solvable in polynomial time if each $M_i[j]$ is either 0 or 1. See reference for additional related results.

[MS15] **DECISION TREE**

INSTANCE: Finite set X of objects, collection $T = \{T_1, T_2, \ldots, T_m\}$ of binary tests $T_i: X \to \{0,1\}$, positive integer K.

QUESTION: Is there a decision tree for X using the tests in T that has total external path length K or less? (A decision tree is a binary tree in which each non-leaf vertex is labelled by a test from T, each leaf is labelled by an object from X, the edge from a non-leaf vertex to its left son is labelled 0 and the one to its right son is labelled 1, and, if $T_{i_1}, O_{i_1}, T_{i_2}, O_{i_2}, \ldots, T_{i_k}, O_{i_k}$ is the sequence of vertex and edge labels on the path from the root to a leaf labelled by $x \in X$, then x is the unique object for which $T_{i_j}(x) = O_{i_j}$ for all j, $1 \leqslant j \leqslant k$. The total external path length of such a tree is the sum, over all leaves, of the number of edges on the path from the root to that leaf.)

Reference: [Hyafil and Rivest, 1976]. Transformation from X3C.

Comment: Remains NP-complete even if for each $T_i \in T$ there are at most three distinct objects $x \in X$ for which $T_i(x) = 1$.

[MS16] MINIMUM WEIGHT AND/OR GRAPH SOLUTION

INSTANCE: Directed acyclic graph $G=(V,A)$ with a single vertex $s \in V$ having in-degree 0, assignment $f(v) \in \{and, or\}$ for each $v \in V$ having nonzero out-degree, weight $w(a) \in Z^+$ for each $a \in A$, and a positive integer K.
QUESTION: Is there a subgraph $G'=(V',A')$ of G such that $s \in V'$, such that if $v \in V'$ and $f(v)=and$ then all arcs leaving v in A belong to A', such that if $v \in V'$ and $f(v)=or$ then at least one of the arcs leaving v in A belongs to A', and such that the sum of the weights of the arcs in A' does not exceed K?

Reference: [Sahni, 1974]. Transformation from X3C.
Comment: Remains NP-complete even if $w(a)=1$ for all $a \in A$ [Garey and Johnson, ——]. The general problem is solvable in polynomial time for rooted directed trees by dynamic programming.

[MS17] FAULT DETECTION IN LOGIC CIRCUITS

INSTANCE: Directed acyclic graph $G=(V,A)$ with a single vertex $v^* \in V$ having out-degree 0, an assignment $f:(V-\{v^*\}) \to \{I, and, or, not\}$ such that $f(v)=I$ implies v has in-degree 0, $f(v)=not$ implies v has in-degree 1, and $f(v)=and$ or $f(v)=or$ implies v has in-degree 2, and a subset $V' \subseteq V$.
QUESTION: Can all single faults occurring at vertices of V' be detected by input-output experiments, i.e., regarding G as a logic circuit with input vertices I, output vertex v^*, and logic gates for the functions "and," "or," and "not" at the specified vertices, is there for each $v \in V'$ and $x \in \{T,F\}$ an assignment of a value to each vertex in I of a value in $\{T,F\}$ such that the output of the circuit for those input values differs from the output of the same circuit with the output of the gate at v "stuck-at" x?

Reference: [Ibarra and Sahni, 1975]. Transformation from 3SAT.
Comment: Remains NP-complete even if $V'=V$ or if V' contains just a single vertex v with $f(v)=I$.

[MS18] FAULT DETECTION IN DIRECTED GRAPHS

INSTANCE: Directed acyclic graph $G=(V,A)$, with $I \subseteq V$ denoting those vertices with in-degree 0 and $O \subseteq V$ denoting those vertices with out-degree 0, and a positive integer K.
QUESTION: Is there a "test set" of size K or less that can detect every "single fault" in G, i.e., is there a subset $T \subseteq I \times O$ with $|T| \leqslant K$ such that, for every $v \in V$, there exists some pair $(u_1,u_2) \in T$ such that v is on a directed path from u_1 to u_2 in G?

Reference: [Ibaraki, Kameda, and Toida, 1977]. Transformation from X3C.
Comment: Remains NP-complete even if $|O|=1$. Variant in which we ask that T be sufficient for "locating" any single fault, i.e., that for every pair $v,v' \in V$ there is some $(u_1,u_2) \in T$ such that v is on a directed path from u_1 to u_2 but v' is on no such path, is also NP-complete for $|O|=1$. Both problems can be solved in polynomial time if $K \geqslant |I| \cdot |O|$.

[MS19] FAULT DETECTION WITH TEST POINTS

INSTANCE: Directed acyclic graph $G=(V,A)$ having exactly one vertex $s \in V$ with in-degree 0 and exactly one vertex $t \in V$ with out-degree 0, and a positive integer K.

QUESTION: Can all "single faults" in G be located by attaching K or fewer "test points" to arcs in A, i.e., is there a subset $A' \subseteq A$ with $|A'| \leqslant K$ such that the test set

$$T = \Big(\{s\} \cup \{u_1 : (u_1,u_2) \in A'\}\Big) \times \Big(\{t\} \cup \{u_2 : (u_1,u_2) \in A'\}\Big)$$

has the property that, for each pair $v,v' \in V-\{s,t\}$, there is some $(u_1,u_2) \in T$ such that v is on a directed path from u_1 to u_2 but v' is on no such path?

Reference: [Ibaraki, Kameda, and Toida, 1977]. Transformation from X3C.

Comment: Variants in which we are asked to locate all single faults by using K or fewer "test connections" or "blocking gates" are also NP-complete, as are the problems of finding a test set T with $|T| \leqslant K$ in the presence of a fixed set of "test points," "test connections," or "blocking gates." See reference for more details.

A13 OPEN PROBLEMS

[OPEN1] GRAPH ISOMORPHISM

INSTANCE: Two graphs $G_1=(V_1,E_1)$ and $G_2=(V_2,E_2)$.
QUESTION: Are G_1 and G_2 isomorphic, i.e., is there a one-to-one onto function $f:V_1 \rightarrow V_2$ such that $\{u,v\} \in E_1$ if and only if $\{f(u),f(v)\} \in E_2$?
Comment: The problem remains open even if G_1 and G_2 are restricted to regular graphs, bipartite graphs, line graphs, comparability graphs, chordal graphs, or undirected path graphs (i.e., intersection graphs for the set of paths in an undirected tree), [Hirschberg and Edelberg, 1973], [Babai, 1976], [Booth, 1978], [Miller, 1977]. Solvable in polynomial time for planar graphs (e.g., see [Hopcroft and Wong, 1974]) and for interval graphs [Booth and Lueker, 1975]. The problem is in NP ∩ co-NP for "arc transitive" cubic graphs [Miller, 1977]. Problems polynomially equivalent to GRAPH ISOMORPHISM include directed graph isomorphism, context-free grammar isomorphism [Hunt and Rosenkrantz, 1977], finitely presented algebra isomorphism [Kozen, 1977a], semi-group isomorphism [Booth, 1978], conjunctive query isomorphism [Chandra and Merlin, 1977], the problem of determining whether a graph is isomorphic to its complement [Colbourne and Colbourne, 1978], and the problem of counting the number of distinct isomorphisms between G_1 and G_2 [Babai, 1977], [Mathon, 1978]. A special case of CLIQUE that is polynomially equivalent to GRAPH ISOMORPHISM is described in [Kozen, 1978]. Isomorphism problems that are perhaps easier than GRAPH ISOMORPHISM include group isomorphism and Latin square isomorphism, both of which can be solved in time $O(n^{\log n})$ [Miller, 1978].

[OPEN2] SUBGRAPH HOMEOMORPHISM (FOR A FIXED GRAPH H)

INSTANCE: Graph $G=(V,E)$.
QUESTION: Does G contain a subgraph homeomorphic to H, i.e., a subgraph $G'=(V',E')$ that can be converted to a graph isomorphic to H by repeatedly removing any vertex of degree 2 and adding the edge joining its two neighbors?
Comment: If H is allowed to vary as part of the instance, the problem is NP-complete, since it contains HAMILTONIAN CIRCUIT as a special case. Solvable in polynomial time for certain fixed graphs H, such as a triangle. Is there any fixed graph H for which this problem is NP-complete? If not, is there any fixed graph $H=(U,F)$ for which the following related problem is NP-complete: Given a graph $G=(V,E)$ and a one-to-one function $f:U \rightarrow V$, is there a subgraph $G'=(V',E')$ that can be converted to a graph isomorphic to H as above and such that f provides the required isomorphism? This latter problem is also known to be NP-complete if H is allowed to vary as part of the instance, since it contains DISJOINT CONNECTING PATHS as a special case. Several complicated polynomial time algorithms have been found for particular values of H, such as a triangle [LaPaugh and Rivest, 1978] and two disjoint edges [Shiloach, 1978]. Is there any fixed integer K such that the problem is NP-complete for H consisting of K disjoint edges?

[OPEN3] **GRAPH GENUS**

INSTANCE: Graph $G=(V,E)$ and a non-negative integer K.

QUESTION: Can G be embedded on a surface of genus K such that no two edges cross one another?

Comment: Solvable in polynomial time for $K=0$, i.e., if the question is whether G is planar (e.g., see [Hopcroft and Tarjan, 1974]). A polynomial time algorithm for $K=1$ and cubic graphs is announced in [Filotti, 1978], and, in [Reif, 1978b], polynomial time algorithms for arbitrary graphs and *any* fixed value of K are presented. In addition, for some restricted classes of graphs, such as cliques, cubes, and complete bipartite graphs, simple closed formulas for the genus have been derived (e.g., see [Harary, 1969]). Although the problem for general G and K is open, the closely related GENUS EXTENSION problem (given G, K, and an embedding of a subgraph of G into a surface of genus K, can the embedding be extended to one for all of G?) is NP-complete [Reif, 1978a]. Open problems for other generalizations of planarity include "Does G have crossing number K or less, i.e., can G be embedded in the plane with K or fewer pairs of edges crossing one another?" and "Does G have thickness K or less, i.e., can E be partitioned into K disjoint sets $E_1,E_2,\ldots,E_k$ such that each subgraph $G_i=(V,E_i)$ is planar?" Related NP-complete problems include PLANAR SUBGRAPH and PLANAR INDUCED SUBGRAPH (see INDUCED SUBGRAPH WITH PROPERTY Π).

[OPEN4] **CHORDAL GRAPH COMPLETION**

INSTANCE: Graph $G=(V,E)$ and a positive integer K.

QUESTION: Is there a superset E' containing E of unordered pairs of vertices from V that satisfies $|E'-E| \leqslant K$ and such that $G'=(V,E')$ is chordal, i.e., such that for every simple cycle of more than 3 vertices in G', there is some edge in E' that is not involved in the cycle but that joins two vertices in the cycle?

Comment: This problem is equivalent to the undirected version of DIRECTED ELIMINATION ORDERING and corresponds to the problem of minimizing "fill-in" when applying Gaussian elimination to symmetric matrices (e.g., see [Rose, Tarjan, and Lueker, 1976]). See [Gavril, 1974b] for an alternative characterization of chordal graphs.

[OPEN5] **CHROMATIC INDEX**

INSTANCE: Graph $G=(V,E)$ and a positive integer K.

QUESTION: Does G have chromatic index K or less, i.e., can E be partitioned into disjoint sets $E_1,E_2,\ldots,E_k$, with $k\leqslant K$, such that, for $1\leqslant i\leqslant k$, no two edges in E_i share a common endpoint in G?

Comment: By Vizing's Theorem (e.g., see [Berge, 1973]), the chromatic index for G is either h or $h+1$, where h is the maximum vertex degree in G, so the above question may be restated as "Given G, is the chromatic index of G equal to its maximum vertex degree?" The answer is always "yes" for bipartite graphs (e.g., see [Berge, 1973]), and there exist polynomial time algorithms for constructing the desired partition in this case (e.g., see [Gabow, 1976]). A particular case that is open is that for cubic graphs (i.e., regular of degree 3), in which case the problem can be restated as "Given G, can the vertices of G be covered by disjoint simple cycles, each involving an even number of vertices?" This latter problem is one of a number of open problems involving parity. Another such problem is: "Given a collection C of subsets of a finite set X, is there a nonempty subcollection $C' \subseteq C$ such that each $x \in X$ belongs to an even number (possibly 0) of sets in C'?" which is equivalent to the open problem mentioned in the comments for DECODING OF LINEAR CODES.

[OPEN6] SPANNING TREE PARITY PROBLEM

INSTANCE: Graph $G=(V,E)$ and a partition of E into disjoint 2-element sets $E_1, E_2, \ldots, E_m$.
QUESTION: Is there a spanning tree $T=(V,E')$ for G such that for each E_i, $1 \leqslant i \leqslant m$, either $E_i \subseteq E'$ or $E_i \cap E' = \emptyset$?
Comment: This is a typical special case of the general "matroid parity problem" (e.g., see [Lawler, 1976a]), which is itself a generalization of graph matching and the two matroid intersection problem, both of which can be solved in polynomial time (assuming, in the matroid case, that there exist polynomial time algorithms for telling whether a set is an independent set of the matroids in question). The related "multiple choice spanning tree" problem, where at *most* one member of each E_i can be in E', is a special case of the two matroid intersection problem and hence can be solved in polynomial time (see MULTIPLE CHOICE BRANCHING).

[OPEN7] PARTIAL ORDER DIMENSION

INSTANCE: Directed acyclic graph $G=(V,A)$ that is transitive, i.e., whenever $(u,v) \in A$ and $(v,w) \in A$, then $(u,w) \in A$, and a positive integer $K \leqslant |V|^2$.
QUESTION: Does there exist a collection of $k \leqslant K$ linear orderings of V such that $(u,v) \in A$ if and only if u is less than v in each of the orderings?
Comment: Solvable in polynomial time for $K=2$ [Lawler, 1976d]. Open for arbitrary K and for any fixed $K \geqslant 3$.

[OPEN8] PRECEDENCE CONSTRAINED 3-PROCESSOR SCHEDULING

INSTANCE: Set T of unit length tasks, partial order $<$ on T, and a deadline $D \in Z^+$.
QUESTION: Can T be scheduled on 3 processors so as to satisfy the precedence constraints and meet the overall deadline D, i.e., is there a schedule $\sigma: T \rightarrow \{0,1, \ldots, D-1\}$ such that $t < t'$ implies $\sigma(t) < \sigma(t')$ and such that for each integer i, $0 \leqslant i \leqslant D-1$, there are at most 3 tasks $t \in T$ for which $\sigma(t) = i$?
Comment: The corresponding problem for 2 processors is solvable in polynomial time [Fujii, Kasami, and Ninomiya, 1969], [Coffman and Graham, 1972], even with individual task deadlines and release times [Garey and Johnson, 1977b]. If the number of processors is allowed to vary as part of the instance, the problem is NP-complete [Ullman, 1975]. See PRECEDENCE CONSTRAINED SCHEDULING for more details. Is there any fixed value of K for which the K-processor version of the above problem is NP-complete?

[OPEN9] LINEAR PROGRAMMING

INSTANCE: Integer-valued vectors $V_i = (v_i[1], v_i[2], \ldots, v_i[n])$, $1 \leqslant i \leqslant m$, $D=(d_1, d_2, \ldots, d_m)$, and $C=(c_1, c_2, \ldots, c_n)$, and an integer B.

QUESTION: Is there a vector $X=(x_1,x_2,\ldots,x_n)$ of rational numbers such that, for $1\leqslant i\leqslant m$, $V_i\cdot X\leqslant d_i$ and such that $C\cdot X\geqslant B$?
Comment: The problem is in NP ∩ co-NP (membership in co-NP follows from the fundamental duality theorem of linear programming). For any fixed value of m, the problem can be solved in polynomial time. There are many variants of LINEAR PROGRAMMING that are polynomially equivalent to it (e.g., see [Reiss and Dobkin, 1976]). One such variant is that in which we drop the vector C from the instance and drop the requirement that $C\cdot X\geqslant B$ (see also [Papadimitriou, 1978b]). Examples of network flow problems polynomially equivalent to LINEAR PROGRAMMING are mentioned in the comments for UNDIRECTED FLOW WITH LOWER BOUNDS, PATH CONSTRAINED NETWORK FLOW, and TWO COMMODITY INTEGRAL FLOW. A generalization of LINEAR PROGRAMMING that is also open though still in NP is the "linear complementarity" problem (see [Murty, 1976]).

[OPEN10] **TOTAL UNIMODULARITY**
INSTANCE: An $m\times n$ matrix M with entries from the set $\{-1,0,1\}$.
QUESTION: Is M *not* totally unimodular, i.e., is there a square submatrix of M whose determinant is *not* in the set $\{-1,0,1\}$?
Comment: The problem remains open even if all entries in M are from $\{0,1\}$. The significance of totally unimodular matrices for integer programming is discussed, for example, in [Lawler, 1976] and [Garfinkel and Nemhauser, 1972].

[OPEN11] **COMPOSITE NUMBER**
INSTANCE: Positive integer N.
QUESTION: Are there positive integers $m,n>1$ such that $N=m\cdot n$?
Comment: The problem is in NP ∩ co-NP [Pratt, 1975]. Although no polynomial time algorithm is known, there is an algorithm for the problem that runs in polynomial time if the "Extended Riemann Hypothesis" holds [Miller, 1976]. However, there is no such algorithm known for determining the prime factors of N, and this latter problem may be harder than the basic decision problem. Of course, all these problems are easily solved in pseudo-polynomial time.

[OPEN12] **MINIMUM LENGTH TRIANGULATION**
INSTANCE: Collection $C=\{(a_i,b_i):1\leqslant i\leqslant n\}$ of pairs of integers, giving the coordinates of n points in the plane, and a positive integer B.
QUESTION: Is there a triangulation of the set of points represented by C that has total "discrete-Euclidean" length B or less? Here a triangulation is a collection of non-intersecting line segments, each joining two points in C, that divides the interior of the convex hull into triangular regions. The discrete-Euclidean length of a line segment joining (a_i,b_i) and (a_j,b_j) is given by $\lceil((a_i-a_j)^2+(b_i-b_j)^2)^{1/2}\rceil$, and the total length of a triangulation is the sum of the lengths of its constituent line segments.
Comment: The analogous problem for the rectilinear metric is also open. [Lloyd, 1977] presents counterexamples to a number of conjectured polynomial time algorithms for the problem and proves that the related CONSTRAINED TRIANGULATION problem is NP-complete.

Symbol Index

Reference and Author Index

ABDEL-WAHAB, H. M. [1976], *Scheduling with Applications to Register Allocation and Deadlock Problems*, Doctoral Thesis, Dept. of Electrical Engineering, University of Waterloo, Waterloo, Ontario. *(A5.1)*

ABDEL-WAHAB, H. M., AND T. KAMEDA [1978], "Scheduling to minimize maximum cumulative cost subject to series-parallel precedence constraints," *Operations Res.* **26**, 141-158. *(A5.1)*

ADLEMAN, L., AND K. MANDERS [1977], "Reducibility, randomness, and intractability (Abstract)," *Proc. 9th Ann. ACM Symp. on Theory of Computing*, Association for Computing Machinery, New York, 151-163. *(7.1)*

ADLEMAN, L. *See also* MANDERS, K.

ADOLPHSON, D. [1977], "Single machine job sequencing with precedence constraints," *SIAM J. Comput.* **6**, 40-54. *(A5.1)*

ADOLPHSON, D., AND T. C. HU [1973], "Optimal linear ordering," *SIAM J. Appl. Math.* **25**, 403-423. *(A1.3; A5.1)*

AHO, A. V., M. R. GAREY, AND F. K. HWANG [1977], "Rectilinear Steiner trees: efficient special case algorithms," *Networks* **7**, 37-58. *(A2.1)*

AHO, A. V., M. R. GAREY, AND J. D. ULLMAN [1972], "The transitive reduction of a directed graph," *SIAM J. Comput.* **1**, 131-137. *(A1.2)*

AHO, A. V., J. E. HOPCROFT, AND J. D. ULLMAN [1968], "Time and tape complexity of pushdown automaton languages," *Information and Control* **13**, 186-206. *(A10.1)*

AHO, A. V., J. E. HOPCROFT, AND J. D. ULLMAN [1974], *The Design and Analysis of Computer Algorithms*, Addison-Wesley, Reading, MA. *(1.3; 2.3; 4.0; 6.1; 7.4)*

AHO, A. V., S. C. JOHNSON, AND J. D. ULLMAN [1977a], "Code generation for expressions with common subexpressions," *J. Assoc. Comput. Mach.* **24**, 146-160. *(A11.1)*

AHO, A. V., S. C. JOHNSON, AND J. D. ULLMAN [1977b], private communication. *(A11.1)*

AHO, A. V., Y. SAGIV, AND J. D. ULLMAN [1978], "Equivalences among relational expressions," unpublished manuscript. *(A4.3)*

AHO, A. V., AND R. SETHI [1977], private communication. *(A1.4)*

AHO, A. V., AND J. D. ULLMAN [1972], *The Theory of Parsing, Translation, and Compiling — Volume 1: Parsing*, Prentice-Hall, Inc., Englewood Cliffs, NJ. *(A10.1)*

AHO, A. V., AND J. D. ULLMAN [1977], private communication. *(A4.2)*

ANGLUIN, D. [1976], *An Application of the Theory of Computational Complexity to the Study of Inductive Inference*, Doctoral Thesis, Dept. of Electrical Engineering and Computer Science, University of California, Berkeley, CA. *(A10.2)*

ANGLUIN, D. [1977], "On the complexity of minimum inference of regular sets," unpublished manuscript. *(A10.1; A10.2)*

ANGLUIN, D., AND L. G. VALIANT [1977], "Fast probabilistic algorithms for Hamiltonian circuits and matchings," *Proc. 9th Ann. ACM Symp. on Theory of Computing*, Association for Computing Machinery, New York, 30-41. *(6.3)*

APPEL, K., AND W. HAKEN [1977a], "Every planar map is 4-colorable — 1: Discharging," *Ill. J. Math.* **21**, 429-490. *(4.1)*

APPEL, K., AND W. HAKEN [1977b], "Every planar map is 4-colorable — 2: Reducibility," *Ill. J. Math.* **21**, 491-567.. *(4.1)*

ARAKI, T., Y. SUGIYAMA, T. KASAMI, AND J. OKUI [1977], "Complexity of the deadlock avoidance problem," *Proc. 2nd IBM Symp. on Mathematical Foundations of Computer Science*, IBM Japan, Tokyo, 229-252. *(A5.3)*

ARAKI, T. *See also* SUGIYAMA, Y.

ARJOMANDI, E. [1977], private communication. *(A1.1)*

BABAI, L. [1976], private communication. *(7.1; A13)*

BABAI, L. [1977], private communication. *(A13)*

BAKER, T., J. GILL, AND R. SOLOVAY [1975], "Relativizations of the P =? NP question," *SIAM J. Comput.* **4**, 431-442. *(7.6)*

BAKER, T. P., AND A. L. SELMAN [1976], "A second step toward the polynomial hierarchy," *Proc. 17th Ann. Symp. on Foundations of Computer Science*, IEEE Computer Society, Long Beach, CA, 71-75. *(7.6)*

BALL, M. O. [1977a], *Network Reliability and Analysis: Algorithms and Complexity*, Doctoral Thesis, Operations Research Dept., Cornell University, Ithaca, NY. *(A2.2)*

BALL, M. O. [1977b], private communication. *(A2.2)*

BARROW, H. G., AND R. M. BURSTALL [1976], "Subgraph isomorphism, matching relational structures and maximal cliques," *Information Processing Lett.* **4**, 83-84. *(A1.4)*

BARTHOLDI, J. J., III, J. B. ORLIN, AND H. D. RATLIFF [1977], "Circular ones and cyclic staffing," Report No. 21, Dept. of Operations Research, Stanford University, Stanford, CA. *(A5.4)*

BARZDIN, Y. M. *See* TRAKHTENBROT, B. A.

BAUER, M., D. BRAND, M. FISCHER, A. MEYER, AND M. PATERSON [1973], "A note on disjunctive form tautologies," *SIGACT News* **5:2**, 17-20. *(7.4)*

BAXTER, L. D. [1976], *The Complexity of Unification*, Doctoral Thesis, Dept. of Computer Science, University of Waterloo, Waterloo, Ontario. *(A9.2)*

BAXTER, L. D. [1977], "The NP-completeness of subsumption," unpublished manuscript. *(A9.2)*

BEERI, C., AND P. A. BERNSTEIN [1978], "Computational problems related to the design of normal form relational schemes," unpublished manuscript. *(A4.3)*

BEERI, C. *See also* BERNSTEIN, P. A.; SHAMIR, E.

BERGE, C. [1973], *Graphs and Hypergraphs*, North-Holland, Amsterdam. *(A13)*

BERGER, R. [1966], *The Undecidability of the Domino Problem* (Mem. Amer. Math. Soc., No. 66), American Mathematical Society, Providence, RI. *(1.4; A8)*

BERLEKAMP, E. R. [1976], private communication. *(A2.1; A8)*

BERLEKAMP, E. R., R. J. MCELIECE, AND H. C. A. VAN TILBORG [1978], "On the inherent intractability of certain coding problems," *IEEE Trans. Information Theory* (to appear). *(A12)*

BERMAN, L., AND J. HARTMANIS [1977], "On isomorphisms and density of NP and other complete sets," *SIAM J. Comput.* **6**, 305-322. *(7.1)*

BERMAN, L. *See also* HARTMANIS, J.

BERNSTEIN, P. A., AND C. BEERI [1976], "An algorithmic approach to normalization of relational database schemas," Report CSRG-73, Computer Systems Research Group, University of Toronto, Canada. *(A4.3)*

BERNSTEIN, P. A. *See also* BEERI, C.; PAPADIMITRIOU, C. H.

BLATTNER, W. O. *See* DANTZIG, G. B.

BLAZEWICZ, J. [1976], "Scheduling dependent tasks with different arrival times to meet deadlines," in H. Beilner and E. Gelenbe (eds.), *Modelling and Performance Evaluation of Computer Systems*, North Holland, Amsterdam, 57-65. *(A5.1)*

BLAZEWICZ, J. [1977a], "Mean flow time scheduling under resource constraints," Report No. PR-19/77, Institute of Control Engineering, Technical University of Poznan, Poland. *(A5.2)*

BLAZEWICZ, J. [1977b], "Scheduling with deadlines and resource constraints," Report PR-25/77, Institute of Control Engineering, Technical University of Poznan, Poland. *(A5.2)*

BLAZEWICS, J. [1978], "Deadline scheduling of tasks with ready times and resource constraints," unpublished manuscript. *(A5.2)*

BOESCH, F. T., S. CHEN, AND J. A. M. MCHUGH [1974], "On covering the points of a graph with point disjoint paths," in *Graphs and Combinatorics* (Proc. Capitol Conf. on Graph Theory and Combinatorics), Lecture Notes in Math., Vol. **46**, Springer, Berlin, 201-212. *(A1.2)*

BOOK, R. V. [1972], "On languages accepted in polynomial time," *SIAM J. Comput.* **1**, 281-287. *(7.5; A10.1)*

BOOK, R. V. [1974], "Comparing complexity classes," *J. Comput. System Sci.* **9**, 213-229. *(7.5)*

BOOK, R. V. [1976], "Translational lemmas, polynomial time, and $(\log n)^j$-space," *Theor. Comput. Sci.* **1**, 215-226. *(7.5)*

BOOK, R. V. [1978], "On the complexity of formal grammars," *Acta Informat.* **9**, 171-182. *(A10.2)*

BOOK, R. V., AND S. GREIBACH [1970], "Quasi-realtime languages," *Math. Systems Theory* **4**, 97-111. *(A10.1)*

BOOTH, K. S. [1975], *PQ Tree Algorithms*, Doctoral Thesis, Dept. of Electrical Engineering and Computer Science, University of California, Berkeley, CA. *(A4.2)*

BOOTH, K. S. [1978], "Isomorphism testing for graphs, semigroups, and finite automata are polynomially equivalent problems," *SIAM J. Comput.* **7**, 273-279. *(7.1; A13)*

BOOTH, K. S., AND G. S. LUEKER [1975], "Linear algorithms to recognize interval graphs and test for the consecutive ones property," *Proc. 7th Ann. ACM Symp. on Theory of Computing*, Association for Computing Machinery, New York, 255-265. *(A13)*

BOOTH, K. S., AND G. S. LUEKER [1976], "Testing for the consecutive ones property, interval graphs, and graph planarity using PQ-tree algorithms," *J. Comput. System Sci.* **13**, 335-379. *(A1.2)*

BOROSH, I., AND L. B. TREYBIG [1976], "Bounds on positive integral solutions of linear Diophantine equations," *Proc. Amer. Math. Soc.* **55**, 299-304. *(A6)*

BRAND, D. *See* BAUER, M.

BROOKS, R. L. [1941], "On coloring the nodes of a network," *Proc. Cambridge Philos. Soc.* **37**, 194-197. *(4.1; A1.1)*

BRUCKER, P. [1978], "On the complexity of clustering problems," in R. Henn, B. Korte, and W. Oletti (eds.), *Optimierung und Operations Research*, Lecture Notes in Economics and Mathematical Systems, Springer, Berlin (to appear). *(A12)*

BRUCKER, P., M. R. GAREY, AND D. S. JOHNSON [1977], "Scheduling equal-length tasks under treelike precedence constraints to minimize maximum lateness," *Math. Oper. Res.* **2**, 275-284. *(A5.2)*

BRUCKER, P. *See also* LENSTRA, J. K.

BRUNO, J., E. G. COFFMAN, JR, AND R. SETHI [1974], "Scheduling independent tasks to reduce mean finishing time," *Comm. ACM* **17**, 382-387. *(A5.2)*

BRUNO, J., AND P. DOWNEY [1978], "Complexity of task scheduling with deadlines, set-up times and changeover costs," *SIAM J. Comput.* (to appear). *(A5.1)*

BRUNO, J., AND R. SETHI [1976], "Code generation for a one-register machine," *J. Assoc. Comput. Mach.* **23**, 502-510. *(A11.1)*

BRUNO, J., AND L. WEINBERG [1970], "A constructive graph-theoretic solution of the Shannon switching game," *IEEE Trans. Circuit Theory* **CT-17**, 74-81. *(A8)*

BURKHARD, W. A. *See* WALSH, A. M.

BURR, S. [1976], private communication. *(A1.1)*

BURR, S., P. ERDÖS, AND L. LOVASZ [1976], "On graphs of Ramsey type," *Ars Combinatorica* **1**, 167-190. *(A1.1)*

BURSTALL, R. M. *See* BARROW, H. G.

CARLIER, J. [1978], "Probleme a une machine," Report No. 78.05, Institut de Programmation, Universite de Pierre et Marie Curie, Paris, France. *(A5.1)*

CHAN, T. [1977], "An algorithm for checking PL/CV arithmetic inferences," Report No. 77-326, Dept. of Computer Science, Cornell University, Ithaca, NY. *(A9.2)*

CHANDRA, A. K., AND P. M. MERLIN [1977], "Optimal implementation of conjunctive queries in relational data bases," *Proc. 9th Ann. ACM Symp. on Theory of Computing*, Association for Computing Machinery, New York, 77-90. *(A4.3; A13)*

CHANDRA, A. K., AND L. J. STOCKMEYER [1976], "Alternation," *Proc. 17th Ann. Symp. on Foundations of Computer Science*, IEEE Computer Society, Long Beach, CA, 98-108. *(7.6)*

CHANDRA, A. K. *See also* STOCKMEYER, L. G.

CHANDY, K. M. *See* VAN SICKLE, L.

CHEN, S. *See* BOESCH, F. T.

CHO, Y., AND S. SAHNI [1978], "Preemptive scheduling of independent jobs with release and due times on open, flow, and job shops," unpublished manuscript. *(A5.3)*

CHO, Y. *See also* SAHNI, S.

CHOMSKY, N., AND G. A. MILLER [1958], "Finite state languages," *Information and Control* **1**, 91-112. *(A10.2)*

CHRISTOFIDES, N. [1976], "Worst-case analysis of a new heuristic for the travelling salesman problem," Technical Report, Graduate School of Industrial Administration, Carnegie-Mellon University, Pittsburgh, PA. *(6.1)*

CHUNG, F. R. K., AND R. L. GRAHAM [1977], private communication. *(A1.2)*

CHVÁTAL, V. [1973], "On the computational complexity of finding a kernel," Report No. CRM-300, Centre de Recherches Mathématiques, Université de Montréal. *(A1.5)*

CHVÁTAL, V. [1975], "On certain polytopes associated with graphs," *J. Combinatorial Theory Ser. B* **18**, 138-154. *(A6)*

CHVÁTAL, V. [1976], private communication. *(A1.2; A1.3)*

CHVÁTAL, V. [1977], "Determining the stability number of a graph," *SIAM J. Comput.* **6**, 643-662. *(7.6)*

CHVÁTAL, V. [1978], private communication. *(A3.1)*

CHVÁTAL, V., AND P. L. HAMMER [1975], "Aggregation of inequalities in integer programming," Report No. STAN-CS-75-518, Computer Science Dept., Stanford University, Stanford, CA. *(A1.5)*

CHVÁTAL, V., AND G. THOMASSEN [1978], "Distances in orientations of graphs," *J. Combinatorial Theory Ser. B* **24**, 61-75. *(A1.5)*

COBHAM, A. [1964], "The intrinsic computational difficulty of functions," in Y. Bar-Hillel (ed.), *Proc. 1964 International Congress for Logic Methodology and Philosophy of Science*, North Holland, Amsterdam, 24-30. *(1.3; 5.2)*

COCKAYNE, E., S. GOODMAN, AND S. HEDETNIEMI [1975], "A linear algorithm for the domination number of a tree," *Information Processing Lett.* **4**, 41-44. *(A1.1)*

COCKAYNE, E. J., AND S. T. HEDETNIEMI [1975], "Optimal domination in graphs," *IEEE Trans. Circuits and Systems* **CAS-22**, 855-857. *(A1.1)*

COCKAYNE, E. J., S. T. HEDETNIEMI, AND P. J. SLATER [1978], private communication. *(A2.5)*

CODY, R. A., AND E. G. COFFMAN, JR [1976], "Record allocation for minimizing expected retrieval costs on drum-like storage devices," *J. Assoc. Comput. Mach.* **23**, 103-115. *(A4.1)*

COFFMAN, E. G., JR, AND R. L. GRAHAM [1972], "Optimal scheduling for two-processor systems," *Acta Informat.* **1**, 200-213. *(A5.2; A13)*

COFFMAN, E. G., JR. *See also* BRUNO, J.; CODY, R. A.; MUNTZ, R. R.

COLBOURN, M. J., AND C. J. COLBOURN [1978], "Graph isomorphism and self-complementary graphs," *SIGACT News* **10:1**, 25-29. *(A13)*

COLBOURNE, C. J. *See* COLBOURNE, M. J.

COMER, D., AND R. SETHI [1976], "Complexity of Trie index construction (Extended abstract)," *Proc. 17th Ann. Symp. on Foundations of Computer Science*, IEEE Computer Society, Long Beach, CA, 197-207. *(A4.1)*

CONSTABLE, R. L., H. B. HUNT, III, AND S. SAHNI [1974], "On the computational complexity of scheme equivalence," Report No. 74-201, Dept. of Computer Science, Cornell University, Ithaca, NY. (Extended abstract appeared in *Proc. 8th Ann. Princeton Conf. on Information Sciences and Systems*, Dept. of Electrical Engineering, Princeton University, Princeton, NJ, 15-20). *(A4.2; A11.2)*

CONWAY, J. H. [1976], *On Numbers and Games*, Academic Press, New York. *(A8)*

CONWAY, R. W., W. L. MAXWELL, AND L. W. MILLER [1967], *Theory of Scheduling*, Addison-Wesley, Reading, MA. *(A5.2)*

COOK, S. A. [1971a], "The complexity of theorem-proving procedures," *Proc. 3rd Ann. ACM Symp. on Theory of Computing*, Association for Computing Machinery, New York, 151-158. *(1.5; 2.6; 3.1.1; 5.2; A1.4; A9.1)*

COOK, S. A. [1971b], "Characterizations of pushdown machines in terms of time-bounded computers," *J. Assoc. Comput. Mach.* **18**, 4-18. *(7.5)*

COOK, S. A. [1973], "A hierarchy for nondeterministic time complexity," *J. Comput. System Sci.* **7**, 343-353. *(7.6)*

COOK, S. A. [1974], "An observation on time-storage trade off," *J. Comput. System Sci.* **9**, 308-316. *(7.5)*

COOK, S., AND R. SETHI [1976], "Storage requirements for deterministic polynomial time recognizable languages," *J. Comput. System Sci.* **13**, 25-37. *(7.5)*

CORNUEJOLS, G., M. L. FISHER, AND G. L. NEMHAUSER [1977], "Location of bank accounts to optimize float: an analytic study of exact and approximate algorithms," *Management Sci.* **23**, 789-810. *(6.1)*

CORNUEJOLS, G., AND G. L. NEMHAUSER [1978], "Tight bounds for Christofides' traveling salesman heuristic," *Math. Programming* **14**, 116-121. *(6.1)*

CRESSWELL, M. J. *See* HUGHES, G. E.

CULIK, K., II *See* OPATRNÝ, J.

DANTZIG, G. B. [1957], "Discrete-variable extremum problems," *Operations Res.* **5**, 266-277. *(4.2.2; A6)*

DANTZIG, G. B. [1960], "On the significance of solving linear programming problems with some integer variables," *Econometrica* **28**, 30-44. *(1.5)*

DANTZIG, G. B., W. O. BLATTNER, AND M. R. RAO [1967], "All shortest routes from a fixed origin in a graph," in *Theory of Graphs: International Symposium*, Gordon and Breach, NY, 85-90. *(1.5)*

DATE, C. J. [1975], *An Introduction to Database Systems*, Addison-Wesley, Reading, MA. *(A4.3)*

DEMERS, A. *See* JOHNSON, D. S.

DEO, N. *See* KRISHNAMOORTHY, M. S.; REINGOLD, E. M.

DOBKIN, D., AND R. E. LADNER [1978], private communication. *(A8)*

DOBKIN, D., R. LIPTON, AND S. REISS [1976], "Linear programming is P-complete," in (same authors), "Excursions into geometry," Report No. 71, Dept. of Computer Science, Yale University, New Haven, CT. *(7.5)*

DOBKIN, D. *See also* REISS, S. P.

DORFMAN, Y. G. *See* ORLOVA, G. I.

DOWNEY, P. J., AND R. SETHI [1976], "Assignment commands and array structures," *Proc. 17th Ann. Symp. on Foundations of Computer Science*, IEEE Computer Society, Long Beach, CA, 57-66. *(A11.2)*

DOWNEY, P. J. *See also* BRUNO, J.

EDELBERG, M. *See* HIRSCHBERG, D.

EDMONDS, J. [1962], "Covers and packings in a family of sets," *Bull. Amer. Math. Soc.* **68**, 494-499. *(1.5)*

EDMONDS, J. [1965a], "Paths, trees, and flowers," *Canad. J. Math.* **17**, 449-467. *(1.3; 5.2)*

EDMONDS, J. [1965b], "Minimum partition of a matroid into independent subsets," *J. Res. Nat. Bur. Standards Sect. B* **69**, 67-72. *(5.2)*

EDMONDS, J., AND E. L. JOHNSON [1970], "Matching: a well-solved class of integer linear programs," in *Combinatorial Structures and their Applications*, Gordon and Breach, New York, 89-92. *(A1.1; A1.2)*

EDMONDS, J., AND E. L. JOHNSON [1973], "Matching, Euler tours, and the Chinese postman," *Math. Programming* **5**, 88-124. *(A2.3)*

EDMONDS, J., AND R. M. KARP [1972], "Theoretical improvements in algorithmic efficiency for network flow problems," *J. Assoc. Comput. Mach.* **19**, 248-264. *(A2.1)*

EDMONDS, J., AND D. W. MATULA [1975], private communication. *(4.2.2; A1.4)*

EHRLICH, G., S. EVEN, AND R. E. TARJAN [1976], "Intersection graphs of curves in the plane," *J. Combinatorial Theory Ser. B* **21**, 8-20. *(A1.1)*

ERDÖS, P. *See* BURR, S.

ESWAREN, K. P., AND R. E. TARJAN [1976], "Augmentation problems," *SIAM J. Comput.* **5**, 653-665. *(A2.2)*

EVEN, S., A. ITAI, AND A. SHAMIR [1976], "On the complexity of timetable and multicommodity flow problems," *SIAM J. Comput.* **5**, 691-703. *(3.1.1; 3.2.3; A2.4; A5.3; A9.1)*

EVEN, S., AND D. S. JOHNSON [1977], unpublished results. *(A2.1; A2.4)*

EVEN, S., D. I. LICHTENSTEIN, AND Y. SHILOACH [1977], "Remarks on Zeigler's method for matrix compression," unpublished manuscript. *(A4.2)*

EVEN, S., A. PNUELI, AND A. LEMPEL [1972], "Permutation graphs and transitive graphs," *J. Assoc. Comput. Mach.* **19**, 400-410. *(A1.1; A1.2)*

EVEN, S., AND Y. SHILOACH [1975], "NP-completeness of several arrangement problems," Report No. 43, Dept. of Computer Science, Technion, Haifa, Israel. *(A1.3)*

EVEN, S., AND R. E. TARJAN [1976], "A combinatorial problem which is complete in polynomial space," *J. Assoc. Comput. Mach.* **23**, 710-719. *(7.4; A8)*

EVEN, S. *See also* EHRLICH, G.

FAGIN, R. [1974], "Generalized first-order spectra and polynomial time recognizable sets," in R. M. Karp (ed.), *Complexity of Computation*, American Mathematical Society, Providence, RI, 43-73. *(A4.2)*

FARLEY, A., S. HEDETNIEMI, S. MITCHELL, AND A. PROSKUROWSKI [1977], "Minimum broadcast graphs," Report No. CS-TR-77-2, Dept. of Computer Science, University of Oregon, Eugene, OR. *(A2.5)*

FISCHER, M. J., AND M. O. RABIN [1974], "Super-exponential complexity of Presburger arithmetic," in R. M. Karp (ed.), *Complexity of Computation*, American Mathematical Society, Providence, RI, 27-41. *(1.4; 7.6)*

FISCHER, M. J. *See also* BAUER, M.; SEIFERAS, J. I.; WAGNER, R. A.

FISHER, M. L. *See* CORNUEJOLS, G.; NEMHAUSER, G. L.

FILOTTI, I. S. [1978], "An efficient algorithm for determining whether a cubic graph is toroidal," *Proc. 10th Ann. ACM Symp. on Theory of Computing*, Association for Computing Machinery, New York, 133-142. *(A13)*

FLORIAN, M., AND M. KLEIN [1971], "Deterministic production planning with concave costs and capacity constraints," *Management Sci.* **18**, 12-20. *(A5.4)*

FLORIAN, M. *See also* LENSTRA, J. K.

FORD, L. R., AND D. R. FULKERSON [1962], *Flows in Networks*, Princeton University Press, Princeton, NJ. *(A2.4)*

FORTUNE, S., J. E. HOPCROFT, AND E. M. SCHMIDT [1977], "The complexity of equivalence and containment for free single variable program schemes," Report No. TR77-310, Dept. of Computer Science, Cornell University, Ithaca, NY. *(A11.2)*

FRAENKEL, A. S., M. R. GAREY, D. S. JOHNSON, T. SCHAEFER, AND Y. YESHA [1978], "The complexity of Checkers on an N×N board — Preliminary report," *Proc. 19th Ann. Symp. on Foundations of Computer Science*, IEEE Computer Society, Long Beach, CA, 55-64. *(7.4; A8)*

FRAENKEL, A. S., AND Y. YESHA [1976], "Theory of annihilation games," *Bull. Amer. Math. Soc.* **82**, 775-777. *(A8)*

FRAENKEL, A. S., AND Y. YESHA [1977], "Complexity of problems in games, graphs, and algebraic equations," unpublished manuscript. *(A1.4; A1.5; A7.2; A8)*

FREDERICKSON, G. N., M. S. HECHT, AND C. E. KIM [1978], "Approximation algorithms for some routing problems," *SIAM J. Comput.* **7**, 178-193. *(A2.3)*

FUJII, M., T. KASAMI, AND K. NINOMIYA [1969], "Optimal sequencing of two equivalent processors," *SIAM J. Appl. Math.* **17**, 784-789. Erratum [1971], *SIAM J. Appl. Math.* **20**, 141. *(A13)*

FULKERSON, D. R., AND D. A. GROSS [1965], "Incidence matrices and interval graphs," *Pacific J. Math.* **15**, 835-855. *(A1.2)*

FULKERSON, D. R. *See also* FORD, L. R.

FULLER, S. H. *See* STONE, H. S.

GABOW, H. N. [1976], "Using Euler partitions to edge color bipartite multigraphs," *Internat. J. Comput. Information Sci.* **5**, 345-355. *(A13)*

GABOW, H. N., S. N. MAHESHWARI, AND L. OSTERWEIL [1976], "On two problems in the generation of program test paths," *IEEE Trans. Software Engrg.* **SE-2**, 227-231. *(A1.5)*

GALIL, Z. [1974], "On some direct encodings of nondeterministic Turing machines operating in polynomial time into P-complete problems," *SIGACT News* **6:1**, 19-24. *(7.3)*

GALIL, Z. [1976], "Hierarchies of complete problems," *Acta Informat.* **6**, 77-88. *(7.5; A10.1)*

GALIL, Z. [1977], "On resolution with clauses of bounded size," *SIAM J. Comput.* **6**, 444-459. *(7.6)*

GALIL, Z., AND N. MEGIDDO [1977], "Cyclic ordering is NP-complete," *Theor. Comput. Sci.* **5**, 179-182. *(A12)*

GAREY, M. R. [1973], "Optimal task sequencing with precedence constraints," *Discrete Math.* **4**, 37-56. *(A5.1)*

GAREY, M. R., F. GAVRIL, AND D. S. JOHNSON [1977], unpublished results. *(A1.2)*

GAREY, M. R., R. L. GRAHAM, AND D. S. JOHNSON [1976], "Some NP-complete geometric problems," *Proc. 8th Ann. ACM Symp. on Theory of Computing*, Association for Computing Machinery, New York, 10-22. *(A2.3)*

GAREY, M. R., R. L. GRAHAM, AND D. S. JOHNSON [1977], "The complexity of computing Steiner minimal trees," *SIAM J. Appl. Math.* **32**, 835-859. *(A2.1)*

GAREY, M. R., R. L. GRAHAM, D. S. JOHNSON, AND D. E. KNUTH [1978], "Complexity results for bandwidth minimization," *SIAM J. Appl. Math.* **34**, 477-495. *(3.2.2; A1.3)*

GAREY, M. R., R. L. GRAHAM, D. S. JOHNSON, AND A. C. YAO [1976], "Resource constrained scheduling as generalized bin packing," *J. Combinatorial Theory Ser. A* **21**, 257-298. *(6.1)*

GAREY, M. R., AND D. S. JOHNSON [1975], "Complexity results for multiprocessor scheduling under resource constraints," *SIAM J. Comput.* **4**, 397-411. *(A3.2; A5.2)*

GAREY, M. R., AND D. S. JOHNSON [1976a], "The complexity of near-optimal graph coloring," *J. Assoc. Comput. Mach.* **23**, 43-49. *(6.2)*

GAREY, M. R., AND D. S. JOHNSON [1976b], "Approximation algorithms for combinatorial problems: an annotated bibliography," in J. F. Traub (ed.), *Algorithms and Complexity: New Directions and Recent Results*, Academic Press, New York, 41-52. *(6.2)*

GAREY, M. R., AND D. S. JOHNSON [1976c], "Scheduling tasks with nonuniform deadlines on two processors," *J. Assoc. Comput. Mach.* **23**, 461-467. *(A5.1; A5.2)*

GAREY, M. R., AND D. S. JOHNSON [1977a], "The rectilinear Steiner tree problem is NP-complete," *SIAM J. Appl. Math.* **32**, 826-834. *(A1.1; A1.2; A2.1)*

GAREY, M. R., AND D. S. JOHNSON [1977b], "Two-processor scheduling with start-times and deadlines," *SIAM J. Comput.* **6**, 416-426. *(A5.1; A5.2; A13)*

GAREY, M. R., AND D. S. JOHNSON [1978], "Strong NP-completeness results: motivation, examples, and implications," *J. Assoc. Comput. Mach.* **25**, 499-508. *(5.2; 6.2)*

GAREY, M. R., AND D. S. JOHNSON [——], unpublished results. *(A1.1; A1.2; A1.4; A1.5; A2.1; A2.2; A2.5; A3.1; A3.2; A5.4; A6; A7.1; A11.1; A12)*

GAREY, M. R., D. S. JOHNSON, G. L. MILLER, AND C. H. PAPADIMITRIOU [1978], unpublished results. *(A1.1; A11.1; A12)*

GAREY, M. R., D. S. JOHNSON, AND C. H. PAPADIMITRIOU [1977], unpublished results. *(A1.3; A8)*

GAREY, M. R., D. S. JOHNSON, AND R. SETHI [1976], "The complexity of flowshop and jobshop scheduling," *Math. Oper. Res.* **1**, 117-129. *(3.2.2; A5.3)*

GAREY, M. R., D. S. JOHNSON, B. B. SIMONS, AND R. E. TARJAN [1978], "Scheduling unit time tasks with arbitrary release times and deadlines," unpublished manuscript. *(A5.1)*

GAREY, M. R., D. S. JOHNSON, AND L. STOCKMEYER [1976], "Some simplified NP-complete graph problems," *Theor. Comput. Sci.* **1**, 237-267. *(3.2.2; 4.1; A1.1; A1.2; A1.3; A2.2; A9.1)*

GAREY, M. R., D. S. JOHNSON, AND R. E. TARJAN [1976a], "The planar Hamiltonian circuit problem is NP-complete," *SIAM J. Comput.* **5**, 704-714. *(3.2.3; A1.3)*

GAREY, M. R., D. S. JOHNSON, AND R. E. TARJAN [1976b], unpublished results. *(A1.1)*

GAREY, M. R. *See also* AHO, A. V.; BRUCKER, P.; FRAENKEL, A. S.; JOHNSON, D. S.

GARFINKEL, R. S. [1977], "Minimizing wallpaper waste, Part 1: a class of traveling salesman problems," *Operations Res.* **25**, 741-751. *(A2.3)*

GARFINKEL, R. S., AND G. L. NEMHAUSER [1972], *Integer Programming*, John Wiley & Sons, New York. *(6.0; A2.3; A13)*

GAVETT, J. [1965], "Three heuristic rules for sequencing jobs to a single production facility," *Management Sci.* **11**, B166-B176. *(6.1)*

GAVRIL, F. [1972], "Algorithms for minimum coloring, maximum clique, minimum covering by cliques, and maximum independent set of a chordal graph," *SIAM J. Comput.* **1**, 180-187. *(A1.1; A1.2; A4.1)*

GAVRIL, F. [1973], "Algorithms for a maximum clique and a maximum independent set of a circle graph," *Networks* **3**, 261-273. *(A1.2)*

GAVRIL, F. [1974a], "Algorithms on circular-arc graphs," *Networks* **4**, 357-369. *(A1.1; A1.2)*

GAVRIL, F. [1974b], "The intersection graphs of subtrees in trees are exactly the chordal graphs," *J. Combinatorial Theory Ser. B* **16**, 47-56. *(A13)*

GAVRIL, F. [1974c], private communication. *(6.1)*

GAVRIL, F. [1977a], "Some NP-complete problems on graphs," *Proc. 11th Conf. on Information Sciences and Systems*, Johns Hopkins University, Baltimore, MD, 91-95. *(A1.1; A1.2; A1.3; A2.5; A4.1)*

GAVRIL, F. [1977b], private communication. *(A1.2)*

GAVRIL, F. *See also* GAREY, M. R.; YANNAKAKIS, M.

GELDMACHER, R. C. *See* LIU, P. C.

GEOFFRION, A. M. [1974], "Lagrangian relaxation and its uses in integer programming," *Math. Prog. Study* **2**, 82-114. *(6.0)*

GILL, J. T., III [1977], "Computational complexity of probabilistic Turing machines," *SIAM J. Comput.* **6**, 675-695. *(7.3)*

GILL, J. T., III. *See also* BAKER, T.

GILMORE, P. C., AND R. E. GOMORY [1964], "Sequencing a one state-variable machine: a solvable case of the traveling salesman problem," *Operations Res.* **12**, 655-679. *(A2.3; A5.3)*

GIMPEL, J. F. [1965], "A method of producing a Boolean function having an arbitrarily prescribed prime implicant table," *IEEE Trans. Computers* **14**, 485-488. *(1.5; A9.1)*

GOLD, E. M. [1974], "Complexity of automaton identification from given data," unpublished manuscript. *(A9.1; A10.1)*

GOLD, E. M. [1978], "Deadlock protection: easy and difficult cases," *SIAM J. Comput.* **7**, 320-336. *(A5.3)*

GOL'DBERG, M. K., AND I. A. KLIPKER [1976], "Minimal placing of trees on a line," Technical Report, Physico-Technical Institute of Low Temeperatures, Academy of Sciences of Ukranian SSR, USSR (in Russian). *(A1.3)*

GOLDSCHLAGER, L. M. [1977], "The monotone and planar circuit value problems are log space complete for P," *SIGACT News* **9:2**, 25-29. *(7.5)*

GOLUMBIC, M. C. [1977], "The complexity of comparability graph recognition and coloring," *Computing* **18**, 199-208. *(A1.1; A1.2)*

GOMORY, R. E. *See* GILMORE, P. C.

GONZALEZ, T. [1977], "Optimal mean finish time preemptive schedules," Report No. 220, Computer Science Dept., Pennsylvania State University, University Park, PA. *(A5.2)*

GONZALEZ, T., E. L. LAWLER, AND S. SAHNI [1978], "Optimal preemptive scheduling of a fixed number of unrelated processors in polynomial time," unpublished manuscript. *(A5.2)*

GONZALEZ, T., AND S. SAHNI [1976], "Open shop scheduling to minimize finish time," *J. Assoc. Comput. Mach.* **23**, 665-679. *(A5.3)*

GONZALEZ, T., AND S. SAHNI [1978a], "Flowshop and jobshop schedules: complexity and approximation," *Operations Res.* **26**, 36-52. *(A5.3)*

GONZALEZ, T., AND S. SAHNI [1978b], "Preemptive scheduling of uniform processor systems," *J. Assoc. Comput. Mach.* **25**, 92-101. *(A5.2)*

GONZALEZ, T. *See also* SAHNI, S.

GOODMAN, S. *See* COCKAYNE, E.; MORROW, C.

GOYAL, D. K. [1976], "Scheduling processor bound systems," Report No. CS-76-036, Computer Science Department, Washington State University, Pullman, WA. *(A5.2)*

GRAHAM, R. L. [1966], "Bounds for certain multiprocessing anomalies," *Bell Syst. Tech. J.* **45**, 1563-1581. *(6.2)*

GRAHAM, R. L., E. L. LAWLER, J. K. LENSTRA, AND A. H. G. RINNOOY KAN [1978], "Optimization and approximation in deterministic sequencing and scheduling: a survey," *Ann. Discrete Math.* (to appear). *(A5.1)*

GRAHAM, R. L. *See also* CHUNG, F. R. K.; COFFMAN, E. G., JR; GAREY, M. R.; JOHNSON, D. S.

GRASSELLI, A., AND F. LUCCIO [1966], "A method for the combined row-column reduction of flow tables," *Proc. 7th Ann. Symp. on Switching and Automata Theory*, IEEE Computer Society, Long Beach, CA, 136-147. *(A10.1)*

GREIBACH, S. [1969], "Checking automata and one-way stack languages," *J. Comput. System Sci.* **3**, 196-217. *(A10.1)*

GREIBACH, S. A. [1973a], "Jump PDA's, deterministic context-free languages, principal AFDL's and polynomial time recognition--Extended abstract," *Proc. 5th Ann. ACM Symp. on Theory of Computing*, Association for Computing Machinery, New York, 20-28. *(A10.2)*

GREIBACH, S. A. [1973b], "The hardest context-free language," *SIAM J. Comput.* **2**, 304-310. *(A10.2)*

GREIBACH, S. *See also* BOOK, R. V.

GRIMMET, G. R., AND C. J. H. MCDIARMID [1975], "On colouring random graphs," *Math. Proc. Cambridge Philos. Soc.* **77**, 313-324. *(6.3)*

GROSS, D. A. *See* FULKERSON, D. R.

GURARI, E. M., AND O. H. IBARRA [1978], "An NP-complete number theoretic problem," *Proc. 10th Ann. ACM Symp. on Theory of Computing*, Association for Computing Machinery, New York, 205-215. *(A7.2)*

HACK, M. [1972], "Analysis of production schemata by Petri nets," Report No. TR-94, Project MAC, Massachusetts Institute of Technology, Cambridge, MA. *(A12)*

HADLOCK, F. O. [1974], "Minimum spanning forests of bounded trees," *Proc. 5th Southeastern Conference on Combinatorics, Graph Theory, and Computing*, Utilitas Mathematica Publishing, Winnipeg, 449-460. *(A2.1; A2.2)*

HADLOCK, F. O. [1975], "Finding a maximum cut of a planar graph in polynomial time," *SIAM J. Comput.* **4**, 221-225. *(4.1; A1.2; A2.2)*

HAKEN, W. *See* APPEL, K.

HAKIMI, S. L. *See* KARIV, O.

HAMMER, P. L. *See* CHVÁTAL, V.

HARARY, F. [1969], *Graph Theory*, Addison-Wesley, Reading, MA. *(A1.2; A3.1; A13)*

HARARY, F., AND R. A. MELTER [1976], "On the metric dimension of a graph," *Ars Combinatorica* **2**, 191-195. *(A1.5)*

HARARY, F., AND E. M. PALMER [1973], *Graphical Enumeration*, Academic Press, New York. *(7.3; A2.1)*

HARRISON, M. A., W. L. RUSSO, AND J. D. ULLMAN [1976], "Protection in operating systems," *Comm. ACM* **19**, 461-471. *(A4.3)*

HARTMANIS, J., AND L. BERMAN [1978], "On polynomial time isomorphisms of some new complete sets," *J. Comput. System Sci.* **16**, 418-422. *(7.1)*

HARTMANIS, J., AND J. E. HOPCROFT [1976], "Independence results in computer science," *SIGACT News* **8:4**, 13-24. *(7.6)*

HARTMANIS, J., AND H. B. HUNT, III [1974], "The LBA problem and its importance in the theory of computing," in R. M. Karp (ed.), *Complexity of Computation*, American Mathematical Society, Providence, RI, 1-26. *(7.4)*

HARTMANIS, J., P. M. LEWIS, AND R. E. STEARNS [1965], "Classification of computations by time and memory requirements," *Proc. IFIP Congress 1965*, Spartan, New York, 31-35. *(7.6)*

HARTMANIS, J., AND R. E. STEARNS [1965], "On the computational complexity of algorithms," *Trans. Amer. Math. Soc.* **117**, 285-306. *(1.4; 7.6)*

HARTMANIS, J. *See also* BERMAN, L.

HASEGAWA, T. *See* IBARAKI, T.

HAVRÁNEK, T. [1975], "Statistical quantifiers in observational calculi: an application in GUHA methods," *Theory and Decision* **6**, 213-230. *(A12)*

HECHT, M. S. *See* FREDERICKSON, G. N.

HEDETNIEMI, S. T. *See* COCKAYNE, E.; FARLEY, A.; MITCHELL, S.

HELD, M., AND R. M. KARP [1971], "The traveling salesman problem and minimum spanning trees: part II," *Math. Programming* **6**, 62-88. *(6.0)*

HELL, P. *See* KIRKPATRICK, D. G.

HERMAN, G. T., AND G. ROZENBERG [1975], *Developmental Systems and Languages*, North-Holland, Amsterdam. *(A10.2)*

HERRMANN, P. P. [1973], "On reducibility among combinatorial problems," Report No. TR-113, Project MAC, Massachusetts Institute of Technology, Cambridge, MA. *(A1.1; A2.4)*

HIRSCHBERG, D., AND M. EDELBERG [1973], "On the complexity of computing graph isomorphism," Report No. TR-130, Computer Science Lab., Dept. of Electrical Engineering, Princeton University, Princeton, NJ. *(A13)*

HIRSCHBERG, D. S., AND C. K. WONG [1976], "A polynomial-time algorithm for the knapsack problem with two variables," *J. Assoc. Comput. Mach.* **23**, 147-154. *(A6)*

HOPCROFT, J. E. [1971], "An $n\log n$ algorithm for minimizing states in a finite automaton," in Z. Kohavi and A. Paz (eds.), *Theory of Machines and Computations*, Academic Press, New York, 189-196. *(A10.1)*

HOPCROFT, J. E., AND R. M. KARP [1973], "An $n^{5/2}$ algorithm for maximum matchings in bipartite graphs," *SIAM J. Comput.* **2**, 225-231. *(3.1.2)*

HOPCROFT, J. E., AND R. E. TARJAN [1974], "Efficient planarity testing," *J. Assoc. Comput. Mach.* **21**, 549-568. *(4.1; A1.2; A13)*

HOPCROFT, J. E., AND J. D. ULLMAN [1967], "Nonerasing stack automata," *J. Comput. System Sci.* **1**, 166-186. *(A10.1)*

HOPCROFT, J. E., AND J. D. ULLMAN [1969], *Formal Languages and their Relation to Automata*, Addison-Wesley, Reading, MA. *(1.3; 2.2; 2.3; 7.4; 7.5; 7.6; A4.2; A10.1; A10.2)*

HOPCROFT, J. E., AND J. K. WONG [1974], "Linear time algorithm for isomorphism of planar graphs (Preliminary report)," *Proc. 6th Ann. ACM Symp. on Theory of Computing*, Association for Computing Machinery, New York, 172-184. *(A13)*

HOPCROFT, J. E. *See also* AHO, A. V.; FORTUNE, S.; HARTMANIS, J.

HORN, W. A. [1972], "Single-machine job sequencing with treelike precedence ordering and linear delay penalties," *SIAM J. Appl. Math.* **23**, 189-202. *(A5.1)*

HORN, W. A. [1973], "Minimizing average flow time with parallel machines," *Operations Res.* **21**, 846-847. *(A5.2)*

HORN, W. A. [1974], "Some simple scheduling algorithms," *Naval Res. Logist. Quart.* **21**, 177-185. *(A5.2)*

HOROWITZ, E., AND S. SAHNI [1974], "Computing partitions with applications to the knapsack problem," *J. Assoc. Comput. Mach.* **21**, 277-292. *(6.0)*

HOROWITZ, E., AND S. SAHNI [1976], "Exact and approximate algorithms for scheduling nonidentical processors," *J. Assoc. Comput. Mach.* **23**, 317-327. *(4.2.2; 6.1)*

HOROWITZ, E., AND S. SAHNI [1978], *Algorithms: Design and Analysis*, Computer Science Press, Potomac, MD. *(6.1)*

HORVATH, E. C., S. LAM, AND R. SETHI [1977], "A level algorithm for preemptive scheduling," *J. Assoc. Comput. Mach.* **24**, 32-43. *(A5.2)*

HOWELL, T. D. [1977], "Grouping by swapping is NP-complete," unpublished manuscript. *(A4.2)*

HU, T. C. [1961], "Parallel sequencing and assembly line problems," *Operations Res.* **9**, 841-848. *(A5.2)*

HU, T. C. [1969], *Integer Programming and Network Flows*, Addison-Wesley, Reading, MA. *(6.0)*

HU, T. C. [1974], "Optimum communication spanning trees," *SIAM J. Comput.* **3**, 188-195. *(A2.1)*

HU, T. C. *See also* ADOLPHSON, D.

HUET, G. P. [1973], "The undecidability of unification in third order logic," *Information and Control* **22**, 257-267. *(A9.2)*

HUGHES, G. E., AND M. J. CRESSWELL [1968], *An Introduction to Modal Logic*, Methuen, London. *(A9.2)*

HUNT, H. B., III [1973a], *On the Time and Tape Complexity of Languages*, Doctoral Thesis, Dept. of Computer Science, Cornell University, Ithaca, NY. *(A10.2)*

HUNT, H. B., III [1973b], "On the time and tape complexity of languages I," *Proc. 5th Ann. ACM Symp. on Theory of Computing*, Association for Computing Machinery, New York, 10-19. *(A10.1; A10.2)*

HUNT, H. B., III [1976], "On the complexity of finite, pushdown, and stack automata," *Math. Systems Theory* **10**, 33-52. *(A10.1)*

HUNT, H. B., III [1977], "A complexity theory of computation structures: preliminary report," unpublished manuscript. *(A9.2; A11.2; A12)*

HUNT, H. B., III [1978], "Uniform lower bounds on scheme equivalence," unpublished manuscript. *(A11.2)*

HUNT, H. B., III, AND D. J. ROSENKRANTZ [1977], "Complexity of grammatical similarity relations: preliminary report," *Proc. Conf. on Theoretical Computer Science*, Dept. of Computer Science, University of Waterloo, Waterloo, Ontario, 139-148. *(A10.2; A13)*

HUNT, H. B., III, AND D. J. ROSENKRANTZ [1978], "Computational parallels between regular and context-free languages," *SIAM J. Comput.* **7**, 99-114. *(A10.2)*

HUNT, H. B., III, D. J. ROSENKRANTZ, AND T. G. SZYMANSKI [1976a], "On the equivalence, containment, and covering problems for the regular and context-free languages," *J. Comput. System Sci.* **12**, 222-268. *(A10.2)*

HUNT, H. B., III, D. J. ROSENKRANTZ, AND T. G. SZYMANSKI [1976b], "The covering problem for linear context-free grammars," *Theor. Comput. Sci.* **2**, 361-382. *(A10.2)*

HUNT, H. B., III, AND T. G. SZYMANSKI [1976a], "Complexity metatheorems for context-free grammar problems," *J. Comput. System Sci.* **13**, 318-334. *(A10.2)*

HUNT, H. B., III, AND T. G. SZYMANSKI [1976b], "Dichotimization, reachability, and the forbidden subgraph problem (Extended abstract)," *Proc. 8th Ann. ACM Symp. on Theory of Computing*, Association for Computing Machinery, New York, 126-134. *(A11.2)*

HUNT, H. B., III, AND T. G. SZYMANSKI [1978], "Lower bounds and reductions between grammar problems," *J. Assoc. Comput. Mach.* **25**, 32-51. *(A10.2)*

HUNT, H. B., III, T. G. SZYMANSKI, AND J. D. ULLMAN [1975], "On the complexity of LR(k) testing," *Comm. ACM* **18**, 707-716. *(A10.2)*

HUNT, H. B., III. *See also* CONSTABLE, R. L.; HARTMANIS, J.

HWANG, F. K. *See* AHO, A. V.

HYAFIL, L., AND R. L. RIVEST [1973], "Graph partitioning and constructing optimal decision trees are polynomial complete problems," Report No. 33, IRIA-Laboria, Rocquencourt, France. *(A2.2)*

HYAFIL, L., AND R. L. RIVEST [1976], "Constructing optimal binary decision trees is NP-complete," *Information Processing Lett.* **5**, 15-17. *(A12)*

IBARAKI, T. [1978], "Approximate algorithms for the multiple-choice continuous knapsack problem," unpublished manuscript. *(A6)*

IBARAKI, T., T. HASEGAWA, K. TERANAKA, AND J. IWASE [1978], "The multiple-choice knapsack problem," *J. Oper. Res. Soc. Japan* **21**, 59-94. *(A6)*

IBARAKI, T., T. KAMEDA, AND S. TOIDA [1977], "NP-complete diagnosis problems on systems graphs," unpublished manuscript. *(A3.1; A12)*

IBARAKI, T. *See also* KISE, H.

IBARRA, O. H., AND C. E. KIM [1975a], "Fast approximation algorithms for the knapsack and sum of subset problems," *J. Assoc. Comput. Mach.* **22**, 463-468. *(6.1)*

IBARRA, O. H., AND C. E. KIM [1975b], "Scheduling for maximum profit," Report No. 75-2, Computer Science Dept., University of Minnesota, Minneapolis, MN. *(A6)*

IBARRA, O. H., AND S. K. SAHNI [1975], "Polynomially complete fault detection problems," *IEEE Trans. Computers* **C-24**, 242-249. *(A12)*

IBARRA, O. H. *See also* GURARI, E. M.

ITAI, A. [1977], "Two commodity flow," Report No. 93, Dept. of Computer Science, Technion, Haifa, Israel. *(7.1; A2.4)*

ITAI, A., Y. PERL, AND Y. SHILOACH [1977], "The complexity of finding maximum disjoint paths with length constraints," Report No. 94, Dept. of Computer Science, Technion, Haifa, Israel. *(A2.4)*

ITAI, A., AND M. RODEH [1977a], "Some matching problems," in *Automata, Languages, and Programming*, Lecture Notes in Computer Science, Vol. 52, Springer, Berlin, 258-268. *(A1.5)*

ITAI, A., AND M. RODEH [1977b], "Finding a minimum circuit in a graph," *Proc. 9th Ann. ACM Symp. on Theory of Computing*, Association for Computing Machinery, New York, 1-10. *(A2.3)*

ITAI, A., M. RODEH, AND S. L. TANIMOTA [1978], "Some matching problems for bipartite graphs," *J. Assoc. Comput. Mach.* (to appear). *(A1.5)*

ITAI, A. *See also* EVEN, S.

IWASE, J. *See* IBARAKI, T.

JACKSON, J. R. [1956], "An extension of Johnson's results on job lot scheduling," *Naval Res. Logist. Quart.* **3**, 201-203. *(A5.3)*

JAZAYERI, M., W. F. OGDEN, AND W. C. ROUNDS [1975], "The intrinsically exponential complexity of the circularity problem for attribute grammars," *Comm. ACM* **18**, 697-706. *(7.6)*

JEROSLOW, R. G. [1973], "There cannot be any algorithm for integer programming with quadratic constraints," *Operations Res.* **21**, 221-224. *(A6)*

JEROSLOW, R. G. [1976], "Bracketing discrete problems by two problems of linear optimization," *Proc. First Symp. on Operations Research (at Heidelberg)*, Verlag Anton Hain, Meisenheim, 205-216. *(A6)*

JOHNSON, D. B., AND S. D. KASHDAN [1976], "Lower bounds for selection in $X+Y$ and other multisets," Report No. 183, Computer Science Department, Pennsylvania State University, University Park, PA (to appear *J. Assoc. Comput. Mach.*). *(5.1; 7.3; A2.1; A2.3; A3.2)*

JOHNSON, D. B., AND T. MIZOGUCHI [1978], "Selecting the Kth element in $X+Y$ and $X_1+X_2+\cdots+X_m$," *SIAM J. Comput.* **7**, 147-153. *(A3.2)*

JOHNSON, D. S. [1973], *Near-Optimal Bin Packing Algorithms*, Doctoral Thesis, Dept. of Mathematics, Massachusetts Institute of Technology, Cambridge, MA. *(6.1; 6.3)*

JOHNSON, D. S. [1974a], "Approximation algorithms for combinatorial problems," *J. Comput. System Sci.* **9**, 256-278. *(6.1)*

JOHNSON, D. S. [1974b], "Worst case behavior of graph coloring algorithms," *Proc. 5th Southeastern Conference on Combinatorics, Graph Theory, and Computing*, Utilitas Mathematica Publishing, Winnipeg, 513-527. *(6.1)*

JOHNSON, D. S., A. DEMERS, J. D. ULLMAN, M. R. GAREY, AND R. L. GRAHAM [1974], "Worst-case performance bounds for simple one-dimensional packing algorithms," *SIAM J. Comput.* **3**, 299-325 *(6.1)*

JOHNSON, D. S., J. K. LENSTRA, AND A. H. G. RINNOOY KAN [1978], "The complexity of the network design problem," *Networks* (to appear). *(A2.1)*

JOHNSON, D. S., AND F. P. PREPARATA [1978], "The densest hemisphere problem," *Theor. Comput. Sci.* **6**, 93-107. *(A6)*

JOHNSON, D. S. *See also* BRUCKER, P.; EVEN, S.; FRAENKEL, A. S.; GAREY, M. R.

JOHNSON, E. L. *See* EDMONDS, J.

JOHNSON, S. C. *See* AHO, A. V.

JOHNSON, S. M. [1954], "Optimal two- and three-stage production schedules with setup times included," *Naval Res. Logist. Quart.* **1**, 61-68. *(A5.3)*

JONES, N. D. [1973], "Reducibility among combinatorial problems in log*n* space," *Proc. 7th Ann. Princeton Conf. on Information Sciences and Systems*, Dept. of Electrical Engineering, Princeton University, Princeton, NJ, 547-551. *(7.5)*

JONES, N. D. [1975], "Space-bounded reducibility among combinatorial problems," *J. Comput. System Sci.* **11**, 68-85. *(7.5)*

JONES, N. D., AND W. T. LAASER [1976], "Complete problems for deterministic polynomial time," *Theor. Comput. Sci.* **3**, 105-117. *(7.5)*

JONES, N. D., L. H. LANDWEBER, AND Y. E. LIEN [1977], "Complexity of some problems in Petri nets," *Theor. Comput. Sci.* **4**, 277-299. *(A12)*

JONES, N. D., Y. E. LIEN, AND W. T. LAASER [1976], "New problems complete for nondeterministic log space," *Math. Systems Theory* **10**, 1-17. *(7.5)*

JONES, N. D., AND S. S. MUCHNIK [1977], "Even simple programs are hard to analyze," *J. Assoc. Comput. Mach.* **24**, 338-350. *(A11.2)*

JONES, N. D., AND S. SKYUM [1976], "Complexity of some problems concerning L systems (preliminary report)," Report No. DAIMI PB-67, University of Aarhus, Aarhus, Denmark. *(A10.2)*

JONES, N. D., AND S. SKYUM [1977], "Recognition of deterministic ETOL languages in logarithmic space," *Information and Control* **35**, 177-181. *(A10.2)*

KAMEDA, T. *See* ABDEL-WAHAB, H. M.; IBARAKI, T.

KANELLAKIS, P. C. *See* PAPADIMITRIOU, C. H.

KARAGANIS, J. J. [1968], "On the cube of a graph," *Canad. Math. Bull.* **11**, 295-296. *(A1.3)*

KARIV, O., AND S. L. HAKIMI [1976a], "An algorithmic approach to network location problems – Part I: the p-centers," unpublished manuscript. *(A2.5)*

KARIV, O., AND S. L. HAKIMI [1976b], "An algorithmic approach to network location problems – Part 2: the p-medians," unpublished manuscript. *(A2.5)*

KARP, R. M. [1972], "Reducibility among combinatorial problems," in R. E. Miller and J. W. Thatcher (eds.), *Complexity of Computer Computations*, Plenum Press, New York, 85-103. *(1.5; 3.1; 5.2; 7.1; 7.4; A1.1; A1.2; A1.3; A2.1; A2.2; A3.1; A3.2; A5.1; A6; A10.1)*

KARP, R. M. [1975a], "On the complexity of combinatorial problems," *Networks* **5**, 45-68. *(A2.4)*

KARP, R. M. [1975b], "The fast approximate solution of hard combinatorial problems," *Proc. 6th Southeastern Conference on Combinatorics, Graph Theory, and Computing*, Utilitas Mathematica Publishing, Winnipeg, 15-31. *(6.3)*

KARP, R. M. [1976], "The probabilistic analysis of some combinatorial search algorithms," in J. F. Traub (ed.), *Algorithms and Complexity: New Directions and Recent Results*, Academic Press, New York, 1-19. *(6.3)*

KARP, R. M. [1977], "Probabilistic analysis of partitioning algorithms for the traveling-salesman problem in the plane," *Math. Oper. Res.* **2**, 209-224. *(6.3)*

KARP, R. M., A. C. MCKELLAR, AND C. K. WONG [1975], "Near-optimal solutions to a 2-dimensional placement problem," *SIAM J. Comput.* **4**, 271-286. *(6.1)*

KARP, R. M. *See also* EDMONDS, J.; HELD, M.; HOPCROFT, J. E.

KASAMI, T. *See* ARAKI, T.; FUJII, M.; SUGIYAMA, Y.

KASHDAN, S. D. *See* JOHNSON, D. B.

KAUFMAN, M. T. [1974], "An almost-optimal algorithm for the assembly line scheduling problem," *IEEE Trans. Computers* **C-23**, 1169-1174. *(6.1)*

KERNIGHAN, B. W. [1971], "Optimal sequential partitions of graphs," *J. Assoc. Comput. Mach.* **18**, 34-40. *(A2.2)*

KIM, C. E. *See* FREDERICKSON, G. N.; IBARRA, O. H.

KIRKPATRICK, D. G., AND P. HELL [1978], "On the complexity of a generalized matching problem," *Proc. 10th Ann. ACM Symp. on Theory of Computing*, Association for Computing Machinery, New York, 240-245. *(A1.1)*

KISE, H., T. IBARAKI, AND H. MINE [1978], "A solvable case of the one-machine scheduling problem with ready and due times," *Operations Res.* **26**, 121-126. *(A5.1)*

KLEE, V. [1978], private communication. *(A6)*

KLEE, V., AND G. J. MINTY [1972], "How good is the simplex algorithm?," in O. Shisha (ed.), *Inequalities III*, Academic Press, New York, 159-175. *(1.3)*

KLEENE, S. C. [1956], "Representation of events in nerve nets and finite automata," in C. E. Shannon and M. McCarthy (eds.), *Automata Studies*, Annals of Math. Studies, No. 34, Princeton University Press, Princeton, NJ, 3-41. *(A10.1)*

KLEIN, M. *See* FLORIAN, M.

KLIPKER, I. A. *See* GOL'DBERG, M. K.

KNUTH, D. E. [1973], private communication. *(A5.1)*

KNUTH, D. E. [1974a], "A terminological proposal," *SIGACT News* **6:1**, 12-18. *(5.2)*

KNUTH, D. E. [1974b], "Postscript about NP-hard problems," *SIGACT News* **6:2**, 15-16. *(5.2)*

KNUTH, D. E. [1974c], private communication. *(A2.4)*

KNUTH, D. E. *See also* GAREY, M. R.

KOU, L. T. [1977], "Polynomial complete consecutive information retrieval problems," *SIAM J. Comput.* **6**, 67-75. *(A4.2)*

KOU, L. T., L. J. STOCKMEYER, AND C. K. WONG [1978], "Covering edges by cliques with regard to keyword conflicts and intersection graphs," *Comm. ACM* **21**, 135-138. *(A1.1; A1.5)*

KOZEN, D. [1976], "Complexity of finitely presented algebras," Report No. 76-294, Dept. of Computer Science, Cornell University, Ithaca, NY. *(A7.3)*

KOZEN, D. [1977a], "Complexity of finitely presented algebras," *Proc. 9th Ann. ACM Symp. on Theory of Computing*, Association for Computing Machinery, New York, 164-177. *(7.1; 7.5; A7.3; A13)*

KOZEN, D. [1977b], "Finitely presented algebras and the polynomial time hierarchy," Report No. 77-303, Dept. of Computer Science, Cornell University, Ithaca, NY. *(7.2; A7.3)*

KOZEN, D. [1977c], "First order predicate logic without negation is NP-complete," Report No. 77-307, Dept. of Computer Science, Cornell University, Ithaca, NY. *(A9.2)*

KOZEN, D. [1977d], "Lower bounds for natural proof systems," *Proc. 18th Ann. Symp. on Foundations of Computer Science*, IEEE Computer Society, Long Beach, CA, 254-266. *(A10.1; A12)*

KOZEN, D. [1978], "A clique problem equivalent to graph isomorphism," unpublished manuscript. *(7.1; A13)*

KRISHNAMOORTHY, M. S. [1975], "An NP-hard problem in bipartite graphs," *SIGACT News* **7:1**, 26. *(A1.3)*

KRISHNAMOORTHY, M. S., AND N. DEO [1977a], "Node deletion NP-complete problems," Technical Report, Computer Centre, Indian Institute of Technology, Kanpur, India. *(A1.2)*

KRISHNAMOORTHY, M. S., AND N. DEO [1977b], "Complexity of the minimum dummy activities problem in a Pert network," Technical Report, Computer Centre, Indian Institute of Technology, Kanpur, India. *(A2.5)*

KRUSKAL, J. B. [1956], "On the shortest spanning subtree of a graph and the traveling salesman problem," *Proc. Amer. Math. Soc.* **7**, 48-50. *(6.1)*

KUCERA, L. [1976], "The complexity of clique finding algorithms," unpublished manuscript. *(6.2)*

KURODA, S. Y. [1964], "Classes of languages and linear-bounded automata," *Information and Control* **7**, 207-223. *(A10.2)*

LAASER, W. T. *See* JONES, N. D.

LABETOULLE, J., E. L. LAWLER, J. K. LENSTRA, AND A. H. G. RINNOOY KAN [1978], "Preemptive scheduling of uniform machines," unpublished manuscript. *(A5.1; A5.2)*

LABETOULLE, J. *See also* LAWLER, E. L.

LADNER, R. E. [1975a], "On the structure of polynomial time reducibility," *J. Assoc. Comput. Mach.* **22**, 155-171. *(7.1)*

LADNER, R. E. [1975b], "The circuit value problem is log space complete for P," *SIGACT News* **7:1**, 18-20. *(7.5)*

LADNER, R. E. [1977], "The computational complexity of provability in systems of modal propositional logic," *SIAM J. Comput.* **6**, 467-480. *(A9.2)*

LADNER, R. E., AND N. LYNCH [1976], "Relativization of questions about log space computability," *Math. Systems Theory* **10**, 19-32. *(7.6)*

LADNER, R. E., N. A. LYNCH, AND A. L. SELMAN [1975], "A comparison of polynomial time reducibilities," *Theor. Comput. Sci.* **1**, 103-123. *(7.1; 7.2)*

LADNER, R. E. *See also* DOBKIN, D.

LAGEWEG, B. J., E. L. LAWLER, J. K. LENSTRA, AND A. H. G. RINNOOY KAN [1978], "Computer aided complexity classification of deterministic scheduling problems," unpublished manuscript, Mathematisch Centrum, Amsterdam. *(5.2)*

LAGEWEG, B. J., AND J. K. LENSTRA [1977], private communication. *(A5.2)*

LAGEWEG, B. J., J. K. LENSTRA, AND A. H. G. RINNOOY KAN [1976], "Minimizing maximum lateness on one machine: computational experience and some applications," *Statistica Neerlandica* **30**, 25-41. *(A5.1)*

LAM, S. *See* HORVATH, E. C.

LANDWEBER, L. H. *See* JONES, N. D.

LANGMAACK, H. [1973], "On correct procedure parameter transmission in higher programming languages," *Acta Informat.* **2**, 110-142. *(A11.2)*

LAPAUGH, A. S., AND R. L. RIVEST [1978], "The subgraph homeomorphism problem," *Proc. 10th Ann. ACM Symp. on Theory of Computing*, Association for Computing Machinery, New York, 40-50. *(A13)*

LAWLER, E. L. [1972], "A procedure for computing the K best solutions to discrete optimization problems and its application to the shortest path problem," *Management Sci.* **18**, 401-405. *(5.1; A2.1; A3.2)*

LAWLER, E. L. [1973], "Optimal sequencing of a single machine subject to precedence constraints," *Management Sci.* **19**, 544-546. *(A5.1)*

LAWLER, E. L. [1976a], *Combinatorial Optimization: Networks and Matroids*, Holt, Rinehart and Winston, New York. *(6.1; A1.1; A1.3; A2.3; A2.4; A3.1; A6; A8; A13)*

LAWLER, E. L. [1976b], "A note on the complexity of the chromatic number problem," *Information Processing Lett.* **5**, 66-67. *(6.0)*

LAWLER, E. L. [1976c], "Sequencing to minimize the weighted number of tardy jobs," *Rev. Francaise Automat. Informat. Recherche Operationnelle Ser. Bleue* **10.5** (suppl.), 27-33. *(A5.1)*

LAWLER, E. L. [1976d], private communication. *(A13)*

LAWLER, E. L. [1977a], "A pseudopolynomial algorithm for sequencing jobs to minimize total tardiness," *Ann. Discrete Math.* **1**, 331-342. *(4.2.2; A5.1)*

LAWLER, E. L. [1977b], "Fast approximation algorithms for knapsack problems," *Proc. 18th Ann. Symp. on Foundations of Computer Science*, IEEE Computer Society, Long Beach, CA, 206-213. *(6.1)*

LAWLER, E. L. [1978], "Sequencing jobs to minimize total weighted completion time subject to precedence constraints," *Ann. Discrete Math.* **2**, 75-90. *(A5.1)*

LAWLER, E. L., AND J. LABETOULLE [1978], "Preemptive scheduling of unrelated parallel processors," *J. Assoc. Comput. Mach.* (to appear). *(A5.2)*

LAWLER, E. L., AND J. M. MOORE [1969], "A functional equation and its applications to resource allocation and sequencing problems," *Management Sci.* **16**, 77-84. *(4.2.2; A5.1)*

LAWLER, E. L. *See also* GONZALEZ, T.; GRAHAM, R. L.; LABETOULLE, J.; LAGEWEG, B. J.

LEGGETT, E. W., JR [1977], *Tools and Techniques for Classifying NP-Hard Problems*, Doctoral Thesis, Dept. of Computer and Information Sciences, Ohio State University, Columbus, OH. *(7.2)*

LEMPEL, A. *See* EVEN, S.

LENSTRA, J. K. [1977], private communication. *(A5.1; A5.3)*

LENSTRA, J. K., AND A. H. G. RINNOOY KAN [1976], "On general routing problems," *Networks* **6**, 273-280. *(A2.3)*

LENSTRA, J. K., AND A. H. G. RINNOOY KAN [1978a], "Complexity of scheduling under precedence constraints," *Operations Res.* **26**, 22-35. *(6.2; A5.1; A5.2)*

LENSTRA, J. K., AND A. H. G. RINNOOY KAN [1978b], "Computational complexity of discrete optimization problems," *Ann. Discrete Math.* (to appear). *(A5.3)*

LENSTRA, J. K., A. H. G. RINNOOY KAN, AND P. BRUCKER [1977], "Complexity of machine scheduling problems," *Ann. Discrete Math.* **1**, 343-362. *(A5.1; A5.2; A5.3)*

LENSTRA, J. K., A. H. G. RINNOOY KAN, AND M. FLORIAN [1978], "Deterministic production planning: algorithms and complexity," unpublished manuscript. *(A5.4)*

LENSTRA, J. K. *See also* GRAHAM, R. L.; JOHNSON, D. S.; LABETOULLE, J.; LAGEWEG, B. J.

LEVIN, L. A. [1973], "Universal sorting problems," *Problemy Peredaci Informacii* **9**, 115-116 (in Russian). English translation in *Problems of Information Transmission* **9**, 265-266. *(5.2; A1.4)*

LEWIS, H. R. [1978], "Satisfiability problems for propositional calculi," unpublished manuscript. *(A9.1)*

LEWIS, H. R., AND C. H. PAPADIMITRIOU [1978], private communication. *(A8)*

LEWIS, J. M. [1976], private communication. *(A1.2)*

LEWIS, J. M. [1978], "On the complexity of the maximum subgraph problem," *Proc. 10th Ann. ACM Symp. on Theory of Computing*, Association for Computing Machinery, New York, 265-274. *(A1.2)*

LEWIS, P. M. *See* HARTMANIS, J.; ROSENKRANTZ, D. J.

LICHTENSTEIN, D. [1977], "Planar satisfiability and its uses," *SIAM J. Comput.* (to appear). *(A2.5; A9.1)*

LICHTENSTEIN, D., AND M. SIPSER [1978], "GO is Pspace hard," *Proc. 19th Ann. Symp. on Foundations of Computer Science*, IEEE Computer Society, Long Beach, CA, 48-54. *(7.4; A8)*

LICHTENSTEIN, D. I. *See also* EVEN, S.

LIEBERHERR, K. [1977], private communication. *(A11.2)*

LIEN, Y. E. *See* JONES, N. D.

LIN, S. [1975], "Heuristic programming as an aid to network design," *Networks* **5**, 33-43. *(6.0)*

LIPSHITZ, L. [1977], "A remark on the Diophantine problem for addition and divisibility," unpublished manuscript. *(A7.1)*

LIPSHITZ, L. [1978], "The Diophantine problem for addition and divisibility," *Trans. Amer. Math. Soc.* **235**, 271-283. *(A7.1)*

LIPSKY, W., JR [1977a], "Two NP-complete problems related to information retrieval," in *Fundamentals of Computation Theory*, Lecture Notes in Computer Science, Springer, Berlin, (to appear). *(A4.3)*

LIPSKY, J., JR [1977b], "One more polynomial complete consecutive retrieval problem," *Information Processing Lett.* **6**, 91-93. *(A4.2)*

LIPSKY, W., JR [1978], private communication. *(A4.2)*

LIPTON, R. J. [1975], "The reachability problem requires exponential space," Report No. 62, Dept. of Computer Science, Yale University, New Haven, CT. *(A12)*

LIPTON, R. J., AND R. E. TARJAN [1977], "Applications of a planar separator theorem," *Proc. 18th Ann. Symp. on Foundations of Computer Science*, IEEE Computer Society, Long Beach, CA, 162-170. *(6.1; 6.2)*

LIPTON, R. J., AND Y. ZALCSTEIN [1977], "Word problems solvable in logspace," *J. Assoc. Comput. Mach.* **24**, 522-526. *(7.5)*

LIPTON, R. J. *See also* DOBKIN, D.

LITVINTCHOUK, S. D., AND V. R. PRATT [1977], "A proof checker for dynamic logic," *Proc. 5th Internat. Joint Conf. on Artificial Intelligence*, International Joint Conferences on Artificial Intelligence, Dept. of Computer Science, Carnegie-Mellon University, Pittsburgh, PA, 552-558. *(A9.2)*

LIU, C. L. [1968], *Introduction to Combinatorial Mathematics*, McGraw-Hill, New York. *(6.1; A1.3)*

LIU, P. C., AND R. C. GELDMACHER [1978], "On the deletion of nonplanar edges of a graph," *SIAM J. Comput.* (to appear). *(3.1.4; 3.2.2; A1.2)*

LLOYD, E. L. [1977], "On triangulations of a set of points in the plane," *Proc. 18th Ann. Symp. on Foundations of Computer Science*, IEEE Computer Society, Long Beach, CA, 228-240. *(A2.5; A13)*

LOVASZ, L. [1973], "Coverings and colorings of hypergraphs," *Proc. 4th Southeastern Conference on Combinatorics, Graph Theory, and Computing*, Utilitas Mathematica Publishing, Winnipeg, 3-12. *(A3.1)*

LOVASZ, L. *See also* BURR, S.

LUCCHESI, C. L. [1976], *A Minimax Equality for Directed Graphs*, Doctoral Thesis, University Of Waterloo, Waterloo, Ontario. *(A1.1)*

LUCCHESI, C. L., AND S. L. OSBORN [1977], "Candidate keys for relations," *J. Comput. System Sci.* (to appear). *(A4.3)*

LUCCIO, F. *See* GRASSELLI, A.

LUCKHAM, D. C., D. M. PARK, AND M. S. PATERSON [1970], "On formalised computer programs," *J. Comput. System Sci.* **4**, 220-249. *(A11.2)*

LUEKER, G. S. [1975], "Two NP-complete problems in nonnegative integer programming," Report No. 178, Computer Science Laboratory, Princeton University, Princeton, NJ. *(A6)*

LUEKER, G. S. *See also* BOOTH, K. S.; ROSE, D. J.

LUKES, J. A. [1974], "Efficient algorithm for the partitioning of trees," *IBM J. Res. Develop.* **18**, 217-224. *(A2.2)*

LYNCH, J. F. [1975], "The equivalence of theorem proving and the interconnection problem," *ACM SIGDA Newsletter* **5:3**. *(A2.4)*

LYNCH, J. F. [1976], private communication. *(A2.4)*

LYNCH, N. [1977], "Log space recognition and translation of parenthesis languages," *J. Assoc. Comput. Mach.* **24**, 583-590. *(7.5)*

LYNCH, N. [1978], "Log space machines with multiple oracle tapes," *Theor. Comput. Sci.* **6**, 25-39. *(7.6)*

LYNCH, N. *See also* LADNER, R. E.

MAHESHWARI, S. [1976], "Traversal marker placement problems are NP-complete," Report No. CU-CS-092-76, Dept. of Computer Science, University of Colorado, Boulder, CO. *(A1.2; A1.5)*

MAHESHWARI, S. N. *See also* GABOW, H. N.

MAIER, D. [1977], "The complexity of some problems on subsequences and supersequences," *J. Assoc. Comput. Mach.* **25**, 322-336. *(A4.2)*

MAIER, D., AND J. A. STORER [1977], "A note on the complexity of the superstring problem," Report No. 233, Computer Science Laboratory, Princeton University, Princeton, NJ. *(A1.2; A4.2)*

MANDERS, K., AND L. ADLEMAN [1978], "NP-complete decision problems for binary quadratics," *J. Comput. System Sci.* **16**, 168-184. *(A7.1; A7.2)*

MANDERS, K. *See also* ADLEMAN, L.

MASEK, W. J. [1978], "Some NP-complete set covering problems," unpublished manuscript. *(a4.2; A9.1)*

MATHON, R. [1978], "A note on the graph isomorphism counting problem," *Information Processing Lett.* (to appear). *(A13)*

MATIJASEVIC, Y. V. [1970], "Enumerable sets are Diophantine," *Dokl. Akad. Nauk SSSR* **191**, 279-282 (in Russian). English translation in *Soviet Math. Dokl.* **11**, 354-357. *(1.4)*

MATIJASEVIC, Y., AND J. ROBINSON [1975], "Reduction of an arbitrary Diophantine equation to one in 13 unknowns," *Acta Arith.* **27**, 521-553. *(A7.2)*

MATULA, D. W. *See* EDMONDS, J.

MAXWELL, W. L. *See* CONWAY, R. W.

MCDIARMID, C. [1976], "Determining the chromatic number of a graph," Report No. STAN-CS-76-576, Computer Science Dept., Stanford University, Stanford, CA. *(7.6)*

McDiarmid, C. *See also* Grimmet, G. R.

McEliece, R. J. *See* Berlekamp, E. R.

McHugh, J. A. M. *See* Boesch, F. T.

McKellar, A. C. *See* Karp, R. M.

McNaughton, R. [1959], "Scheduling with deadlines and loss functions," *Management Sci.* **6**, 1-12. *(A5.2)*

Megiddo, N. [1977], private communication. *(A2.3)*

Megiddo, N. *See also* Galil, Z.

Melter, R. A. *See* Harary, F.

Merlin, P. M. *See* Chandra, A. K.

Meyer, A. R. [1975], "Weak monadic second order theory of successor is not elementary recursive," in R. Parikh (ed.), *Logic Colloquium*, (Proc. Symposium on Logic, Boston, 1972), Lecture Notes In Mathematics, Vol. 453, Springer, Berlin, 132-154. *(7.6)*

Meyer, A. R., and D. M. Ritchie [1967], "The complexity of loop programs," *Proc. 22nd Natl. Conf. of the ACM*, Thompson Book Co., Washington, DC, 465-469. *(A11.2)*

Meyer, A. R., and M. I. Shamos [1977], "Time and Space," in A. K. Jones (ed.), *Perspectives on Computer Science*, Academic Press, New York, 125-146. *(7.5)*

Meyer, A. R., and L. J. Stockmeyer [1972], "The equivalence problem for regular expressions with squaring requires exponential time," *Proc. 13th Ann. Symp. on Switching and Automata Theory*, IEEE Computer Society, Long Beach, CA, 125-129. *(1.4; 7.2; 7.4; 7.6)*

Meyer, A. R. *See also* Bauer, M.; Seiferas, J. I.; Stockmeyer, L. J.

Miller, G. A. *See* Chomsky, N.

Miller, G. L. [1976], "Riemann's Hypothesis and tests for primality," *J. Comput. System Sci.* **13**, 300-317. *(7.1; A13)*

Miller, G. L. [1977], "Graph isomorphism, general remarks," *Proc. 9th Ann. ACM Symp. on Theory of Computing*, Association for Computing Machinery, New York, 143-150. *(7.1; A13)*

Miller, G. L. [1978], "On the $n^{\log n}$ isomorphism technique: A preliminary report," *Proc. 10th Ann. ACM Symp. on Theory of Computing*, Association for Computing Machinery, New York, 51-58. *(A13)*

Miller, G. L. *See also* Garey, M. R.

Miller, L. W. *See* Conway, R. W.

Mine, H. *See* Kise, H.

Minsky, M. [1967], *Computation: Finite and Infinite Machines*, Prentice Hall, Englewood Cliffs, NJ. *(2.2)*

Minty, G. J. [1977], "On maximal independent sets of vertices in claw-free graphs," unpublished manuscript. *(A1.2)*

Minty, G. J. *See also* Klee, V.

MITCHELL, S., AND S. HEDETNIEMI [1977], "Edge domination in trees," *Proc. 8th Southeastern Conference on Combinatorics, Graph Theory, and Computing*, Utilitas Mathematica Publishing, Winnipeg, 489-509. *(A1.1)*

MITCHELL, S. *See also* FARLEY, A.

MIZOGUCHI, T. *See* JOHNSON, D. B.

MONMA, C. L., AND J. B. SIDNEY [1977], "A general algorithm for optimal job sequencing with series-parallel precedence constraints," Report No. 347, School of Operations Research, Cornell University, Ithaca, NY. *(A5.1)*

MOORE, J. M. [1968], "An *n* job, one machine sequencing algorithm for minimizing the number of late jobs," *Management Sci.* **15**, 102-109. *(A5.1)*

MOORE, J. M. *See also* LAWLER, E. L.

MORROW, C., AND S. GOODMAN [1976], "An efficient algorithm for finding a longest cycle in a tournament," *Proc. 7th Southeastern Conference on Combinatorics, Graph Theory, and Computing*, Utilitas Mathematica Publishing, Winnipeg, 453-462. *(A1.3; A2.3)*

MUCHNIK, S. S. *See* JONES, N. D.

MUNRO, I. *See* ROBERTSON, E.

MUNTZ, R. R., AND E. G. COFFMAN, JR [1969], "Optimal preemptive scheduling on two-processor systems," *IEEE Trans. Computers* **C-18**, 1014-1020. *(A5.2)*

MUNTZ, R. R., AND E. G. COFFMAN, JR [1970], "Preemptive scheduling of real-time tasks on multiprocessor systems," *J. Assoc. Comput. Mach.* **17**, 324-338. *(A5.2)*

MURTY, K. G. [1972], "A fundamental problem in linear inequalities with applications to the traveling salesman problem," *Math. Programming* **2**, 296-308. *(A6)*

MURTY, K. G. [1976], *Linear and Combinatorial Programming*, John Wiley and Sons, Inc., New York. *(A13)*

NELSON, G., AND D. C. OPPEN [1977], "Fast decision algorithms based on union and find," *Proc. 18th Ann. Symp. on Foundations of Computer Science*, IEEE Computer Society, Long Beach, CA, 114-119. *(A9.2)*

NEMHAUSER, G. L., L. A. WOLSEY, AND M. L. FISHER [1978], "An analysis of approximations for maximizing submodular set functions — I," *Math. Programming* **14**, 265-294. *(6.1)*

NEMHAUSER, G. L. *See also* CORNUEJOLS, G.; GARFINKEL, R. S.

NESETRIL, J., AND A. PULTR [1977], "The complexity of a dimension of a graph," *Proc. Wroclaw Conf. on Foundations of Computer Science* (to appear). *(A1.5)*

NESETRIL, J., AND V. RÖDL [1977], "A simple proof of Galvin-Ramsey properties of finite graphs and a dimension of a graph," unpublished manuscript. *(A1.5)*

NIEVERGELT, J. *See* REINGOLD, E. M.

NIGMATULLIN, R. G. [1975], "Complexity of the approximate solution of combinatorial problems," *Dokl. Akad. Nauk. SSSR* **224**, 289-292 (in Russian). English translation in *Soviet Math. Dokl.* **16**, 1199-1203. *(6.2)*

NINOMIYA, K. *See* FUJII, M.

OGDEN, W. F. *See* JAZAYERI, M.

OKUI, J. *See* ARAKI, T.; SUGIYAMA, Y.

OPATRNÝ, J. [1978], "Total ordering problem," unpublished manuscript. *(A12)*

OPATRNÝ, J., AND K. CULIK, II [1975], "Time complexity of L languages," *Abstracts of Papers: Conference on Formal Languages, Automata, and Development*, University of Utrecht, Netherlands. *(A10.2)*

OPPEN, D. C. *See* NELSON, G.

ORLIN, J. [1976], "Contentment in graph theory: covering graphs with cliques," unpublished manuscript. *(A1.1)*

ORLIN, J. B. *See also* BARTHOLDI, J. J., III.

ORLOVA, G. I., AND Y. G. DORFMAN [1972], "Finding the maximum cut in a graph," *Engrg. Cybernetics* **10**, 502-506. *(4.1; A1.2; A2.2)*

OSBORN, S. L. *See* LUCCHESI, C. L.

OSTERWEIL, L. *See* GABOW, H. N.

PALMER, E. M. *See* HARARY, F.

PAPADIMITRIOU, C. H. [1976a], "The NP-completeness of the bandwidth minimization problem," *Computing* **16**, 263-270. *(A1.3; A4.2)*

PAPADIMITRIOU, C. H. [1976b], "On the complexity of edge traversing," *J. Assoc. Comput. Mach.* **23**, 544-554. *(A2.3)*

PAPADIMITRIOU, C. H. [1976c], "The complexity of the capacitated tree problem," Report No. TR-21-76, Center for Research in Computing Technology, Harvard University, Cambridge, MA. *(A2.1)*

PAPADIMITRIOU, C. H. [1977], "The Euclidean traveling salesman problem is NP-complete," *Theor. Comput. Sci.* **4**, 237-244. *(A2.3)*

PAPADIMITRIOU, C. H. [1978a], "The adjacency relation on the traveling salesman polytope is NP-complete," *Math. Programming* **14**, 312-324. *(A6)*

PAPADIMITRIOU, C. H. [1978b], "Efficient search for rationals," Report No. TR-01-78, Center for Research in Computing Technology, Harvard University, Cambridge, MA. *(A13)*

PAPADIMITRIOU, C. H. [1978c], "Serializability of concurrent updates," Report No. TR-14-78, Center for Research in Computing Technology, Harvard University, Cambridge, MA. *(A4.3)*

PAPADIMITRIOU, C. H. [1978d], private communication. *(A1.1)*

PAPADIMITRIOU, C. H., P. A. BERNSTEIN, AND J. B. ROTHNIE [1977], "Some computational problems related to database concurrency control," *Proc. Conf. on Theoretical Computer Science*, University of Waterloo, Waterloo, Ontario, 275-282. *(A4.3)*

PAPADIMITRIOU, C. H., AND P. C. KANELLAKIS [1978], "Flowshop scheduling with limited temporary storage," unpublished manuscript. *(A5.3)*

PAPADIMITRIOU, C. H., AND K. STEIGLITZ [1976], "Some complexity results for the traveling salesman problem," *Proc. 8th Ann. ACM Symp. on Theory of Computing*, Association for Computing Machinery, New York, 1-9. *(A1.3)*

PAPADIMITRIOU, C. H., AND M. YANNAKAKIS [1978a], "On the complexity of minimum spanning tree problems," unpublished manuscript. *(A2.1)*

PAPADIMITRIOU, C. H., AND M. YANNAKAKIS [1978b], "Scheduling interval-ordered tasks," Report No. TR-11-78, Center for Research in Computing Technology, Harvard Universty, Cambridge, MA. *(A5.2)*

PAPADIMITRIOU, C. H. *See also* GAREY, M. R.; LEWIS, H. R.

PARK, D. M. *See* LUCKHAM, D. C.

PATERSON, M. S. [1967], *Equivalence Problems in a Model of Computation*, Doctoral Thesis, Cambridge University, Cambridge, England. *(A11.2)*

PATERSON, M. S., AND M. N. WEGMAN [1978], "Linear unification," *J. Comput. System Sci.* **16**, 158-167. *(A7.3; A9.2)*

PATERSON, M. S. *See also* BAUER, M.; LUCKHAM, D. C.

PAUL, W. J. [1977], "A $2.5n$ lower bound on the combinational complexity of Boolean functions," *SIAM J. Comput.* **6**, 427-443. *(7.6)*

PAULL, M., AND S. UNGER [1959], "Minimizing the number of states in incompletely specified sequential switching functions," *IRE Trans. Electron. Comput.* **EC-8**, 356-367. *(A10.1)*

PERL, Y., AND Y. SHILOACH [1978], "Finding two disjoint paths between two pairs of vertices in a graph," *J. Assoc. Comput. Mach.* **25**, 1-9. *(A2.4)*

PERL, Y., AND S. ZAKS [1978], private communication. *(A1.5)*

PERL, Y. *See also* ITAI, A.

PFLEEGER, C. F. [1973], "State reduction in incompletely specified finite-state machines," *IEEE Trans. Computers* **C-22**, 1099-1102. *(A10.1)*

PFLEEGER, C. F. [1974], *Complete Sets and Time and Space Bounded Computation*, Doctoral Thesis, Computer Science Dept., Pennsylvania State University, University Park, PA. *(A10.1)*

PIPPENGER, N., AND L. G. VALIANT [1976], "Shifting graphs and their applications," *J. Assoc. Comput. Mach.* **23**, 423-432. *(7.6)*

PLAISTED, D. [1976], "Some polynomial and integer divisibility problems are NP-hard," *Proc. 17th Ann. Symp. on Foundations of Computer Science*, IEEE Computer Society, Long Beach, CA, 264-267. *(A3.1; A6; A7.1; A7.2; A7.3)*

PLAISTED, D. [1977a], "Sparse complex polynomials and polynomial reducibility," *J. Comput. System Sci.* **14**, 210-221. *(A7.1; A7.2)*

PLAISTED, D. [1977b], "New NP-hard and NP-complete polynomial and integer divisibility problems," *Proc. 18th Ann. Symp. on Foundations of Computer Science*, IEEE Computer Society, Long Beach, CA, 241-253. *(7.1; A7.1; A7.2)*

PLESNÍK, J. [1978], "The NP-completeness of the Hamiltonian cycle problem in planar digraphs with degree bound two," unpublished manuscript. *(A1.3)*

PNUELI, A. *See* EVEN, S.

POLJAK, S. [1974], "A note on stable sets and colorings of graphs," *Comment. Math. Univ. Carolinae* **15**, 307-309. *(A1.2)*

PRATT, V. [1975], "Every prime has a succinct certificate," *SIAM J. Comput.* **4**, 214-220. *(7.1; A1.3)*

PRATT, V. [1977], "Two easy theories whose combination is hard," unpublished manuscript. *(A9.2)*

PRATT, V. *See also* LITVINTCHOUK, S. D.

PREPARATA, F. P. *See* JOHNSON, D. S.

PRÖMEL, H. J. [1978], private communication. *(A2.4)*

PROSKUROWSKI, A. *See* FARLEY, A.

PUDLÁK, P. [1975], "Polynomially complete problems in the logic of automated discovery," in *Mathematical Foundations of Computer Science*, Lecture Notes in Computer Science, Vol. 32, Springer, Berlin, 358-361. *(A9.2)*

PUDLÁK, P., AND F. N. SPRINGSTEEL [1975], "Complexity in mechanized hypothesis formation," unpublished manuscript. *(A12)*

PULTR, A. *See* NESETRIL, J.

QUEYRANNE, M. [1976], private communication. *(A1.5)*

RABIN, M. O. [1958], "Recursive unsolvability of group theoretic problems," *Ann. of Math.* **67**, 172-194. *(1.4)*

RABIN, M. O. [1976], "Probabilistic algorithms," in J. F. Traub (ed.), *Algorithms and Complexity: New Directions and Recent Results*, Academic Press, New York, 21-39. *(6.3)*

RABIN, M. O. *See also* FISCHER, M. J.

RAFSKY, L. [1977], private communication. *(A12)*

RAO, M. R. *See* DANTZIG, G. B.

RATLIFF, H. D. *See* BARTHOLDI, J. J., III.

REIF, J. H. [1978a], "A note on the complexity of imbedding extension problems," unpublished manuscript. *(A13)*

REIF, H. J. [1978b], "Polynomial time recognition of graphs of fixed genus," unpublished manuscript. *(A13)*

REINGOLD, E. M., J. NIEVERGELD, AND N. DEO [1977], *Combinatorial Algorithms: Theory and Practice*, Prentice-Hall, Inc., Englewood Cliffs, NJ. *(4.0)*

REISS, S. P. [1977a], *Inverse Translation: the Theory of Practical Automatic Programming*, Doctoral Thesis, Dept. of Computer Science, Yale University, New Haven, CT. *(A10.2)*

REISS, S. P. [1977b], "Statistical database confidentiality," Report No. 25, Dept. of Statistics, University of Stockholm, Stockholm, Sweden. *(A4.3)*

REISS, S. P., AND D. P. DOBKIN [1976], "The complexity of linear programming," Report No. 69, Dept. of Computer Science, Yale University, New Haven, CT. *(7.1; A6; A13)*

REISS, S. P. *See also* DOBKIN, D.

REYNER, S. W. [1977], "An analysis of a good algorithm for the subtree problem," *SIAM J. Comput.* **6**, 730-732. *(4.2.2; A1.4)*

RINNOOY KAN, A. H. G. *See* GRAHAM, R. L.; JOHNSON, D. S.; LABETOULLE, J.; LAGEWEG, B. J.; LENSTRA, J. K.

RITCHIE, D. M. *See* MEYER, A. R.

RIVEST, R. L. *See* HYAFIL, L.; LAPAUGH, A. S.

ROBERTSON, E. L. [1977], "Code generation for short/long address machines," Report No. 1779, Mathematics Research Center, University of Wisconsin, Madison, WI. *(A11.1)*

ROBERTSON, E. L. [1978], "Microcode bit optimization is NP-complete," *IEEE Trans. Computers* (to appear). *(A11.1)*

ROBERTSON, E., AND I. MUNRO [1978], "NP-completeness, puzzles, and games," *Utilitas Math.* **13**, 99-116. *(7.4; A8)*

ROBINSON, J. *See* MATIJASEVIC, Y.

RODEH, M. *See* ITAI, A.

RÖDL, V. *See* NESETRIL, J.

ROGERS, H., JR [1967], *Theory of Recursive Functions and Effective Computability*, McGraw-Hill, New York. *(7.2)*

ROSE, D. J., AND R. E. TARJAN [1978], "Algorithmic aspects of vertex elimination on directed graphs," *SIAM J. Appl. Math.* **34**, 176-197. *(A1.3)*

ROSE, D., R. TARJAN, AND G. LUEKER [1976], "Algorithmic aspects of vertex elimination on graphs," *SIAM J. Comput.* **5**, 266-283. *(A13)*

ROSENKRANTZ, D. J. [1969], "Programmed grammars and classes of formal languages," *J. Assoc. Comput. Mach.* **16**, 107-131. *(A10.2)*

ROSENKRANTZ, D. J., R. E. STEARNS, AND P. M. LEWIS [1977], "An analysis of several heuristics for the traveling salesman problem," *SIAM J. Comput.* **6**, 563-581. *(6.1)*

ROSENKRANTZ, D. J. *See also* HUNT, H. B., III.

ROSENTHAL, A. [1974], *Computing Reliability of Complex Systems*, Doctoral Thesis, Dept. of Electrical Engineering and Computer Science, University of California, Berkeley, CA. *(A2.2)*

ROSENTHAL, A. [1977], "Computing the reliability of a complex network," *SIAM J. Appl. Math.* **32**, 384-393. *(6.2)*

ROTHNIE, J. B. *See* PAPADIMITRIOU, C. H.

ROUNDS, W. C. [1973], "Complexity of recognition in intermediate level languages," *Proc. 14th Ann. Symp. on Switching and Automata Theory*, IEEE Computer Society, Long Beach, CA, 145-158. *(A10.1; A10.2)*

ROUNDS, W. C. *See also* JAZAYERI, M.

ROZENBERG, G. *See* HERMAN, G. T.

RUSSO, W. L. *See* HARRISON, M. A.

RUTLEDGE, J. [1964], "On Ianov's program schemata," *J. Assoc. Comput. Mach.* **11**, 1-9. *(A11.2)*

SAGIV, Y., AND M. YANNAKAKIS [1978], "Equivalencce among relational expressions with the union and difference operations," Report No. 241, Dept. of Electrical Engineering and Computer Science, Princeton University, Princeton, NJ. *(A4.3)*

SAGIV, Y. *See also* AHO, A. V.

SAHNI, S. [1974], "Computationally related problems," *SIAM J. Comput.* **3**, 262-279. *(5.2; A1.2; A2.4; A6; A7.3; A12)*

SAHNI, S. [1975], "Approximate algorithms for the 0/1 knapsack problem," *J. Assoc. Comput. Mach.* **22**, 115-124. *(6.1)*

SAHNI, S. [1976], "Algorithms for scheduling independent tasks," *J. Assoc. Comput. Mach.* **23**, 116-127. *(4.2.2; 6.1)*

SAHNI, S. [1977], "General techniques for combinatorial approximation," *Operations Res.* **25**, 920-936. *(6.3)*

SAHNI, S., AND Y. CHO [1977a], "Scheduling independent tasks with due times on a uniform processor system," Report No. 77-7, Computer Science Dept., University of Minnesota, Minneapolis, MN. *(A5.2)*

SAHNI, S., AND Y. CHO [1977b], "Complexity of scheduling shops with no wait in process," Report No. 77-20, Computer Science Dept., University of Minnesota, Minneapolis, MN. *(A5.3)*

SAHNI, S., AND T. GONZALEZ [1976], "P-complete approximation problems," *J. Assoc. Comput. Mach.* **23**, 555-565. *(6.2; A2.3)*

SAHNI, S. *See also* CHO, Y.; CONSTABLE, R. L.; GONZALEZ, T.; HOROWITZ, E.; IBARRA, O. H.

SAVITCH, W. J. [1970], "Relationship between nondeterministic and deterministic tape complexities," *J. Comput. System Sci.* **4**, 177-192. *(7.4)*

SAVITCH, W. J. [1974], "Nondeterministic log *n* space," *Proc. 8th Ann. Princeton Conf. on Information Sciences and Systems*, Dept. of Electrical Engineering, Princeton University, Princeton, NJ, 21-23. *(7.5)*

SCHAEFER, T. J. [1974], private communication. *(A1.1)*

SCHAEFER, T. J. [1978a], "Complexity of some two-person perfect-information games," *J. Comput. System Sci.* **16**, 185-225. *(7.4; A8)*

SCHAEFER, T. J. [1978b], "The complexity of satisfiability problems," *Proc. 10th Ann. ACM Symp. on Theory of Computing*, Association for Computing Machinery, New York, 216-226. *(A1.1; A9.1; A9.2)*

SCHAEFER, T. *See also* FRAENKEL, R. S.

SCHMIDT, E. M. *See* FORTUNE, S.

SEIFERAS, J. I., M. J. FISCHER, AND A. R. MEYER [1978], "Separating nondeterministic time complexity classes," *J. Assoc. Comput. Mach.* **25**, 146-167. *(7.6)*

SELMAN, A. L. *See* BAKER, T. P.; LADNER, R. E.

SETHI, R. [1973], "A note on implementing parallel assignment instructions," *Information Processing Lett.* **2**, 91-95. *(A11.1)*

SETHI, R. [1975], "Complete register allocation problems," *SIAM J. Comput.* **4**, 226-248. *(3.2.3; A11.1)*

SETHI, R. [1977a], "On the complexity of mean flow time scheduling," *Math. Oper. Res.* **2**, 320-330. *(A5.2)*

SETHI, R. [1977b], private communication. *(A7.3)*

SETHI, R., AND J. D. ULLMAN [1970], "The generation of optimal code for arithmetic expressions," *J. Assoc. Comput. Mach.* **17**, 715-728. *(A11.1)*

SETHI, R. *See also* AHO, A. V.; BRUNO, J.; COMER, D.; COOK, S.; DOWNEY, P. J.; GAREY, M. R.; HORVÁTH, E. C.

SHAMIR, A. [1977], "Finding minimum cutsets in reducible graphs," Report No. MIT/LCS/TM-85, Laboratory for Computer Science, Massachusetts Institute of Technology, Cambridge, MA. *(A1.1)*

SHAMIR, A. *See also* EVEN, S.

SHAMIR, E., AND C. BEERI [1974], "Checking stacks and context-free programmed grammars accept P-complete languages," in J. Loeckx (ed.), *Proc. 2nd Colloq. on Automata, Languages, and Programming*, Lecture Notes in Computer Science, Vol. 14, Springer, Berlin, 27-33. *(A10.1; A10.2)*

SHAMOS, M. I. [1976], "Geometry and statistics: problems at the interface," in J. F. Traub (ed.), *Algorithms and Complexity: New Directions and Recent Results*, Academic Press, New York, 251-280. *(A12)*

SHAMOS, M. I. *See also* MEYER, A. R.

SHAPIRO, S. D. [1977], "Performance of heuristic bin packing algorithms with segments of random length," *Information and Control* **35**, 146-158. *(6.3)*

SHAPLEY, L. S., AND M. SHUBIK [1954], "A method of evaluating the distribution of power in a committee system," *Amer. Pol. Sci. Rev.* **48**, 787-792. *(A12)*

SHILOACH, Y. [1976], "A minimum linear arrangement algorithm for undirected trees," Report, Dept. of Applied Mathematics, Weizmann Institute, Rehovot, Israel. *(A1.3)*

SHILOACH, Y. [1978], "The two paths problem is polynomial," Report No. STAN-CS-78-654, Computer Science Department, Stanford University, Stanford, CA. *(A2.4; A13)*

SHILOACH, Y. *See also* EVEN, S.; ITAI, A.; PERL, Y.

SHUBIK, M. *See* SHAPLEY, L. S.

SIDNEY, J. B. [1973], "An extension of Moore's due date algorithm," in S. E. Elmaghraby (ed.), *Symposium on the Theory of Scheduling and its Applications*, Lecture Notes in Economics and Mathematical Systems, Vol. 86, Springer, Berlin, 393-398. *(A5.1)*

SIDNEY, J. B. [1975], "Decomposition algorithms for single-machine sequencing with precedence relations and deferral costs," *Operations Res.* **23**, 283-298. *(A5.1)*

SIDNEY, J. B. *See also* MONMA, C. L.

SIMON, J. [1975], *On Some Central Problems in Computational Complexity*, Doctoral Thesis, Dept. of Computer Science, Cornell University, Ithaca, NY. *(7.3)*

SIMON, J. [1977], "On the difference between the one and the many (preliminary version)," in *Automata, Languages, and Programming*, Lecture Notes in Computer Science, Vol. 52, Springer, Berlin, 480-491. *(7.3)*

SIMONS, B. [1978], "A fast algorithm for single processor scheduling," *Proc. 19th Ann. Symp. on Foundations of Computer Science*, IEEE Computer Society, Long Beach, CA, 246-252. *(A5.1)*

SIMONS, B. *See also* GAREY, M. R.

SIPSER, M. *See*

SKYUM, S. *See* JONES, N. D.

SLATER, P. J. [1976], "*R*-domination in graphs," *J. Assoc. Comput. Mach.* **23**, 446-450. *(A2.5)*

SLATER, P. J. *See also* COCKAYNE, E.

SMITH, W. E. [1956], "Various optimizers for single-state production," *Naval Res. Logist. Quart.* **3**, 59-66. *(A5.1)*

SOLOVAY, R. *See* BAKER, T.

SPRINGSTEEL, F. N. *See* PUDLÁK, P.

STATMAN, R. [1976], private communication. *(A1.4; A9.1)*

STEARNS, R. E. *See* HARTMANIS, J.; ROSENKRANTZ, D. J.

STEIGLITZ, K. *See* PAPADIMITRIOU, C. H.

STOCKMEYER, L. J. [1973], "Planar 3-colorability is NP-complete," *SIGACT News* **5:3**, 19-25. *(3.2.3; 4.1)*

STOCKMEYER, L. J. [1974a], *The Complexity of Decision Problems in Automata Theory and Logic*, Doctoral Thesis, Dept. of Electrical Engineering, Massachusetts Institute of Technology, Cambridge, MA. *(A10.2)*

STOCKMEYER, L. J. [1974b], private communication. *(A1.3)*

STOCKMEYER, L. J. [1975], "The set basis problem is NP-complete," Report No. RC-5431, IBM Research Center, Yorktown Heights, NY. *(A3.1)*

STOCKMEYER, L. J. [1976a], "The polynomial-time hierarchy," *Theor. Comput. Sci.* **3**, 1-22. *(7.2; 7.4; 7.5; A7.3)*

STOCKMEYER, L. J. [1976b], private communication. *(A4.1)*

STOCKMEYER, L. J., AND A. K. CHANDRA [1978], "Provably difficult combinatorial games," Report No. RC-6957, IBM Thomas J. Watson Research Center, Yorktown Heights, NY. *(7.6)*

STOCKMEYER, L. J., AND A. R. MEYER [1973], "Word problems requiring exponential time," *Proc. 5th Ann. ACM Symp. on Theory of Computing*, Association for Computing Machinery, New York, 1-9. *(7.4; 7.5; A7.1; A7.3; A8; A9.2; A10.2)*

STOCKMEYER, L. J. *See also* CHANDRA, A. K.; GAREY, M. R.; KOU, L. T.; MEYER, A. R.

STONE, H. S., AND S. H. FULLER [1973], "On the near-optimality of the shortest-latency-time-first drum scheduling discipline," *Comm. ACM* **16**, 352-353. *(6.1)*

STORER, J. A. [1977], "NP-completeness results concerning data compression," Report No. 234, Dept. of Electrical Engineering and Computer Science, Princeton University, Princeton, NJ. *(A4.2)*

STORER, J. A., AND T. G. SZYMANSKI [1978], "The macro model for data compression (Extended abstract)," *Proc. 10th Ann. ACM Symp. on Theory of Computing*, Association for Computing Machinery, New York, 30-39. *(A4.2)*

STORER, J. A. *See also* MAIER, D.

SUDBOROUGH, I. H. [1975], "A note on tape-bounded complexity classes and linear context-free languages," *J. Assoc. Comput. Mach.* **22**, 499-500. *(7.5)*

SUGIYAMA, Y., T. ARAKI, J. OKUI, AND T. KASAMI [1977], "Complexity of the deadlock avoidance problem," *Trans. IECE Japan* **60-D**, 251-258 (in Japanese). *(A5.3)*

SUGIYAMA, Y. *See also* ARAKI, T.

SUURBALLE, J. W. [1975], "Minimal spanning trees subject to disjoint arc set constraints," unpublished manuscript. *(A2.1)*

SYSLO, M. M. [1973], "A new solvable case of the traveling salesman problem," *Math. Programming* **4**, 347-348. *(A2.3)*

SZYMANSKI, T. G. [1978], "Assembling code for machines with span-dependent instructions," *Comm. ACM* **21**, 300-308. *(3.2.2; A11.1)*

SZYMANSKI, T. G. *See also* HUNT, H. B., III; STORER, J. A.

TANIMOTA, S. L. *See* ITAI, A.

TARJAN, R. E. [1977], "Finding optimum branchings," *Networks* **7**, 25-35. *(A2.1)*

TARJAN, R. E., AND A. E. TROJANOWSKI [1977], "Finding a maximum independent set," *SIAM J. Comput.* **6**, 537-546. *(6.0; 7.6)*

TARJAN, R. E. *See also* EHRLICH, G.; ESWAREN, K. P.; EVEN, S.; GAREY, M. R.; HOPCROFT, J. E.; LIPTON, R. J.; ROSE, D. J.

TERANAKA, K. *See* IBARAKI, T.

THOMASSEN, G. *See* CHVÁTAL, V.

TOIDA, S. *See* IBARAKI, T.

TRAKHTENBROT, B. A., AND Y. M. BARZDIN [1973], *Finite Automata*, North-Holland, Amsterdam. *(A10.1)*

TREYBIG, L. B. *See* BOROSH, I.

TROJANOWSKI, A. E. *See* TARJAN, R. E.

TSEITIN, G. S. [1970], "On the complexity of derivation in propositional calculus," in A. O. Slisenko (ed.), *Studies in Constructive Mathematics and Mathematical Logic — Part II*, Consultants Bureau, New York, 115-125. *(7.6)*

TSICHRITZIS, D. [1970], "The equivalence problem of simple programs," *J. Assoc. Comput. Mach.* **17**, 729-738. *(A11.2)*

TURING, A. [1936], "On computable numbers, with an application to the Entscheidungsproblem," *Proc. London Math. Soc. Ser. 2* **42**, 230-265 and **43**, 544-546. *(1.4)*

ULLMAN, J. D. [1975], "NP-complete scheduling problems," *J. Comput. System Sci.* **10**, 384-393. *(4.1; A5.2; A13)*

ULLMAN, J. D. [1976], "Complexity of sequencing problems," in E. G. Coffman, Jr., (ed.), *Computer and Job/Shop Scheduling Theory*, John Wiley & Sons, New York, 139-164. *(A5.2)*

ULLMAN, J. D. *See also* AHO, A. V.; HARRISON, M. A.; HOPCROFT, J. E.; HUNT, H. B., III; JOHNSON, D. S.; SETHI, R.

UNGER, S. *See* PAULL, M.

VALIANT, L. G. [1976a], "Relative complexity of checking and evaluating," *Information Processing Lett.* **5**, 20-23. *(5.1)*

VALIANT, L. G. [1976b], "A polynomial reduction of satisfiability to Hamiltonian circuits that preserves the number of solutions," unpublished manuscript. *(7.3)*

VALIANT, L. G. [1977a], "The complexity of computing the permanent," Report No. CSR-14-77, Computer Science Department, University of Edinburgh, Edinburgh, Scotland, (to appear *Theor. Comput. Sci.*). *(7.3; A1.1; A7.3)*

VALIANT, L. G. [1977b], "The complexity of enumeration and reliability problems," Report No. CSR-15-77, Computer Science Dept., University of Edinburgh, Edinburgh, Scotland. *(7.3; A2.2)*

VALIANT, L. G. [1977c], private communication. *(A1.5; A7.2)*

VALIANT, L. G. *See also* ANGLUIN, D.; PIPPENGER, N.

VAN LEEUWEN, J. [1975], "The membership question for ETOL-languages is polynomially complete," *Information Processing Lett.* **3**, 138-143. *(A10.2)*

VAN LEEUWEN, J. [1976a], "Having a Grundy-numbering is NP-complete," Report No. 207, Computer Science Dept., Pennsylvania State University, University Park, PA. *(A1.5)*

VAN LEEUWEN, J. [1976b], "Variations on a new machine model," *Proc. 17th Ann. Symp. on Foundations of Computer Science*, IEEE Computer Society, Long Beach, CA, 228-235. *(7.5)*

VAN LEEUWEN, J. [1977], "Inequivalence of program-segments and NP-completeness," Report No. 216, Computer Science Dept., Pennsylvania State University, University Park, PA. *(A11.2)*

VAN SICKLE, L., AND K. M. CHANDY [1977], "The complexity of computer network design problems," unpublished manuscript. *(A4.1)*

VAN TILBORG, H. C. A. *See* BERLEKAMP, E. R.

WAGNER, R. A. [1975], "On the complexity of the extended string-to-string correction problem," *Proc. 7th Ann. ACM Symp. on Theory of Computing*, Association for Computing Machinery, New York, 218-223. *(A4.2)*

WAGNER, R. A., AND M. J. FISCHER [1974], "The string-to-string correction problem," *J. Assoc. Comput. Mach.* **21**, 168-173. *(A4.2)*

WALSH, A. M., AND W. A. BURKHARD [1977], "Efficient algorithms for (3,1) graphs," *Information Sci.* **13**, 1-10. *(A1.1)*

WEGMAN, M. N. *See* PATERSON, M. S.

WEINBERG, L. *See* BRUNO, J.

WINKLMANN, K. A. [1977], *A Theoretical Study of Some Aspects of Parameter Passing in ALGOL-60 and in Similar Programming Languages*, Doctoral Thesis, Purdue University, Lafayette, IN. *(A11.2)*

WITSENHAUSEN, H. S. [1978], "Information aspects of stochastic control," unpublished manuscript. *(A3.2)*

WOLSEY, L. A. *See* NEMHAUSER, G. L.

WONG, C. K., AND A. C. YAO [1976], "A combinatorial optimization problem related to data set allocation," *Rev. Francaise Automat. Informat. Recherche Operationnelle Ser. Bleue* **10.5** (suppl.), 83-95. *(A3.2)*

WONG, C. K. *See also* HIRSCHBERG, D. S.; KARP, R. M.; KOU, L. T.

WONG, J. K. *See* HOPCROFT, J. E.

WRATHALL, C. [1976], "Complete sets and the polynomial-time hierarchy," *Theor. Comput. Sci.* 3, 23-33. *(7.2; 7.4)*

YANNAKAKIS, M. [1978a], "The node deletion problem for hereditary properties," Report No. TR-240, Computer Science Laboratory, Princeton University, Princeton, NJ. *(A1.2)*

YANNAKAKIS, M. [1978b], "Node- and edge-deletion NP-complete problems," *Proc. 10th Ann. ACM Symp. on Theory of Computing*, Association for Computing Machinery, New York, 253-264. *(A1.1; A1.2; A2.2)*

YANNAKAKIS, M. [1978c], private communication. *(A1.2)*

YANNAKAKIS, M., AND F. GAVRIL [1978], "Edge dominating sets in graphs," unpublished manuscript. *(A1.1; A1.2; A12)*

YANNAKAKIS, M. *See also* PAPADIMITRIOU, C. H.; SAGIV, Y.

YAO, A. C. [1976], private communication. *(6.3)*

YAO, A. C. [1978a], "New algorithms for bin packing," Report No. STAN-CS-78-662, Computer Science Department, Stanford University, Stanford, CA. *(6.1)*

YAO, A. C. [1978b], private communication. *(A3.2)*

YAO, A. C. *See also* GAREY, M. R.; WONG, C. K.

YESHA, Y. *See* FRAENKEL, A. S.

ZADEH, N. [1973], "A bad network problem for the simplex method and other minimum cost flow algorithms," *Math. Programming* 5, 255-266. *(1.3)*

ZAKS, S. *See* PERL, Y.

ZALCSTEIN, Y. *See* LIPTON., R. J.

Subject Index

UPDATE

In this addendum we briefly survey some recent advances that have been made on the open problems in A13 and also correct several errors and omissions that occur in the body of the text. Of the twelve open problems, eight have been resolved.

The most famous of the open problems to be resolved is LINEAR PROGRAMMING, which is now known to be solvable in polynomial time by the ellipsoid method and by interior point approaches [Khachian, 1979], [Karmarkar, 1984]. Also solvable in polynomial time are SUBGRAPH HOMEOMORPHISM (FOR ANY FIXED GRAPH H) [Robertson and Seymour, 1986], SPANNING TREE PARITY [Lovasz, 1980] and TOTAL UNIMODULARITY [Seymour,1980]. Open problems that have been proved to be NP-complete are GRAPH GENUS [Thomassen, 1990], CHORDAL GRAPH COMPLETION [Yannakakis,1981], CHROMATIC INDEX [Holyer,1981], and PARTIAL ORDER DIMENSION [Yannakakis,1982], along with the following three problems, mentioned in passing in A13: CROSSING NUMBER [Garey and Johnson, 1983], GRAPH THICKNESS [Mansfield,1983], and LINEAR COMPLEMENTARITY [Chung,1979].

For a more detailed (and continuing) update, the reader is referred to the second author's column in the *Journal of Algorithms*, "The NP-Completeness Column: An Ongoing Guide." The first edition of this column (December, 1981) is a greatly expanded (and only slightly out-of-date) version of the above paragraph, with the June, 1987 and September, 1988 columns covering the more recent results.

As to the main body of our list (A1 through A12), several significant errors have been pointed out to us. There was a misstatement in the comments to DOMINATING SET [GT2]; the cases actually proved NP-complete are planar cubic graphs and, for CONNECTED DOMINATING SET, regular graphs of degree 4 and planar graphs of *maximum* degree 4. HAMILTONIAN CIRCUIT [GT37] is NP-complete for edge graphs, our claim of polynomial time solvability having been based on a faulty analogy with Euler tours. In the definition of K-CLOSURE [GT57], we meant to require that $|V'| = K$. In the first comment on MINIMUM EDGE-COST FLOW [ND32], we must allow $c(a) \leq 2$ for the claimed NP-completeness result to hold. MINIMUM TEST SET [SP6] is NP-complete even when restricted to the case where each subset has at most two elements. SUBSET PRODUCT [SP14], although not solvable in time polynomial in $|A|$ and $\max\{s(a): a \in A\}$, is solvable in pseudo-polynomial time due to the presence of B in the input. In the last comment on INTEGER PROGRAMMING [MP1], we meant to require that $\overline{x} \cdot \overline{y} \geq b$. In the definition of COMPARATIVE VECTOR INEQUALITIES [MP13], "at least as large as" should be replaced by "strictly larger than." In our comments on QUADRATIC DIOPHANTINE EQUATIONS [AN8], our claim that $\Sigma_{i=1}^{k} a_i x_i = c$ is solvable in polynomial time holds only if arbitrary integer solutions are allowed. If only non-negative solutions are allowed, the problem is of course NP-complete. Finally, the direction of the inequality in MATRIX COVER [MS13] should be reversed. Our thanks to D. Denning, J. Feigenbaum, T. Ibaraki, J. K. Lenstra, D. Richards, and J. Shepherdson, among others, for pointing these out to us.

With respect to the body of the text, three omissions should be mentioned: the omission of a dot from the third circle in the top row of Figure 4.3, the omission of the qualifier ''directed'' for graphs when discussing Eulerian paths in the first paragraph on p. 168, and the omission on the third line from the bottom of p. 112 of the proviso that the string *y* extend only as far to the right as the current position of the oracle head. (Exercise: show that there would be NP-hard problems that can be solved in polynomial time if the original definition of *y* were used.)

Finally, we would like to apologize to K. Steiglitz, W. Lipski, Jr. and C. P. Pfleeger for misspelling their names.

March, 1991 M.R.G. D.S.J.

REFERENCES

1. S. J. CHUNG, A note on the complexity of LCP: the LCP is NP-complete, Report No. 79-2, Department of Industrial and Operations Engineering, University of Michigan, Ann Arbor, 1979.
2. M. R. GAREY AND D. S. JOHNSON, Crossing number is NP-complete, *SIAM J. Algebraic and Discrete Methods* **4** (1983), 312-316.
3. I. HOLYER, The NP-completeness of edge-coloring, *SIAM J. Comput.* **10** (1981), 718-720.
4. N. KARMARKAR, A new polynomial-time algorithm for linear programming, *Combinatorica* **4** (1984), 373-395.
5. L. G. KHACHIAN, A polynomial algorithm in linear programming, *Dokl. Akad. Nauk. SSSR* **244** (1979), 1093-1096 (in Russian). English translation in *Soviet Math. Dokl.* **20** (1979), 191-194.
6. L. LOVÁSZ, Matroid matching and some applications, *J. Combinatorial Theory Ser. B* **28** (1980), 208-236.
7. A. MANSFIELD, Determining the thickness of graphs is NP-hard, *Math. Proc. Camb. Phil. Soc.* **93** (1983), 9-23.
8. N. ROBERTSON AND P. D. SEYMOUR, Graph minors XIII: The disjoint paths problem, manuscript (1986).
9. P. D. SEYMOUR, Decomposition of regular matroids, *J. Combinatorial Theory Ser. B* **28** (1980), 305-359.
10. C. THOMASSEN, The graph genus problem is NP-complete, *J. Algorithms* **10** (1989), 568-576.
11. M. YANNAKAKIS, Computing the minimum fill-in is NP-complete, *SIAM J. Algebraic and Discrete Methods* **2** (1981), 77-79.
12. M. YANNAKAKIS, The complexity of the partial order dimension problem, *SIAM J. Algebraic and Discrete Methods* **3** (1982), 351-358.